X-RAY DIFFRACTION METHODS

X-RAY DIFFRACTION METHODS

E. W. NUFFIELD

PROFESSOR OF GEOLOGY
UNIVERSITY OF TORONTO

JOHN WILEY & SONS

New York • Chichester • Brisbane • Toronto

ISBN 0 471 65170 2

Library of Congress Catalog Card Number: 66-22843
Printed in the United States of America

5 6 7 8 9 10

PREFACE

This book is an attempt to condense the basic concepts and modern methods of x-ray crystallography into one small volume. It is meant to be a practical book in the sense that students, I hope, will be able to use it as a laboratory manual. It is directed particularly at students in the earth sciences and others whose mathematical background is generally below the level attained by students in the physical sciences. The mathematical treatments are therefore not always so elegant as they might be. Readers are expected to have had an elementary course in crystallography; nevertheless, I have included a summary in Chapter 1 to serve as a guide for extracurricular reading.

The plan of the book follows the scheme of instruction used in the Department of Geology at the University of Toronto. A course constituting the substance of the first five chapters (i.e., excluding single-crystal methods) is offered in the second year of instruction in geology. This is a terminal course for many students, and therefore the reciprocal-lattice concept, which is not needed to gain a working knowledge of polycrystal methods, is left until later. A course in single-crystal methods, based on the remainder of the book, is offered as an option in the succeeding undergraduate years.

No book on x-ray diffraction methods could be considered adequate without special mention of Professor M. J. Buerger's contributions. His *X-Ray Crystallography*, published by John Wiley and Sons in 1942, is a classic. This and his other books, particularly *The Powder Method* with L. V. Azaroff (McGraw-Hill, 1958), *Crystal-Structure Analysis* (Wiley, 1960), and *The Precession Method* (Wiley, 1964), and many articles have made him the leading figure in the field with which this book is concerned. Professor Buerger has improved the designs of standard diffraction instruments and has provided a deeper understanding of their use. He is responsible for a number of new instruments, of which the precession goniometer is the most successful. This instrument already ranks with the powerful equi-inclination Weissenberg camera, and its full potential has not yet been realized. Everyone who writes on diffraction methods draws on the work of Professor Buerger. Certainly my debt is a heavy one.

I am grateful for permission to reproduce the many figures I have taken from the literature. Professor Dame Kathleen Lonsdale kindly gave me an opportunity to learn something of how the Laue and oscillation methods are used for orienting crystals in her laboratory. I am indebted to my colleague, Professor E. J. Brooker, for many discussions about the material in the book. The substance of Chapter 13, which deals with a general method for orienting a randomly set crystal, was extracted from his Ph.D. thesis. Dr. F. W. Matthews allowed me to read an unpublished manuscript on the use of the A.S.T.M. Index and File. Individual references in Chapter 5 to figures, tables, and other data do not properly acknowledge the use that I have made of papers published by members of the Philips Laboratories, New York, N.Y.

Toronto, Canada
June 1966

E. W. NUFFIELD

CONTENTS

1. Introduction

HISTORICAL BACKGROUND 1
The Laue Experiments and the Bragg Equation 1
Wavelength units, 5
ELEMENTARY CRYSTALLOGRAPHY 7
The Classification of Crystals 8
The translation lattices, 9 · Point groups, 14 · Space groups, 18 · Crystal classes, 26
Choice of the Triclinic cell, Its Orientation, and Axial Labels 26

2. The Nature and Generation of X-Rays

The continuous spectrum (white radiation), 30 · Characteristic or line spectra, 31 · Fluorescence, 35 · β filters, 39 · X-ray diffraction tubes, 40 · X-ray generators, 45

3. The Diffraction of X-Rays

THE GEOMETRY OF DIFFRACTION 46
Diffraction by a row of identical, equally spaced atoms, 48 · Diffraction by a plane lattice-array of atoms, 50 · Diffraction by a three-dimensional lattice-array of atoms, 52 · The "reflection" of x-rays, 55
THE INTENSITY OF DIFFRACTION 62
Scattering by an electron, 62 · Scattering by an atom, 64 · Scattering by a unit cell, 66 · The temperature factor, 69 · The absorption factor, 71 · The Lorentz factor, 72

THE INTEGRATED REFLECTION 73
INTEGRATED REFLECTIONS FROM REAL CRYSTALS 75
The effect of extinction, 75
THE DETERMINATION OF CRYSTAL STRUCTURES 79
DIFFRACTION OR LAUE SYMMETRY 85
REFERENCES 86

4. Systematically Absent X-Ray Reflections and the Determination of Space Group

SYSTEMATIC ABSENCES DUE TO LATTICE CENTERING 87
One face-centered lattice (*C*, *B*, *A*), 87 · All-face-centered lattice (*F*), 90 · Body-centered lattice (*I*), 90 · The rhombohedral lattice (*R*), 91
SYSTEMATIC ABSENCES DUE TO SPACE-GROUP SYMMETRY ELEMENTS 93
Glide planes, 93 · Screw axes, 96
THE DETERMINATION OF SPACE GROUP 98

5. Polycrystal or Powder Methods

PART 1. FILM TECHNIQUES 105
THE DEBYE–SCHERRER METHOD 105
Camera sizes, 107 · The relative intensities of the reflections, 107 · The sample, 111 · The choice of radiation, 114 · Measurement of the film, 116
THE INTERPRETATION OF X-RAY POWDER DATA 118
Indexing the reflections and deriving the dimensions of the unit cell, 118 · Calculation of the unit cell contents, 125 · The determination of lattice type and space group, 128 · The identification of crystalline substances, 130 · The quantitative analysis of mixtures, 137 · The determination of crystallite size, 147
THE ATTAINMENT OF PRECISE LATTICE DIMENSIONS WITH THE DEBYE–SCHERRER METHOD 149
The source and correction of errors, 150
SPECIAL POWDER TECHNIQUES AND APPARATUS 166
Parafocusing cameras, 166 · Pinhole techniques, 169 · High-temperature cameras, 171
REFERENCES 171
PART 2. THE DIFFRACTOMETER 172
The Geometry and Optical System of the Goniometer 172

X-Ray Counters 177
Gas-filled detectors, 177 · Photoelectric detectors, 184 · A comparison of the three types of counters, 185
The Counting Circuits 196
The Reliability of Diffractometer Powder Data ... 201
Specimen preparation, 204
A Comparison of the Debye–Scherrer and Diffractometer Methods 206

6. Orientation and Projection of Morphological Crystals

Orientation of Crystals by Optical Methods 208
Optical goniometers, 208
Methods of Projection 212
The stereographic projection, 214 · The gnomonic projection, 221

7. The Laue Method

The transmission method, 223 · The back-reflection method, 226
The Projection of Laue Photographs.................... 226
Stereographic projection of transmission Laue photographs, 226 · Stereographic projection of back-reflection Laue photographs, 230
Orienting a Crystal with the Laue Method 236
Correction of a small error in setting a mirror plane normal to the axis of the cylindrical film, 236 · Orientation of a plane, which reflects in the equatorial plane, precisely parallel to the beam, 237

8. The Reciprocal Lattice

The reciprocal lattice, 239 · Formation of a lattice-array by reciprocal points, 240 · Relation between reciprocal-lattice planes and direct-lattice rows, 241 · The centered lattices, 242 · Relation between the reciprocal lattice and the gnomonic projection, 243
The Reflection of X-Rays in Terms of the Reciprocal Lattice.. 244
The sphere of reflection, 244 · Diffraction from planes of reciprocal points, 247

9. The Buerger Precession Method

Motion of the crystal, 253 · Motion of the zero-level plane, 253 · Upper-level reciprocal planes, 257 · The precession instrument, 257 · Photographing reciprocal levels, 258 · Reciprocal-level screen setting, 260 · Instrumental settings, 260 · The determination of *d**, 261 · The effect of the wavelength of incident radiation, 264

ORIENTATION PHOTOGRAPHS 266

The preparation of orientation photographs, 269

THE CHOICE OF THE CRYSTAL'S ORIENTATION IN THE BEAM 272

Function of the various precession films, 272 · The choice of precession axis, 274 · Relation between the axis of crystal holder and precession axes, 276 · Orienting a crystal for precession with two axes, 277

MEASUREMENT OF ZERO-LEVEL FILMS 278

A measuring instrument, 278 · The determination of lattice constants, 278 · Correction for film shrinkage, 279 · Calibrating a precession instrument, 280

10. The Rotation and Oscillation Methods

ROTATION PHOTOGRAPHS 283

The method, 283 · Recording the reflections, 284 · Determination of the direct-lattice period in the rotation axis, 285 · Indexing rotation photographs, 288 · Summary of the rotation method, 293

OSCILLATION PHOTOGRAPHS...................................... 294

Orienting a crystal for rotation about a direct-lattice row, 296

11. The Equi-inclination Weissenberg Method

The apparatus, 302 · Central and noncentral rows, 304

DIFFRACTION IN TERMS OF CENTRAL ROWS 305

Weissenberg projection of a central line in the zero level, 305 · Weissenberg projection of the angular relations between central-lattice rows in the zero level, 308 · Weissenberg projection of upper-level central rows, 309 · Reconstructing the reciprocal lattice from equi-inclination Weissenberg photographs, 315

DIFFRACTION IN TERMS OF NONCENTRAL ROWS........... 324

Reconstructing the reciprocal lattice with a lattice line template, 328

THE RECOGNITION OF DIFFRACTION SYMMETRY ON EQUI-INCLINATION WEISSENBERG PHOTOGRAPHS................. 330

THE DETERMINATION OF LATTICE CONSTANTS.............. 332

Linear constants, 332 · Angular constants, 336

12. Single-Crystal X-Ray Diffractometry

The Buerger equi-inclination diffractometer, 343 · The Eulerian cradle, 345 · The linear diffractometer, 347

13. A General Method for Orienting a Crystal

An expression for the orientation of a randomly set reflecting plane, 352 · Procedure for determining a plane's orientation, 356 · Application of the method, 357 · Setting the crystal, 359 · The recognition of crystal symmetry, 362

Appendix

1. THE 230 SPACE GROUPS 365
2. THE SIX PERMUTATION OF THE ORTHORHOMBIC SPACE GROUPS 370
3. SPACE-GROUP DETERMINATIVE TABLES 372
4. TABLE OF $\frac{1}{2}\left(\frac{\cos^2\theta}{\sin\theta}+\frac{\cos^2\theta}{\theta}\right)$ 385
5. DATA FOR CONSTRUCTING STEREOGRAPHIC SCALES 388
6. MATHEMATICAL RELATIONS BETWEEN RECIPROCAL- AND DIRECT-LATTICE CONSTANTS..................................... 391

   The Volume of a Triclinic Cell.................................. 391

   The Volume of a Triclinic Cell in Terms of Its Own Elements.. 393

   A Unit Translation in Terms of Its Corresponding Reciprocal Constant.. 395

   An Angle Entirely in Terms of the Angles of Its Reciprocal Cell 398

The Volume Entirely in Terms of Its Own Elements 400
Equations Relating Triclinic-, Reciprocal-, and Direct-Lattice Constants 402

Index 405

X-RAY DIFFRACTION METHODS

CHAPTER ONE

INTRODUCTION

HISTORICAL BACKGROUND

The Laue Experiments and the Bragg Equation

The investigation of the geometrical theory of crystal structure was completed by the end of the nineteenth century with the publication of the 230 space groups by Federov, Schoenflies, and Barlow. A number of English scientists then turned to problems connected with the determination of crystal structures. They devised three-dimensional models of possible structures, considered the packing of spheres, and even deduced the probable space groups of some minerals. However, the investigations were not systematic and attracted little attention, partly because the reality of the atom was still not universally accepted, but largely because there was no way to determine a space group and test the theory that the atoms of crystal structures were actually arranged in patterns predicted by space-group theory. Further progress toward the goal of a firm basis for the theory of crystal structure awaited advances in the studies of the x-rays discovered by Roentgen in 1895.

Wiechert and Stokes had concluded in 1896 that x-rays must be short waves consisting of electromagnetic pulses. Their wavelengths were thought to be of the order of 1 Å, based on estimates of 7×10^{-9} cm (Wien in 1907) and 4×10^{-9} cm (Sommerfeld in 1912). Nevertheless, a number of eminent physicists of the time, including W. H. Bragg, believed in the corpuscular theory of x-rays.

Perrin and Millikan had measured Avogadro's number [the number of molecules (or atoms) in a gram-molecular (or atomic) weight] and had obtained a value of 6.2×10^{23}. This provided the key for calculating the approximate volume occupied by a molecule or an atom:

$$\text{volume of one molecule} = \frac{\text{molecular weight}}{\text{Avogadro's number} \times \text{density}}$$

The compound NaCl, for example, with a molecular weight of 58.5 and a density of 2.2, gave a value of

$$\frac{58.5}{6.2 \times 10^{23} \times 2.2} = 42.9 \times 10^{-24}\ \text{cm}^3 = 42.9\ \text{Å}^3$$

for the volume of one molecule of NaCl. From this it followed that the atoms are spaced at intervals of about 2 Å in the NaCl structure.

The phenomenon of diffraction of visible light by ruled gratings was well known. Lord Rayleigh had frequently emphasized that the essential characteristic of a grating is the periodic repetition of its elements and not the nature of these elements. About 1910 M. von Laue elaborated on Lord Rayleigh's idea and arrived at an equation for the position of the diffraction maxima which could be extended without difficulty to the case of double periodicity as it exists in cross-gratings. Shortly thereafter it occurred to von Laue that crystals were supposed to be regular arrangements of matter on the same scale as the estimated order of wavelength of x-rays. He reasoned that if the assumptions concerning the internal structure of crystals were correct, and if x-rays were electromagnetic waves, crystals should act as three-dimensional diffraction gratings for the penetrating x-rays. At this time von Laue was an assistant lecturer in the physics department at Munich headed by Professor Summerfeld. It was arranged that Friedrich and Knipping, who had just completed their doctorates under Professor Roentgen, would undertake experiments to test the hypothesis.

A copper sulfate crystal, chosen for the first experiment, was placed in the path of a fine beam of x-rays; a photographic plate was used to record the effects. The first trials were unsuccessful because the plate was improperly positioned. However, when the plate was placed behind the crystal and normal to the beam, a pattern of spots in addition to the direct-beam spot was revealed when the plate was developed (Fig. 1-1). They obtained an identical but enlarged pattern when the crystal-to-plate distance was increased which indicated that the x-rays emanating from the crystal traveled in straight lines. They also found that the pattern was changed when the orientation of the crystal was altered. The choice of copper sulfate for the first experiment was unfortunate, because at this stage it was virtually impossible to interpret the diffraction effects from a triclinic crystal. When a crystal of cubic ZnS was substituted, it was found that the symmetry of the arrangement of spots on the plate was directly related to the orientation of the crystal in the beam. This finally appeared to prove the periodic arrangement of atoms in crystals, the wave nature of x-rays, and von Laue's diffraction theory. Admittedly, the ZnS crystal gave only a selection of the diffraction spots that von Laue had expected, but he explained this by ascribing to the atoms a strongly selective scattering power for x-rays and was thus led to believe in the existence of five characteristic wavelengths in the radiation.

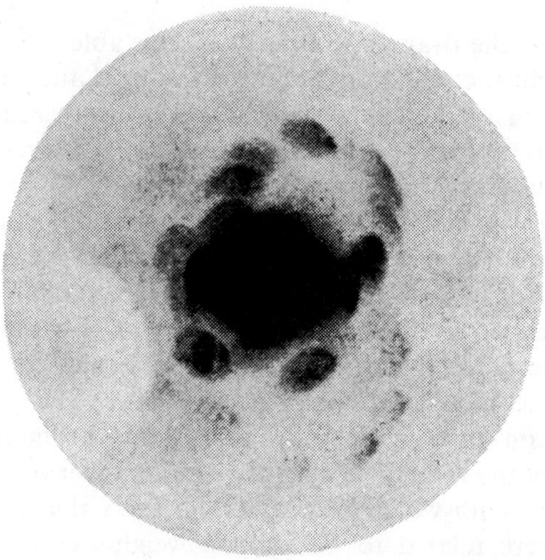

Fig. 1-1. The first x-ray diffraction photograph [P. P. Ewald, *Fifty Years of X-ray Diffraction*, N. V. A. Oosthoek, Utrecht, The Netherlands (1962)].

The experimental results and von Laue's interpretations were published in 1912 and aroused much interest. In England W. H. Bragg and his son W. L. Bragg immediately undertook experiments to test von Laue's conclusions. The younger Bragg, working at Cambridge, was soon satisfied that von Laue's photographs were due to diffraction, but concluded that von Laue's explanation for the spots had to be modified. He observed in his own experiments that the diffracted beam responsible for any spot was slightly converging—an observation that was easily explained, if the slightly diverging incident beam was being reflected by the crystal. This led him to picture the diffraction effect as a reflection of x-rays from the lattice planes of the crystal. He tested this by reflecting white x-rays from a mica plate set at a series of angles and obtained in every case a spot in the reflection position. This indicated that all wavelengths over a certain range were represented in the x-ray beam and that the reflection effect was equivalent to a selection from the continuous spectrum of a wavelength determined by the spacing of the lattice planes parallel to the plate surface of the mica crystal. From this Bragg derived a mathematical relation connecting the wavelength λ of the reflected x-radiation, the spacing d of the reflecting planes, and the angle of reflection θ. This relation

$$n\lambda = 2d \sin \theta$$

is now known as the Bragg law. Bragg was also able to explain why only a selection of diffraction spots appeared in von Laue's photographs of ZnS by ascribing to this compound an all-face-centered lattice instead of the primitive lattice von Laue had assumed. These results were communicated to the Cambridge Philosophical Society in November of 1912.

W. H. Bragg was much interested in the reflection concept and set up at Leeds the first x-ray spectrometer for accurately measuring the position and intensity of x-ray beams reflected from crystal planes. He was primarily interested in the nature of x-rays at this time. He satisfied himself that the reflected rays were really x-rays and confirmed that the spectrum of the incident beam was continuous. However, he found superimposed on this continuous range monochromatic components characteristic of the target. He constructed a number of x-ray tubes with different targets, with which he was able to show that the characteristic wavelengths were related to the atomic weights of the target metals. This was, in part, the groundwork for Moseley's discovery, published in 1914, of the mathematical relation between the characteristic wavelengths and the atomic number of the chemical element of which the target was composed.

W. L. Bragg had been able to deduce the atomic arrangements in NaCl and KCl from Laue photographs, but he had not been able to solve the structure of the diamond. W. H. Bragg found little difficulty with this structure when it was investigated with the x-ray spectrometer. Father and son then joined forces and collaborated in working out a number of structures, including ZnS, CaF_2, and FeS_2.

The Braggs used the NaCl structure to make still another important advance. X-ray diffraction photographs provided n and θ for the various reflections from a crystal, but could not evaluate λ without a knowledge of d. It was possible to establish the relative wavelength values of x-rays from different targets by reflection from a standard crystal (same value of d). Before absolute values of the wavelengths could be deduced, however, it was necessary to know the absolute values of the lattice constants of one compound. W. L. Bragg had shown that the Laue photographs of NaCl (and KCl) could be explained by two interpenetrating all-face-centered structures, one composed of Na atoms and the other of Cl atoms (Fig. 1-2)—in agreement with one of the structures postulated for this compound before the Laue experiments by Pope and

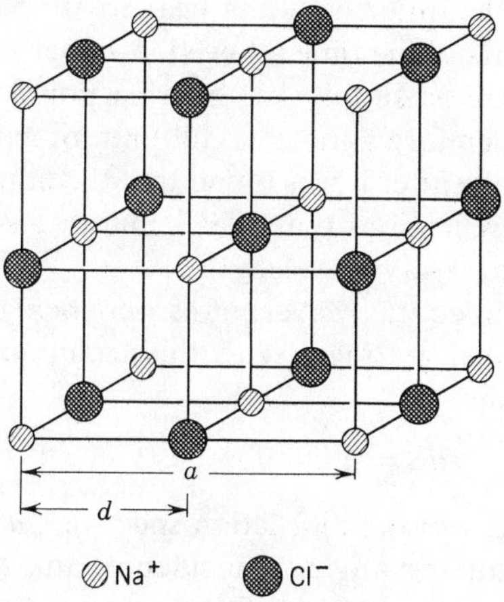

Fig. 1-2. The crystal structure of NaCl.

Barlow. The unit cell of this structure contained four molecules. By using an improved value for Avogadro's number (6.06×10^{23}), the volume of the cell was calculated to be

$$4 \times \frac{58.46}{6.06 \times 10^{23} \times 2.167} = 178.40 \times 10^{-24}\ \text{cm}^3 = 178.40\ \text{Å}^3$$

This corresponded to a cubic cell edge of 5.63 Å and, because of the all-face-centered nature of the structure, to a spacing of 2.815 Å for the $\{100\}$ planes. This first absolute spacing value provided an absolute value for the wavelengths of radiation reflected from the $\{100\}$ planes of NaCl. Alternatively, a known wavelength could be used to calculate the absolute lattice constants of an unknown crystal.

Wavelength Units

Soon after the first crystal structures had been determined, it became evident that x-ray measurements were more precise than measurements of constants, such as Avogadro's number, on which the absolute value of the $\{100\}$ spacing of NaCl was based. It seemed certain that the values of the constants would be progressively refined and would result in a changing series of values for the {100} spacing. To avoid the consequent confusion in the literature, Siegbahn, one of the foremost x-ray spectroscopists, proposed in 1925 to express x-ray wavelengths in terms of a new unit of measurement—the x unit, abbreviated xu. At this time the best value of the NaCl spacing was 2.814 Å. Siegbahn defined the x unit by assigning a value of 2814.00 xu to the fundamental NaCl spacing. This gave the unit a value as near to 10^{-11} cm, or 1/1000 Å, as the physical constants of the time allowed. A larger unit, equal to 1000 xu and abbreviated kxu or simply kx, became popular because its value was near 1 Å and therefore a more useful unit for expressing interatomic distances. Shortly thereafter it was found that the mineral calcite ($CaCO_3$) gave much sharper reflections than NaCl, and it came to be the standard crystal for measuring x-ray wavelengths.

Meanwhile, absolute x-ray wavelengths were being measured by means of diffraction by ruled gratings. With this technique, the wavelength is given by the relation

$$n\lambda = a[\cos\theta - \cos(\theta + \alpha)]$$

where n is of the order of the diffraction spectrum, a is the spacing of the grating, θ is the glancing angle of incidence, and $(\theta + \alpha)$ the glancing angle of diffraction. The values of a, θ, and α are capable of being measured with great accuracy, and today the ruled-grating method yields the best absolute wavelength values. As the technique developed to a high degree of precision, it became evident that absolute wavelengths determined by diffraction from crystals were lower by 0.2 per cent than the grating-determined values. This led to a thorough reevaluation of the constants involved in calculating the absolute value of the {100} spacing of NaCl—namely, the molecular weight, the density, and Avogadro's number. The investigation suggested that the only possible error of this magnitude lay in the determination of Avogadro's number. Eventually the error was traced to one of the constants—e, the charge on the electron—on which the number depended. A redetermination of e removed the discrepancy between crystal and ruled-grating values. However, the value of e, deduced by combining the two x-ray methods, was recognized as more accurate than the direct determination. Thus x-ray diffraction is responsible for finding errors in two fundamental physical constants, N and e, and for refining their values.

The new wavelength values came into general use in the early nineteen forties. Unfortunately, the literature for this period is somewhat confusing. Many crystallographers who used the Siegbahn kx values, wrongly stated their measurements in angstrom units. Consequently, the literature for the change-over period contains measurements both in angstrom units and in kx units but stated to be in angstrom units. Most of the data available before 1951 have been reviewed by Donnay and Nowacki.[1] It is advisable to consult the second edition[2] of this reference if any doubt exists. Values of wavelengths and lattice constants expressed in kx units are multiplied by the generally accepted factor 1.00202 to correct the 0.2 per cent discrepancy and to convert to angstrom units.

ELEMENTARY CRYSTALLOGRAPHY

The word *crystal* was used by the Greeks as early as 400 B.C. At first it referred specifically to transparent, crystallized quartz that was found in fine specimens in the high regions of the Alps. This mineral was believed to be ice—cooled and hardened to a permanent solid by the intense cold of the Alps. As a result of this belief, which persisted into the seventeenth century, the mineral became known as rock crystal from the Greek word *kyros*, meaning ice. There was much speculation about the origin of its hexagonal form, and by a natural association of ideas other minerals which showed faces, including those with metallic luster, were gradually spoken of as crystals. Eventually the word became a general term to describe any crystallized compound—that is, one bounded by faces—although "rock crystal" persists to this day as a varietal name for clear, crystallized quartz. Toward the end of the eighteenth century Rome de l'Isle used the term *crystallography* for the science concerned with the study of crystal forms of all substances, both transparent and opaque, that had formed by solidification in nature and in the laboratory.

The contemporary crystallographer is more concerned with the study of atomic arrangements, of which the crystal faces are one external expression. Because of this, the definition of the word crystal is changing and varies with the discipline. In physics and metallurgy, for example, "a region of matter within which the atoms are arranged in a three-dimensional translationally periodic pattern" (Buerger[3]) is a crystal, whether faces are present or not. The usage is not consistent in the earth sciences, even in the same field of concentration. Petrologists find the terms euhedral and anhedral useful in describing the fabric of rocks. Both words refer to grains that show optical continuity under the microscope. Euhedral implies a characteristic geometrical outline, anhedral the absence of crystal form. Some petrologists (Tyrrell[4]) speak

[1] J. D. H. Donnay and W. Nowacki, "Crystal Data," *Geol. Soc. Am. Mem.*, **60** (1954).

[2] *Crystal Data*, Second Edition, American Crystallographic Association, Monograph 5.

[3] M. J. Buerger, *Elementary Crystallography*, John Wiley and Sons, New York (1956).

[4] G. W. Tyrrel, *Principles of Petrology*, Methuen and Co. London (1929).

of both euhedral and anhedral grains as crystals, whereas others make the distinction of euhedral crystals and anhedral grains. The presence of faces is implicit in the definition of a crystal in descriptive mineralogy—"a crystal is a solid body bounded by plane natural surfaces which are the external expression of a regular internal arrangement of constituent atoms or ions" (Berry and Mason[5]). The word crystal is used in a curious way by many x-ray crystallographers. A specimen without faces but with continuous atomic arrangements qualifies as a crystal, if a crystallographic direction can be located and oriented in an x-ray beam—that is, if the specimen is amenable to the single crystal x-ray methods described in this book.

The orderly arrangement of atoms that constitute a crystal is known as the crystal structure. Some materials are obviously crystalline in composition but others, like the mineral limonite, are not. The ability to detect crystallinity improved enormously with the Laue experiments in 1912, which showed that x-rays are diffracted by crystals to give a characteristic x-ray pattern. Limonite was formerly classed as a mineral colloid, mainly because its earthy nature defied the usual microscopic tests for crystallinity. X-ray studies have shown that most specimens of this mineral produce a diffraction pattern characteristic of the mineral goethite ($HFeO_2$). These specimens of limonite might therefore be described as cryptocrystalline varieties of goethite. Some solids, such as glasses, appear to be obviously noncrystalline (amorphous), but they give weak and diffuse diffraction effects which indicate that their structures are at least partly ordered. Certain compounds give a range of patterns that vary from the diffraction effects of truly crystalline solids to the indistinct effects associated with amorphous materials. In these compounds, at least, there is no sharp distinction between the crystalline and amorphous states. It may be surmised that in the amorphous types there is a short-order range of crystallinity (individual crystals are only a few atoms in size), whereas in the highly ordered types the crystallinity is of a long-range order.

The Classification of Crystals

A crystal structure is like a three-dimensional wallpaper design in that it is an infinite repetition of some motif (a group of atoms). It is a periodic space pattern. Study has shown that there are 230 kinds of space pattern. Each crystal belongs to one of the 230 types, hence elementary crystallography is vitally concerned with the characteristics of

[5] L. G. Berry and B. Mason, *Mineralogy*, W. H. Freeman and Co., San Franscisco and London (1959).

the patterns. Because repetition is a fundamental property of the patterns, it is reasonable to base their classification on the repetition (symmetry) operations that yield them.

It is convenient, in developing the classification of crystals, to subdivide the elements of symmetry into three categories:

(1) translation (parallel periodic displacement);
(2) point group symmetry (rotation and rotation-inversion axes, reflection planes);
(3) space-group symmetry (screw axes, glide planes).

The Translation Lattices

The lattice is derived from the crystal structure by replacing each repeat motif with a point. The lattice is therefore a collection of equipoints that portrays the translational periodicity of the structure (hence the term translation lattice). This fundamental property of all crystalline structures limits the number of distinct two-dimensional lattices to the five shown in Fig. 1-3. These plane lattices can be systematically stacked

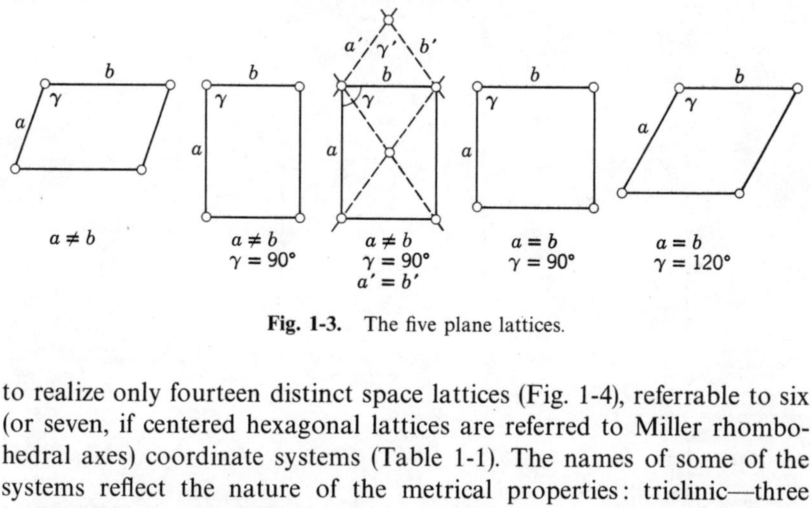

Fig. 1-3. The five plane lattices.

to realize only fourteen distinct space lattices (Fig. 1-4), referrable to six (or seven, if centered hexagonal lattices are referred to Miller rhombohedral axes) coordinate systems (Table 1-1). The names of some of the systems reflect the nature of the metrical properties: triclinic—three inclined axes; monoclinic—one inclined axis; orthorhombic—axes mutually perpendicular; isometric (cubic)—three mutually perpendicular equal axes. The remaining names, tetragonal and hexagonal, refer to the dominant symmetry of crystals belonging to these systems. A repeat unit of a lattice is known as the unit cell.

NOTATION OF LATTICE POINTS, ROWS, AND PLANES. Figure 1-5 illustrates the method of notating lattice points, rows, and planes on the

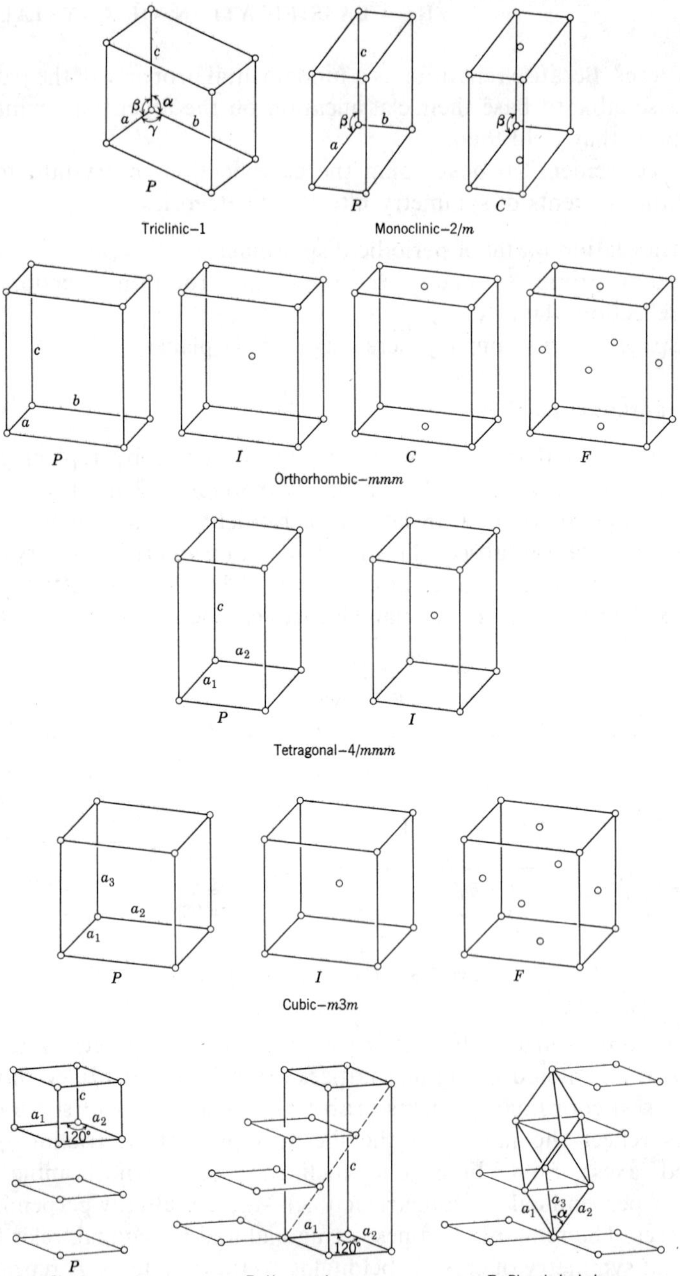

Fig. 1-4. The fourteen space lattices.

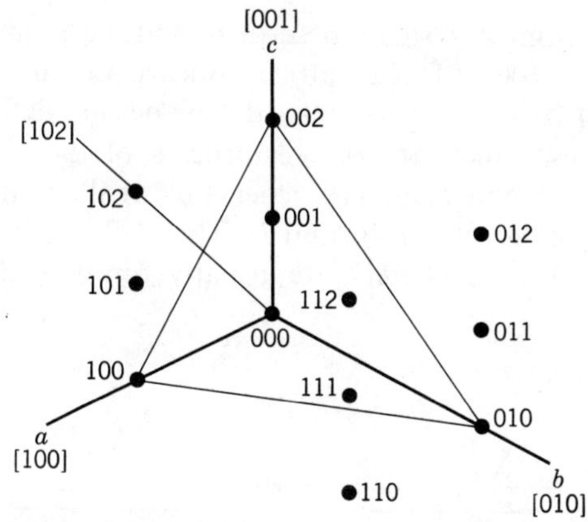

Fig. 1-5. Notation of lattice points, rows, and planes.

Table 1-1. The Fourteen Bravais Lattices

System	Type	Nature of Edges and Angles	Symmetry
Triclinic	P	$a \neq b \neq c$ $\alpha \neq \beta \neq \gamma$	$\bar{1}$
Monoclinic	P, C (b = twofold axis)	$a \neq b \neq c$ $\alpha = \gamma = 90° \neq \beta$	$2/m$
	P, B (c = twofold axis)	$a \neq b \neq c$ $\alpha = \beta = 90° \neq \gamma$	
Orthorhombic	P C (or A, B) I F	$a \neq b \neq c$ $\alpha = \beta = \gamma = 90°$	mmm
Tetragonal	P I	$a_1 = a_2 \neq c$ $\alpha = \beta = \gamma = 90°$	$4/mmm$
Hexagonal	R[a] P	$a_1 = a_2 \neq c$ $\alpha = \beta = 90°, \gamma = 120°$	$\bar{3}m$ $6/mmm$
Cubic	P I F	$a_1 = a_2 = a_3$ $\alpha = \beta = \gamma = 90°$	$m3m$

[a] Alternatively, a rhombohedral unit cell, characterized by edges and angles $a_1 = a_2 = a_3$; $\alpha < 120° \neq 90°$, can be chosen (see Fig. 1–3). This cell has one third the volume of the cell outlined by hexagonal axes.

basis of the coordinate systems described. Lattice points are specified without brackets—100, 101, 102; lattice rows are identified by brackets—[100] (the *a* axis), [010] (the *b* axis). Lattice planes are defined in terms of the Miller indices, which are the reciprocals of the intercepts of the planes on the coordinate axes. The general plane illustrated in Fig. 1-5*c*, for example, has intercepts 1*a*, 1*b*, and 2*c*. The Miller indices are therefore $\frac{1}{1}a, \frac{1}{1}b, \frac{1}{2}c = 2a\,2b\,1c$. The letters are usually omitted, and the indices

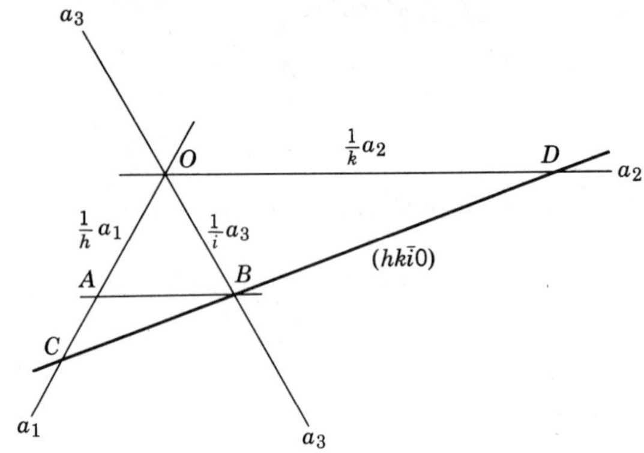

Fig. 1-6. Hexagonal indices.

are enclosed in parentheses; thus (221). If the calculations result in indices that have a common factor—for example, (422)—the factor is removed to give the simplest set of integers: (221). The symbol (221) therefore applies equally to all individuals of a stack of identical, parallel planes related by a simple translation operation. Braces { } are used to designate a family of planes related by the symmetry of the lattice. Thus, in the tetragonal system, {110} refers to the four planes (110), $(\bar{1}10)$, $(\bar{1}\bar{1}0)$, and $(1\bar{1}0)$; in the cubic system the same symbol stands for the twelve planes (110), $(\bar{1}10)$, $(\bar{1}\bar{1}0)$, $(1\bar{1}0)$, (101), $(10\bar{1})$, $(\bar{1}01)$, $(\bar{1}0\bar{1})$, (011), $(01\bar{1})$, $(0\bar{1}1)$, and $(0\bar{1}\bar{1})$.

The notation of hexagonal planes requires special attention. Hexagonal crystals are usually referred to the Bravais–Miller axes—a_1, a_2, a_3, and *c*. The three *a* axes are equivalent and, as a consequence, there should be a simple relation between the corresponding three indices. Figure 1-6 shows a plane with indices $(hk\bar{i}0)$ intersecting a set of hexagonal axes. The intercepts of the plane are the reciprocals of the indices, multiplied by the lattice period, *a*: $(1/h)a_1(1/k)a_2(1/i)a_3$. *AB* is drawn parallel to *OD* to construct two similar triangles *OCD* and *ACB*. *OAB* is an equilateral triangle. Then

$$\frac{OC}{OD} = \frac{AC}{AB} = \frac{OC - OA}{AB}$$

and

$$\frac{a_1/h}{a_2/k} = \frac{a_1/h - a_3/i}{a_3/i}$$

Because the axes are equivalent, this reduces to

$$\frac{1}{hi} = \frac{1}{hk} - \frac{1}{ki}$$

After multiplying by hki

$$k = i - h$$

The plane intersects the negative end of the a_3 axis and consequently, having regard for the sign of the i index,

$$h + k + i = 0$$

In other words, the sum of the indices is zero. For the sake of simplicity the i index is frequently replaced by a dot: $(hk \cdot l)$.

THE SPACING OF LATTICE PLANES. The perpendicular distance between the individuals of a stack of lattice planes is known as the *spacing* of the planes and is denoted by the letter d (Fig. 1-7). The value

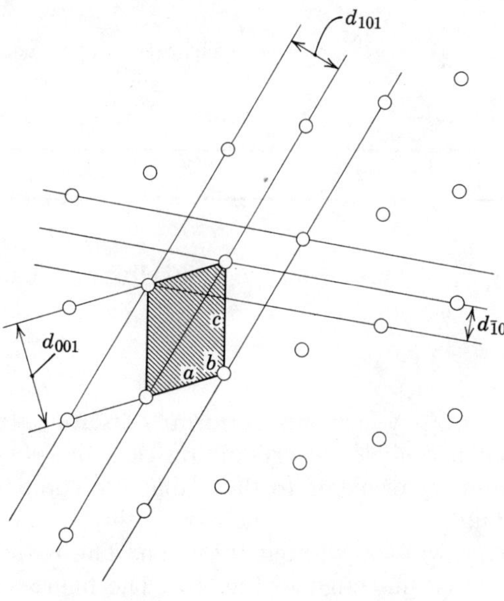

Fig. 1-7. Interplanar spacings.

Table 1-2. Values of the Interplanar Spacing (d_{hkl}) in the Six Crystal Systems

System	d_{hkl}
Cubic	$\left[\frac{1}{a^2}(h^2+k^2+l^2)\right]^{-1/2}$
Tetragonal	$\left[\frac{h^2+k^2}{a^2}+\frac{l^2}{c^2}\right]^{-1/2}$
Orthorhombic	$\left[\frac{h^2}{a^2}+\frac{k^2}{b^2}+\frac{l^2}{c^2}\right]^{-1/2}$
Hexagonal	$\left[\frac{4}{3a^2}(h^2+hk+k^2)+\frac{l^2}{c^2}\right]^{-1/2}$ hexagonal indexing $\left[\frac{1}{a^2}\frac{(h^2+k^2+l^2)\sin^2\alpha+2(hk+kl+lh)(\cos^2\alpha-\cos\alpha)}{1-2\cos^3\alpha+3\cos^2\alpha}\right]^{-1/2}$ rhombohedral indexing
Monoclinic	$\left[\frac{\frac{h^2}{a^2}+\frac{l^2}{c^2}-\frac{2hl\cos\beta}{ac}}{\sin^2\beta}+\frac{k^2}{b^2}\right]^{-1/2}$
Triclinic	$\left[\frac{\frac{h^2}{a^2}\sin^2\alpha+\frac{k^2}{b^2}\sin^2\beta+\frac{l^2}{c^2}\sin^2\gamma+\frac{2hk}{ab}(\cos\alpha\cos\beta-\cos\gamma)+\frac{2kl}{bc}(\cos\beta\cos\gamma-\cos\alpha)+\frac{2lh}{ca}(\cos\gamma\cos\alpha-\cos\beta)}{1-\cos^2\alpha-\cos^2\beta-\cos^2\gamma+2\cos\alpha\cos\beta\cos\gamma}\right]^{-1/2}$

of d depends on the dimensions of the unit cell in the manner shown in Table 1-2.

Point Groups

Point-group symmetry elements perform translation-free operations. The translational periodicity of crystals restricts the possible kinds of point-group symmetry elements to those that are compatible with the lattices. This limits us to axes of one-, two-, three-, four-, and sixfold rotation symmetry with or without inversion. The nature of the permissible operators is illustrated in Fig. 1-8. The figure shows that the onefold axis signifies asymmetry and $\bar{1}$ indicates centrosymmetry. The

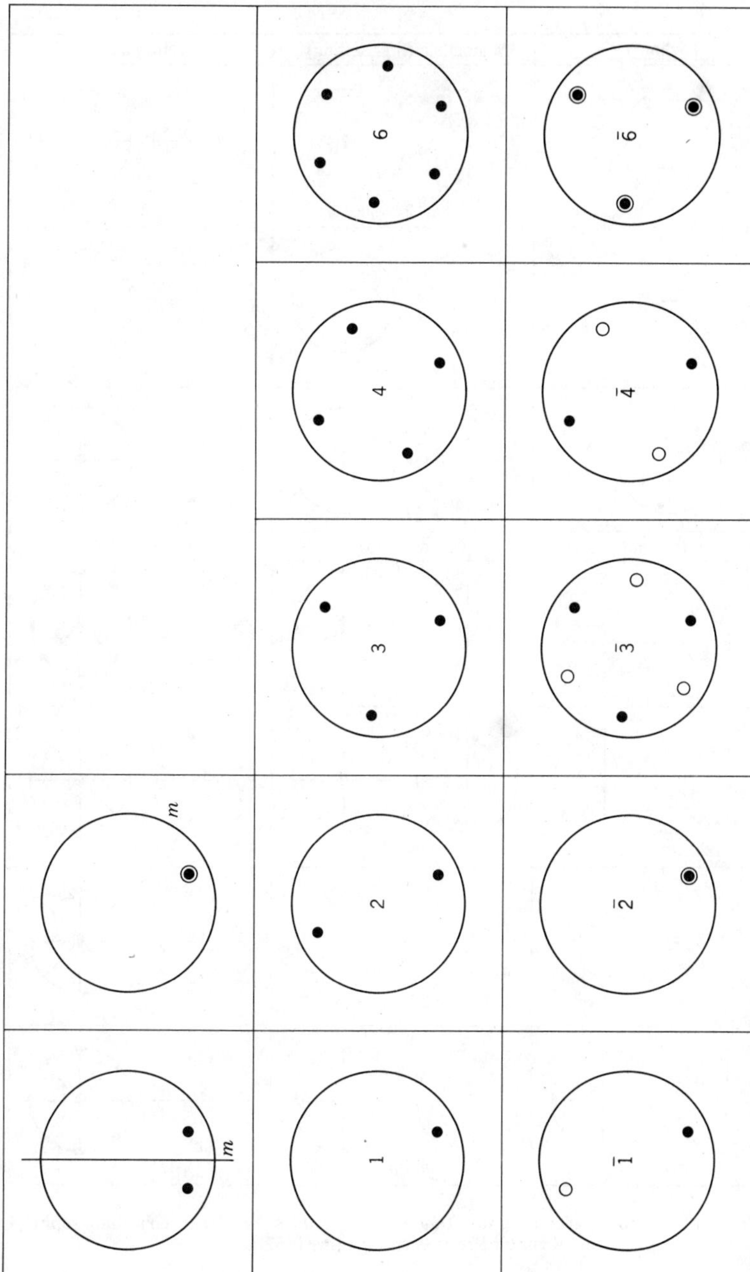

Fig. 1-8. The translation-free symmetry operations.

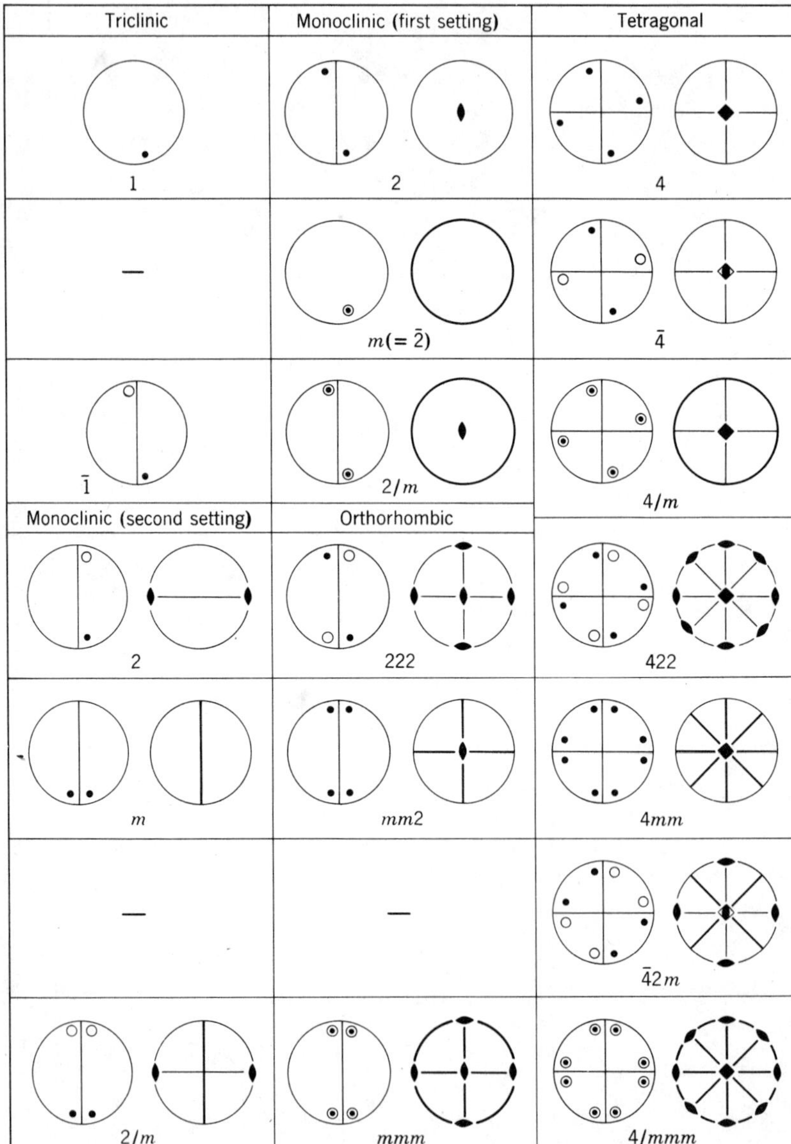

Fig. 1-9. The 32 point groups (after *International Tables for X-ray Crystallography*, **1**, Kynoch Press, Birmingham (1952)].

3 $\overline{3}$	3 $\overline{3}$	$\overline{3}$	Trigonal	A single 3 or $\overline{3}$ symmetry axis without a mirror plane normal to it; lattice *P* or *R*	$a_1 = a_2 = a_3$ $\alpha = \beta = \gamma < 120° \neq 90°$[c] or $a_1 = a_2 \neq c$ $\alpha = \beta = 90°;\ \gamma = 120°$[d]	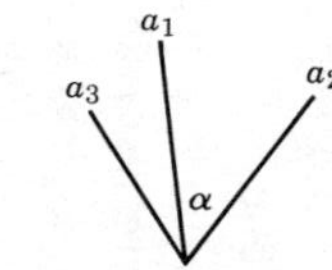
32 3*m* $\overline{3}$ 2/*m*	32 3*m* $\overline{3}m$	$\overline{3}m$				
6 $\overline{6}$ 6*m*	6 $\overline{6}$ 6*m*	6/*m*	Hexagonal	A single 6 or $\overline{6}$ (3/*m*) symmetry axis; lattice *P*	$a_1 = a_2 \neq c$ $\alpha = \beta = 90°;\ \gamma = 120°$[d]	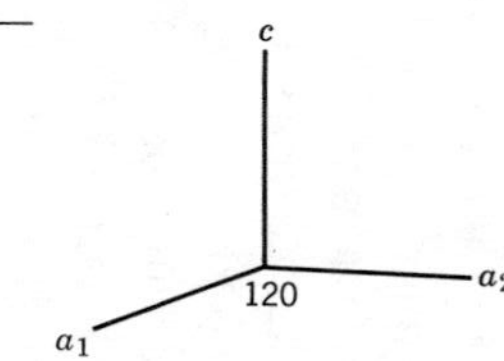
622 6*mm* $\overline{6}$m2 6/*m* 2/*m* 2/*m*	622 6*mm* $\overline{6}$*m*2 6/*mmm*	6/*mmm*				
23 2/*m* $\overline{3}$	23 *m*3	*m*3	Cubic	Four equivalent 3 or $\overline{3}$ symmetry axes	$a = a_2 = a_3$ $\alpha = \beta = \gamma = 90°$	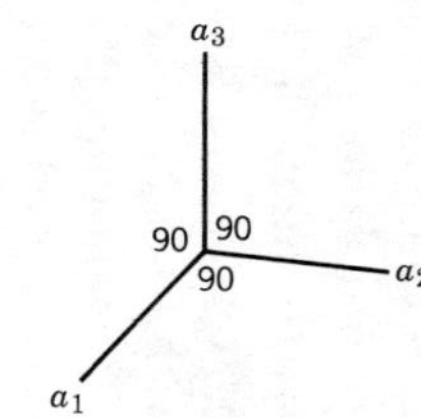
432 $\overline{4}3m$ 4/*m* $\overline{3}$ 2/*m*	432 $\overline{4}3m$ *m*3*m*	*m*3*m*				

The symbol $\neq$ implies nonequality by reason of the symmetry; accidental equality may occur.

[a] With the twofold axis taken as the *b* axis.

[b] With the twofold axis taken as the *c* axis.

[c] Miller or rhombohedral axes; threefold axis along [111].

[d] Bravais–Miller or hexagonal axes; it is customary to take three axes—a_1, a_2, and a_3—at 120° to each other.

Fourfold screw axes combine rotations of 90° counterclockwise (4_1) or clockwise (4_3) with translations of one quarter to give enantiomorphous arrangements, or a rotation of 90° with a translation of one half the repeat dimension (4_2). Sixfold screw axes combine 60° rotations in the opposite senses with translations of one sixth and one third, respectively, to give two sets of enantiomorphous arrangements—6_1, 6_5, and 6_2, 6_4. A fifth arrangement is produced by a rotation of 60° combined with a translation of one half (6_3). The various operations are illustrated in Fig. 1-10.

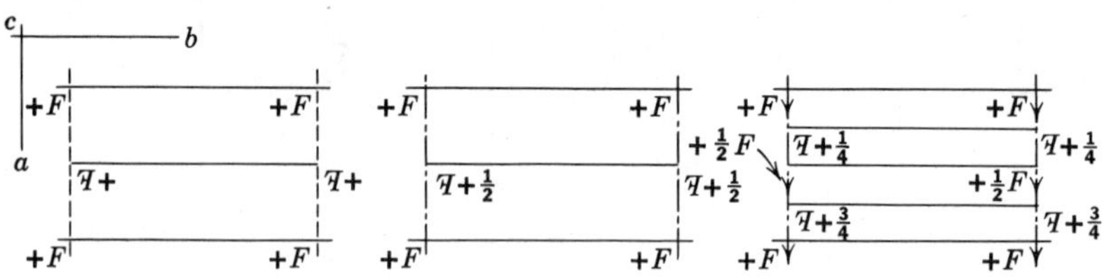

Fig. 1-11. Glide planes parallel to (010): (*a*) axial glide, translation $a/2$; (*b*) diagonal glide, translation $a/2 + c/2$; (*c*) diamond glide, translation $a/4 + c/4$.

The translation component of a glide plane is a simple fraction ($\frac{1}{2}$ or $\frac{1}{4}$) of a repeat period of the motif. The plane is described as an axial glide (*a*, *b*, or *c*) if the glide is parallel to the edge of the unit cell. If the glide is along a diagonal of the cell, the plane is described as a diagonal glide (*d*) if the component is one half the repeat period and as a diamond glide (*n*) if the component is one quarter. Because translation takes place parallel to the plane, the orientation of the plane limits the direction of the glide. For example, an axial glide situated parallel to (100) cannot have an $a/2$ glide component. The components of diagonal and diamond glides are even more restricted—the component of a diagonal glide parallel to (010) can be only $a/2 + c/2$. The various kinds of glide are illustrated in Fig. 1-11 and described in Table 1-4.

Point groups become infinite groups (space groups) if we add the element of translation—that is, if we apply the point groups to infinite lattices. The (lattice) translations multiply a symmetry element into an infinite number of parallel, identical elements. A reflection plane, for example, becomes a range of equally spaced reflection planes if a translation is applied normal to the plane.

If we take into account the possibility that the symmetry operations may include translation components, we increase the number of space groups to 230 (including eleven enantiomorphously related pairs), each

Table 1-4. Glide Planes

Glide Component	Kind of Glide and Designation		Orientation
$\frac{a}{2}$		a	(010), (001)
$\frac{b}{2}$	axial	b	(100), (001)
$\frac{c}{2}$		c	(100), (010), $(1\bar{1}0)$, $(11\bar{2}0)$, $(1\bar{1}00)$
$\frac{a}{2}+\frac{b}{2}$			(001)
$\frac{b}{2}+\frac{c}{2}$			(100)
$\frac{c}{2}+\frac{a}{2}$	diagonal	n	(010)
$\frac{a}{2}+\frac{b}{2}+\frac{c}{2}$			$(1\bar{1}0)$
$\frac{a}{4}+\frac{b}{4}$			(001), $(1\bar{1}0)$
$\frac{b}{4}+\frac{c}{4}$			(100)
$\frac{c}{4}+\frac{a}{4}$	diamond	d	(010)
$\frac{a}{4}+\frac{b}{4}+\frac{c}{4}$			$(1\bar{1}0)$

characterized by a particular framework of symmetry elements. An initial point in such a framework is multiplied into an infinite number of equivalent points.

It is interesting to note that the space groups were established late in the nineteenth century, independently and almost simultaneously, by a Russian crystallographer Federov, a German mathematician Schoenflies, and a British businessman-turned-scientist Barlow.

Appendix 1 lists the 230 space groups with symbols and space-group numbers as they appear in the *International Tables for X-ray Crystallography.*[6] The arrangement is according to crystal system and point

[6] *International Tables for X-ray Crystallography*, **1**, Kynoch Press, Birmingham (1952).

group. Space groups are designated by both the Schoenflies and the Hermann–Mauguin symbols. The Schoenflies symbols are based on the Schoenflies notation for the point groups. The symbols D_2, C_{2v}, and D_{2h} refer, respectively, to the orthorhombic point groups 222, *mm*2, and *mmm*. The space group symbol $D_2^{\,5}$ refers to the fifth space group (number 20 of Appendix 1) in a list of space groups that are isomorphous with the point group 222. A Schoenflies symbol immediately gives the point group to one familiar with the notation, but provides no information about the lattice type or the presence and orientation of space-group symmetry elements. The Hermann–Mauguin symbols, on the other hand, clearly set forth the lattice type and, to one familiar with the Hermann–Mauguin point group symbols, the point group and the presence and orientation of diagnostic space-group symmetry elements. Space group 20, already referred to, is designated as $C222_1$ in the Hermann–Mauguin notation. The symbol indicates that the lattice is *C*- (or base-) centered, the point group is 222, and the twofold axis parallel to [001] is a screw axis. The Hermann–Mauguin symbols are widely used today; the Schoenflies notation is retained mainly as an aid to consulting older descriptions in the literature.

The space-group symbols used in Appendix 1 are based on certain arbitrary rules set up by the International Union of Crystallography to achieve a uniformity in the presentation of space-group data. The symbols are described as the standard symbols and refer to a setting for the unit cell which has been adopted as the standard. The unit cell is chosen in accordance with the symmetry, but additional rules are required to fix the setting in the orthorhombic system because the cell can be oriented in six different ways (Fig. 1-12) which result in six permutations for some of the space-group symbols. Because many crystallographers are guided in their selection of a setting for an orthorhombic crystal by rules that label the unit-cell edges on the basis of their relative lengths or on the habit of the crystal, the literature includes "nonstandard" descriptions of crystal structures. Appendix 2, which lists the standard space-group symbols and their permutations, may be useful for interpreting the settings.

EQUIVALENT POSITIONS. A point placed in a space group can be regarded as multiplied by the symmetry elements to give a set of points which conforms to the symmetry of the space group. Consequently, the points occupy *equivalent positions*. The number of points in a set of equivalent positions depends on the space group and the coordinates of the representative point. If the point has coordinates *x*, *y*, *z*, the equivalent positions are said to be *general positions*, and, if the coordinates are

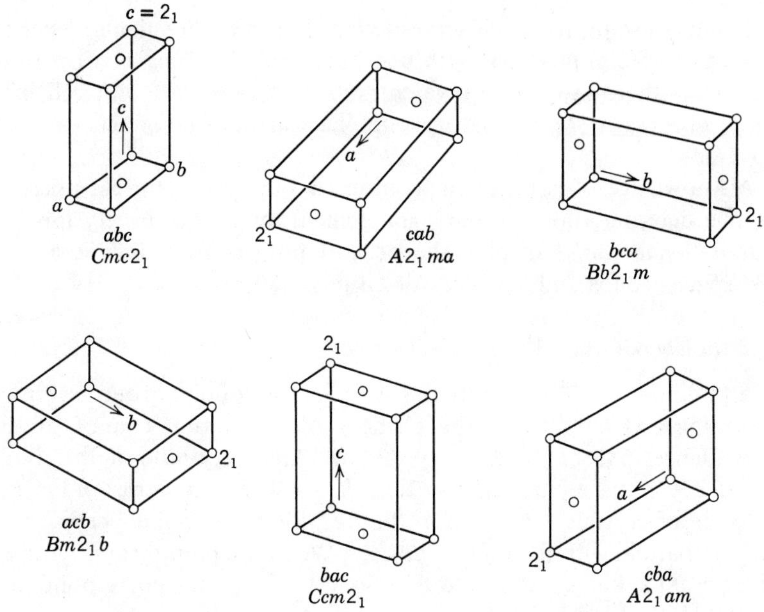

Fig. 1-12. The six orientations of the space group $Cmc2_1$.

specialized, the points are situated at *special positions.* Figure 1-13 shows points in the general positions and in the special $x, 0, z$ positions in the space group *Cm.* Clearly there are four equivalent general positions per unit cell. Two of the special positions occur in planes that bound the

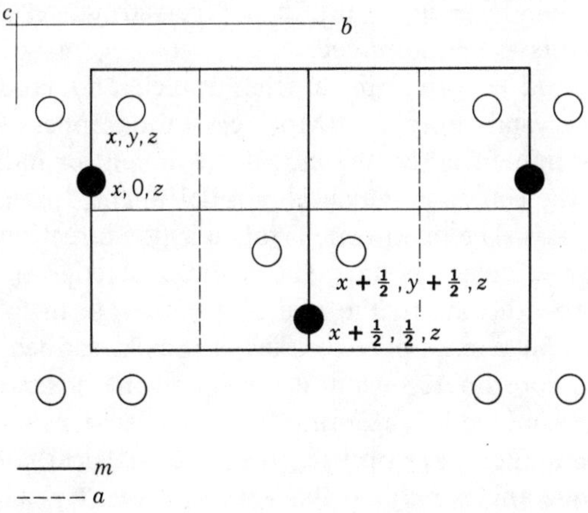

Fig. 1-13. Equivalent special and general positions in the space group *Cm.*

cell, and consequently each serves two cells and has a value of $\frac{1}{2}$ per cell. The third special position, with coordinates $x + \frac{1}{2}, \frac{1}{2}, z$, serves only one cell. Thus there are two equivalent $x, 0, z$ positions per unit cell in the space group *Cm*. Most space groups have a number of sets of special positions.

A knowledge of equivalent positions is obviously of great assistance in the determination of the arrangement of atoms in crystals. The *International Tables* display the general positions on diagrams of the 230 space groups and list data for all positions.

Crystal Classes

A finite body can have only the symmetry of a point group; symmetry operations which include the element of translation cannot produce coincidence. Morphological crystals are finite crystallographic bodies with faces that are parallel to lattice planes. The symmetries of idealized versions of crystals are therefore limited to the 32 point groups which are compatible with the infinite lattices. When the point-group symmetry of a crystal is known, it is said to belong to one of the corresponding 32 crystal classes.

Choice of the Triclinic Cell, its Orientation, and Axial Labels

The choice of the conventional unit cell is fixed by the symmetry in all crystal systems, except the triclinic. A cell can be selected from a triclinic space lattice in an infinite number of ways (Fig. 1-14), but one, in which the three shortest noncoplanar translations are chosen to define the edges of the cell, is unique and has received universal acceptance. Buerger[3] calls this the *reduced* cell.

No one method of orienting a triclinic cell and labeling its edges appeals to all crystallographers. Most crystallographers in the field of mineralogy are influenced by the face development or habit of a crystal and designate the cell edge which is parallel to the "main zone" of the crystal as the c axis. The main zone is taken as the direction of elongation in prismatic or acicular crystals, the prominent edge in blade-shaped crystals, or perpendicular to the plane of flattening in tabular crystals. The shorter of the remaining two axes is then described as the a axis. A right-handed coordinate system is used, and the positive directions of the axes are determined by orienting the cell so that α (between b and c) and β (between a and c) are obtuse. In the geometrical sense all triclinic cells with oblique angles may be classified into two types on the basis of whether two of the eight solid angles have three obtuse or three acute plane (interedge) angles. The two types of cell are, respectively, obtuse and acute parallelepipeds. The remaining six solid angles consist of two acute and one obtuse plane angles in the first case and two obtuse and one acute in the second case. Thus, if α and β are taken to be obtuse, γ is obtuse in the first cell and acute in the second. The axis of crystal elongation frequently coincides with the shortest lattice period, whereas the normal to flattening in tabular crystals is usually parallel to the longest lattice period. Consequently, the unit cells of many triclinic crystals chosen by the conventions described conform to one or other of the rules: $c < a < b$ and $a < b < c$.

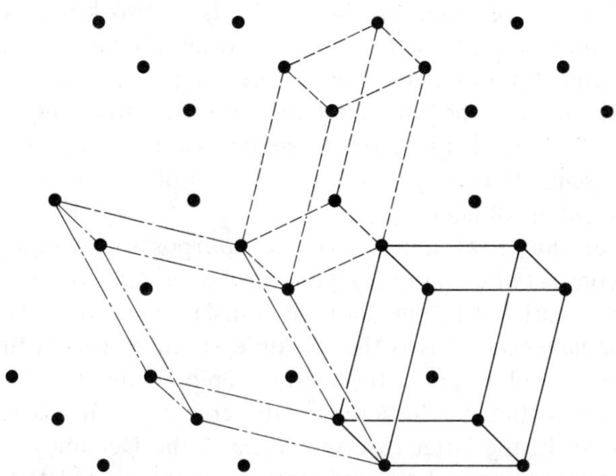

Fig. 1-14. Primitive triclinic cells (adapted from W. L. Bond, *Univ. Toronto Studies, Geol. Ser.*, No. 50, 9 (1947).

Rules based on a consideration of the morphology of crystals cannot always lead to a unique description of the unit cell, mainly because the choice of the main zone is often subject to interpretation. Many crystals have more than one prominent zone. Furthermore, the rules are based on the assumption that a crystal has a characteristic habit. If we grant that the assumption is valid, the morphological characteristics may become evident only after statistical studies have been undertaken of the external development of crystals grown under a variety of laboratory or geological conditions. The original descriptions of crystals are often based on a single laboratory synthesis or occurrence in nature, in which the compound may exhibit an unusual face development.

Rules of orientation based on morphology are of interest chiefly to mineralogists. Workers in other fields are generally less concerned about relations between the external form development and the crystal structure. It is not surprising therefore that rules have been formulated on geometrical grounds. The most generally accepted proposal of this kind was advanced by Donnay,[7] who suggested abandoning the rule to define the c edge of the unit cell as the axis of the main zone in favor of designating the edges of the reduced cell in accordance with the scheme $c < a < b$ and taking α and β to be obtuse. Because this condition often results from morphological considerations, Donnay's proposal in effect forges a common convention into a rule.

Buerger[3] suggested a useful geometrical method for orienting the reduced cell. The method is based on the consideration that oblique parallelepipeds may be classified as obtuse or acute. As already stated, the usual practice of orienting the cell leads to two kinds of cell. The first corresponds to a cell with α, β, and γ all obtuse and the second to a cell with α and β obtuse and γ acute. Buerger pointed out that the two representations treat the interaxial angles differently—the first is all-obtuse and the second mixed obtuse-acute. An equal-treatment scheme would represent the first type of crystal as all-obtuse, as is the custom, and the second as all-acute.

A practice that has gained favor for purposes of cataloguing cell constants consists of defining crystallographic descriptions in terms of a primitive cell outlined by the shortest translations that *enable all three angles to be nonacute.* This is the Delauney reduced cell. If the conventional reduced cell is an obtuse parallelepiped, the Delauney cell is identical with it, but it is different if the conventional cell is an acute parallelepiped. In the latter case the edges of the Delauney cell are not the shortest three noncoplanar translations in the space lattice.

It is likely that crystallographers will eventually follow rules proposed by an international commission on crystallographic procedures. Meanwhile, it is reasonable to follow the generally accepted practice of referring a triclinic compound to the primitive cell whose edges are the three shortest noncoplanar translations in the lattice. The cell is correctly chosen if the edges of each face are shorter than the corresponding face diagonals. If referencing is restricted to a right-handed coordinate system, the cell can be oriented in 24 ways. However, its dimensions are constant and its interedge angles can differ only by becoming the supplements of those in another orientation. Hence the cell data are readily recognized and can be successfully catalogued.

[7] J. D. H. Donnay, *Am. Mineralogist*, **28**, 313 (1943).

CHAPTER TWO

THE NATURE AND GENERATION OF X-RAYS

X-rays, along with visible light, form a part of the electromagnetic spectrum. The various types of radiation that are grouped under this heading are similar in nature and may be roughly distinguished or classified on the basis of wavelength (Table 2–1). The members of the spectrum have a dual nature. Under some conditions they act like waves—the study of the diffraction of x-rays is based on their wavelike nature. Under other conditions they behave like particles: their energy, for example, is radiated not continuously, as we would expect from a pure wave motion, but rather as bundles of waves, called quanta or photons. The energy of a quantum is proportional to the wavelength of the radiation in accordance with the Einstein equation

$$E = \frac{hc}{\lambda}$$

where E is the energy of each quantum, c is the speed of light, λ is the wavelength of the radiation, and h is a proportionality constant known as Planck's constant.

X-rays are generated when electrons, traveling at high speeds, collide with the atoms of an obstacle. Broadly speaking, two types of interaction occur, and these result in two types of x-ray spectra—continuous and characteristic.

An x-ray tube is a device for promoting the interaction between high-speed electrons and matter. Its essential features are a source of electrons and two electrodes. A high voltage, applied across the tube, accelerates the electrons toward the anode, which is aptly called the target. X-rays are generated at the point of impact with this obstacle and radiate in all directions.

Table 2-1. Members of the Electromagnetic Spectrum

	Wavelength[a]
Gamma rays	0.0056–0.1 Å
X-rays	0.1–45 Å
Ultraviolet rays	45–3500 Å
Visible light	3500–7000 Å
Infrared rays	7000–2,000,000 Å
Etc.	

[a] After K. Krauskopf, *Fundamentals of Physical Science*, Third Edition, McGraw-Hill Co., New York (1953).

The Continuous Spectrum (White Radiation)

An electron is abruptly slowed and suffers a loss of energy as it enters the electric field surrounding the nucleus of an atom of the target. This results in the emission of a quantum of x-radiation whose wavelength is given by the expression

$$\Delta E = \frac{hc}{\lambda}$$

where ΔE is the loss in energy. An electron, accelerated by a difference in potential of V volts, has the energy eV (where e is the charge on the electron). If the electron is instantaneously stopped, the quantum energy is equal to the energy of the electron and the radiation has the wavelength

$$\lambda = \frac{hc}{eV} = \frac{(6.624 \times 10^{-27}\text{ erg-sec}) \times (2.998 \times 10^{10}\text{ cm/sec})}{(1.602 \times 10^{-20}\text{ emu}) \times (V\text{ volts})} = \frac{12{,}400}{V}\text{ Å} \qquad (1)$$

Normally an electron's energy is lost in a stepwise fashion by a series of encounters, giving rise to a number of quanta whose wavelengths are longer than if the energy were lost instantaneously. As a result, the spectrum from a stream of electrons exhibits a continuous range of wavelengths to a short wavelength limit, which represents the maximum electron energy transformed to x-radiation in a single encounter. The plot of intensity against wavelength (Fig. 2-1) is a smooth curve which begins abruptly at a short wavelength limit (maximum quantum energy), rises to a maximum (average quantum energy), and extends toward infinity (minimum quantum energy). An increase in the voltage across the x-ray tube increases the speed of the bombarding electrons; the electrons arrive at the target with greater energy, hence the average energy of the quanta is greater. As a consequence, the total intensity of the x-radiation is increased, and both the short wavelength limit and the intensity maximum of the spectrum shift to shorter wavelengths (higher energy). The wavelength distribution of the spectrum is independent

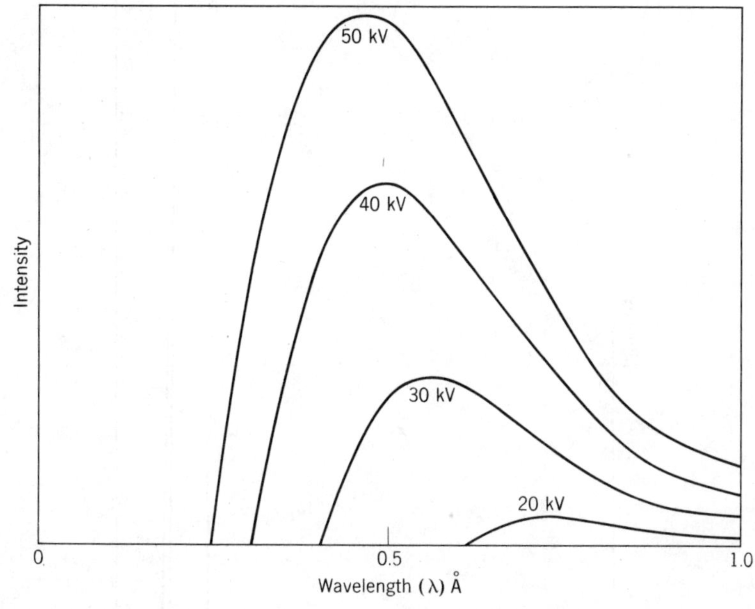

Fig. 2-1. Nature of the variation of the continuous spectrum with voltage (after C. T. Ulrey).

of the target material, but the total intensity of the radiation is proportional to the atomic number of the material. The target of an x-ray tube which is meant to function as an emitter of the continuous spectrum is, therefore, always manufactured from a heavy element.

Characteristic or Line Spectra

If the voltage across an x-ray tube is raised to a critical level, which depends on the element of the target, sharp peaks are superimposed on the smooth curve that represents the continuous spectrum (Fig. 2-2). The peaks occur in groups known as series and are designated K, L, M and N in order of increasing wavelength. In general, each series has the same appearance from one element to the next, but the given peak shifts to a shorter wavelength in passing from one element to the next heavier one. The peaks for a given target element always occur at the same wavelengths and are, therefore, *characteristic* spectra.

Characteristic spectra are produced when the bombarding electrons have sufficient energy to penetrate to the interior of the atoms of the

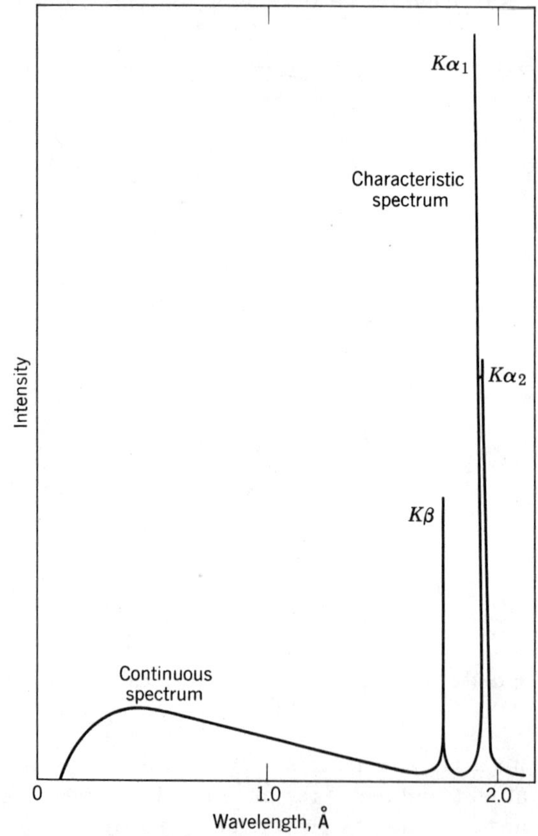

Fig. 2-2. The *K* characteristic peaks of iron superimposed on the continuous spectrum curve.

target and to dislodge electrons from their innermost shells. An atom may be pictured as a nucleus surrounded by shells of electrons. The inner or *K* shell can have two electrons, the *L* shell eight electrons, the *M* shell eighteen, and so on. The *K* shell electrons are the most difficult to dislodge. An expelled electron must be replaced, and the probability is that the vacancy will be filled from the next outer shell rather than from a more remote shell. This creates a new vacancy, which is filled from the next shell, or less probably from a more distant shell, and thus the atom returns to a normal state in a series of steps.

An electron moves to a lower energy level when it "falls in" to an inner shell. The transition is accompanied by the emission of energy in the form of x-radiation, the wavelength of which is precisely fixed by the difference in energy between the shells. Falling-in from the various levels

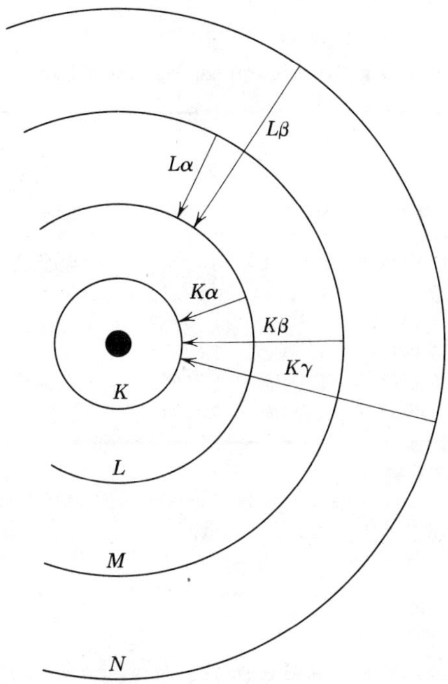

Fig. 2-3. Schematic drawing to illustrate the production of K and L characteristic peaks.

to replace K electrons generates the K series of peaks (Fig. 2-3). Falling-in of L electrons gives rise to the $K\alpha$ peaks. L electrons are distributed over three sublevels—L_i with two electrons, L_{ii} with two electrons, and L_{iii} with four electrons. Transition does not occur from the L_i sublevel. The L_{ii} and L_{iii} sublevels represent slightly different energy levels, and as a result the $K\alpha$ peak is actually a doublet whose individual peaks are designated, respectively, as α_2 and α_1. Similarly, the $K\beta$ peak is a closely spaced doublet, which arises from the transfer of electrons from two of the five M sublevels. A third doublet, the γ peak, is formed by electrons falling in from two of the five N sublevels. The total of such transfers gives rise to the K characteristic spectrum. The L spectrum is produced by electrons falling in from more distant shells to replace electrons expelled from the L shell. About 30 L peaks have been identified.

If the electron stream from the cathode of an x-ray tube has the energy to expel K electrons, obviously the more lightly held electrons in the other shells will also be expelled. Thus, if the K series of peaks is generated, as many additional series of peaks are generated as the electron configuration of the atoms in the target allows.

Table 2-2. Wavelengths, Absorption Edges, and Excitation Potentials

	$K\alpha_2$	$K\alpha_1$	$K\alpha$[a]	$K\beta$	$K_{\text{abs edge}}$	Excitation Potential $\frac{12.4}{K_{\text{abs edge}}}$
Ag	0.56378 Å	0.55936 Å	0.5608 Å	0.49701 Å	0.4858 Å	25.5 kV
Mo	0.71354	0.70926	0·7107	0.63225	0.6197	20.0
Cu	1.54433	1.54051	1.5418	1.39217	1.380	8.98
Co	1.79278	1.78892	1.7902	1.62075	1.608	7.71
Fe	1.93991	1.93597	1.9373	1.75653	1.743	7.10
Cr	2.29351	2.28962	2.2909	2.08480	2.070	5.98

[a] Diffraction effects due to $K\alpha_1$ and $K\alpha_2$ are frequently not resolved on x-ray photographs. In this case an average wavelength, weighted in accordance with the ratio of the intensities of the peaks, is used:

$$\frac{2\alpha_1 + \alpha_2}{3}$$

The relative intensities of the peaks in a series are governed by the relative probabilities of falling in from the various shells. This depends partly on the number of electrons in each sublevel. Thus the chance of an electron falling in from the L_{iii} sublevel with four electrons should be twice that from the L_{ii} sublevel with two electrons. This is confirmed by an observed $K\alpha_1 : K\alpha_2$ intensity ratio of about 2:1 for all the elements. The intensity of the $K\beta$ component is not so readily predicted; the ratio, $\alpha_1 : \alpha_2 : \beta$, is observed to be about 10:5:2 for the radiation from a copper target. The $K\gamma$ component of the spectrum is much weaker and does not produce detectable diffraction effects.

Only the K spectrum, and in particular the intense $K\alpha$ and $K\beta$ peaks, is of significance in x-ray diffraction. The useful and practical wavelength range extends from about 0.7 to 2.3 Å. This radiation (Table 2-2) is provided by the K spectra of elements from atomic number 42 (molybdenum) down to 24 (chromium). Longer-wavelength radiation is too easily absorbed by the x-ray tube windows and by the air, and shorter-wavelength radiation—of wavelength values below 0.5 Å—is accompanied by a serious measure of white radiation. L peaks which fall into the useful range are, by comparison, much weaker.

The minimum voltage that must be applied to an x-ray tube to give the electron stream the energy to excite the K spectrum is given by the energy equation (1), in which λ corresponds to the wavelength of the K absorption edge of the target element

$$V_{\text{min}} = \frac{12{,}400}{\lambda_{K_{\text{abs edge}}}} \text{ (volts)}$$

This is known as the *excitation potential.* Table 2-2 lists the values of the K absorption edges and the excitation potentials for six useful K spectra. In contrast to the continuous spectrum, an adjustment of the voltage across the tube has no effect on the wavelengths of the characteristic peaks. The adjustment does affect the intensities, provided that the supply of bombarding electrons is not limited. An increase in the voltage causes an increase in the rate at which target electrons are expelled from their shells and therefore an increase in peak intensity. The continuous spectrum is simultaneously intensified. An operating voltage of several times the excitation potential provides a good ratio of characteristic-to-continuous radiation.

Fluorescence

The K characteristic spectrum of an element can also be excited by a beam of x-rays incident on the element, provided that the x-rays have the energy to eject K electrons from its atoms. The minimum energy (eV) depends on the element and corresponds to the wavelength of the absorption edge of the element

$$eV_{\text{min}} = \frac{hc}{\lambda_{K_{\text{abs edge}}}}$$

The absorption edge marks an abrupt change in the absorption coefficient of the element (Fig. 2-4). At the critical value, most of the energy of the incident radiation is utilized in ejecting K electrons, and consequently the x-rays are heavily absorbed. X-rays of shorter wavelengths (higher energy) have the energy to expel electrons from the K shell, but the probability of collision decreases rapidly with decreasing wavelength and the absorption effect falls off sharply. X-rays of longer wavelengths lack the energy to excite the K spectrum and are readily transmitted by the element. The absorption edge marks the boundary between transmission of incident radiation on the long-wavelength side and absorption of radiation with consequent generation of secondary radiation on the short-wavelength side. The secondary radiation is characteristic of the absorber, and therefore its wavelength is always longer than that of the exciter. When an element is generating x-radiation in this manner, it is said to be fluorescing; the secondary radiation is usually called fluorescent radiation.

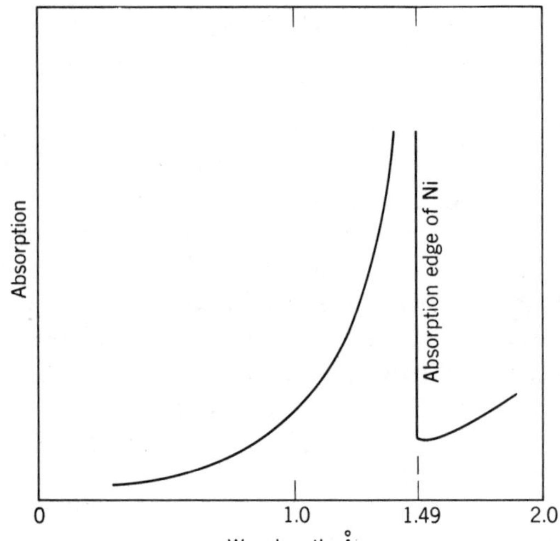

Fig. 2-4. Absorption edge of nickel.

The importance of fluorescence in x-ray diffraction studies can be illustrated by the effect of the $K\alpha$ component of an x-ray beam from a copper target on the element cobalt. Table 2-2 shows that the wavelength of the $K\alpha$ radiation ($\lambda = 1.5418$ Å) lies close to and on the short-wavelength side of the absorption edge of cobalt ($K_{\text{abs edge}} = 1.61$ Å). As a result, the element fluoresces strongly in a Cu$K\alpha$ beam. If appreciable amounts of cobalt are present in a sample that is being photographed with this radiation, two undesirable effects will be observed. First, the diffraction pattern will be weakened because of the absorption of the incident radiation, and second, the fluorescent radiation generated in the sample will scatter in all directions in the camera, thereby increasing the background and obscuring the diffraction pattern. The elements iron and manganese (K absorption edges 1.74 and 1.90, respectively) behave in a similar manner with copper radiation.

Table 2-3. Fluorescent X-Rays

Element	Atomic Number	$K_{\text{abs edge}}$	Target	$\lambda_{K\alpha}$	Fluoresce Strongly
Ca	20	3.07 Å			
Sc	21	2.76			
Ti	22	2.50			
			Cr	2.2909 Å	Ti, Sc, Ca
V	23	2.27			
Cr	24	2.07			
			Fe	1.9373	Cr, V, Ti
Mn	25	1.90			
			Co	1.7902	Mn, Cr, V
Fe	26	1.74			
Co	27	1.61			
			Cu	1.5418	Co, Fe, Mn
Ni	28	1.49			
Cu	29	1.38			
Zn	30	1.28			
Ga	31	1.20			
Ge	32	1.12			
As	33	1.04			
Se	34	0.980			
Br	35	0.920			
Kr	36	0.865			
Rb	37	0.815			
Sr	38	0.770			
Y	39	0.728			
			Mo	0.7107	Y, Sr, Rb
Zr	40	0.689			
Nb	41	0.654			
Mo	42	0.621			
Tc	43	0.589			
			Ag	0.5608	Tc, Mo, Nb
Ru	44	0.560			

Table 2-3 lists the elements which, as experience has shown, fluoresce strongly in the $K\alpha$ radiation from common target materials. The table indicates that fluorescence occurs if the atomic number of the sample element is lower by at least two than that of a light target element (Cr to Cu) and lower by at least three when the target is made of molybdenum. Although the absorption effect falls off rapidly on the short-wavelength side of the absorption edge (Fig. 2-4), in general three elements fluoresce strongly with each target element. Thus titanium, vanadium, and chromium (atomic numbers 22 to 24) fluoresce strongly with iron radiation (atomic number 26). The radiation for a particular specimen must therefore be selected with some care. Table 2-3 shows that an element is transparent to x-rays that are excited by a target manufactured from this element or a lighter one. If it is necessary to use a heavier target element, fluorescence will not be serious if a large interval separates the *K* absorption edge of the element and the wavelength of the incident radiation. Titanium compounds, for example, give sharp patterns with little background when photographed with copper radiation.

The fluorescence phenomenon is turned to a useful purpose in x-ray spectroscopy. In this method of elemental analysis the sample is bathed in a beam of x-radiation. The *K* and/or *L* spectra of the elements are thereby excited, and the sample fluoresces. It remains only to determine the wavelengths of the characteristic peaks in the radiation emitted by

Table 2-4. β Filters

Element	Atomic Number	*K* Absorption Edge
Ti	22	2.50 Å
V	23	2.27
Cr	24	2.07
Mn	25	1.90
Fe	26	1.74
Co	27	1.61
Ni	28	1.49
Cu	29	1.38
Zn	30	1.28
Ga	31	1.20
Ge	32	1.12
As	33	1.04
Se	34	0.980
Br	35	0.920
Kr	36	0.865
Rb	37	0.815
Sr	38	0.770
Y	39	0.728
Zr	40	0.689
Nb	41	0.654
Mo	42	0.621
Tc	43	0.589
Ru	44	0.560
Rh	45	0.533
Pd	46	0.509
Ag	47	0.486

			Filter Data		
Target	$K\alpha$	$K\beta$	Filter	Thickness[a]	Loss of $K\alpha_1$
Cr	2.2909 Å	2.0848 Å	V (V_2O_5)	0.036 mm	48 %
Fe	1.9373	1.7565	Mn (MnO_2)	0.026	45
Co	1.7902	1.6208	Fe	0.012	39
Cu	1.5418	1.3922	Ni	0.015	45
Mo	0.7107	0.6323	Zr	0.081	57
Ag	0.5608	0.49701	Rh	0.062	59
			Pd	0.062	60

[a] To achieve a $K\alpha_1/K\beta$ ratio of 100/1. From *International Tables for X-ray Crystallography*, **3**, Kynoch Press, Birmingham (1962).

the sample and their intensities, to identify the elements, and to estimate their abundance.

β *Filters*

A monochromatic beam of x-rays is desirable for certain diffraction studies—especially when the sample is a powder. The reason for this is that the $K\beta$ component of the beam is sufficiently intense to produce its own diffraction pattern. When this is superimposed on the $K\alpha$ pattern, the difficulty of interpreting the diffraction effects is materially increased.

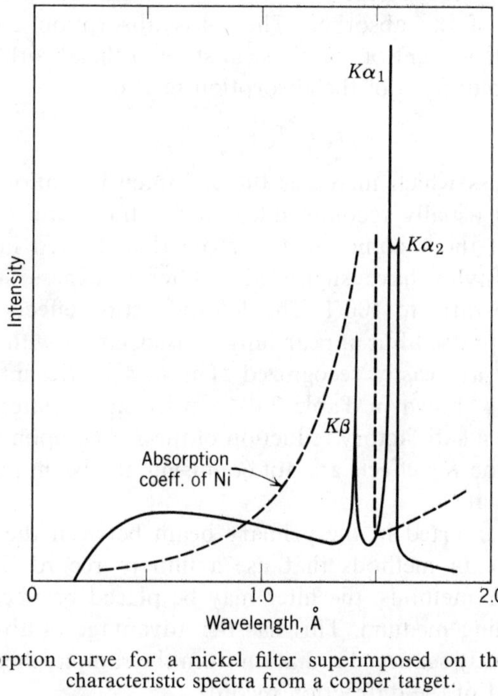

Fig. 2-5. Absorption curve for a nickel filter superimposed on the continuous and characteristic spectra from a copper target.

The absorption effect described in a preceding section and illustrated in Fig. 2-4 offers the means of removing the unwanted radiation to leave a beam that is essentially monochromatic. The element nickel, for example, has an absorption edge of 1.49 Å, intermediate between the wavelengths of $Cu_{K\alpha}$ (1.5418 Å) and $Cu_{K\beta}$ (1.3922 Å) radiation (Fig. 2-5). The element is therefore relatively transparent to the α radiation, but it strongly absorbs the β component of an x-ray beam from a copper target. Thus it can act as a β filter and, as indicated by Fig. 2-5, it can also reduce the intensity of the continuous spectrum. An examination of Table 2-4 shows that an element with the proper absorption edge is available for each of the commonly used target materials. The β filter naturally also reduces the intensity of the α component of the beam. X-radiation is always partly absorbed when it passes through matter—experiment shows that the intensity of the transmitted beam I_x, after passing through a thickness x, of an absorber is

$$I_x = I_0 e^{-\mu x}$$

in which I_0 is the intensity of the incident beam and μ is the linear absorption coefficient of the absorber. This coefficient is dependent on the density ρ of the absorber. The mass absorption coefficient μ/ρ, however, is independent of the physical state of the absorber, and therefore a more useful form of the absorption relation[1] is

$$I_x = I_0 e^{-(\mu/\rho)\rho x}$$

A filter thickness which increases the α/β intensity ratio to a value of about 600/1 is usually recommended in the literature. Unfortunately, the intensity of the α component is more than halved in the process. Parrish and Taylor have suggested a filter thickness sufficient only to increase the ratio to 100/1. The $K\beta$ diffraction effects are then not completely suppressed but appear only in association with the strongest $K\alpha$ effects and are easily recognized. The loss of $K\alpha$ intensity is still considerable, as shown in Table 2-4—a price in $K\alpha$ intensity must be paid to achieve a satisfactory reduction of the $K\beta$ component. If the loss is serious and the $K\beta$ effects are not confusing, the beam may be used in the unfiltered state.

The filter is inserted in the primary beam between the x-ray source and the sample in methods that use a film to record the diffraction effects. In other methods, the filter may be placed between the sample and the recording medium. This has the advantage of absorbing some of the incoherently scattered radiation from the sample, thereby reducing the background of the diffraction record.

X-Ray Diffraction Tubes

An x-ray tube was described as consisting essentially of a source of electrons and two electrodes. The electrons are produced in two ways, which gives rise to two kinds of tubes—cold-cathode or gas tubes and hot-cathode or filament tubes.

[1] The mass absorption coefficient of a compound M, with n elements is $(\mu/\rho)_M = \Sigma(\mu/\rho)_n W_n$, in which W_n is the weight fraction of the nth element.

GAS TUBES. The space between the electrodes in a gas tube is evacuated to a pressure of about 0.01 mm Hg and is maintained with the aid of a vacuum pump and a controlled air leak. When a voltage is applied to the tube, ions, which are present in the residual air, are attracted to the cathode, where they release electrons. The electrons, driven toward the anode or target, ionize the air in the process to provide more ions which in turn release more electrons from the cathode. X-rays are generated when the electrons collide with the atoms of the target. The generation of x-rays is inefficient in all diffraction tubes. About 99 per cent of the electron energy is converted to heat, and consequently the target must be constantly cooled by a flow of cold water to its underside.

A certain minimum voltage (the ionization potential) is required to give the electrons, released by the cathode, enough energy to ionize the residual air in their passage to the target. The ionization potential varies with the pressure of air in the tube. When the voltage is increased beyond the ionization potential, the speed of the electrons is increased. This results in a higher degree of ionization and consequently more electrons to carry the current. Hence the tube current (and therefore the intensity of the x-ray beam) depends on both the applied voltage and the gas pressure in the tube.

The cathode of a gas tube soon becomes pitted from the impact of ions and must be polished at frequent intervals. For this reason the tube is made demountable. The demountable feature has the advantage that targets of different metals may be easily interchanged to give a source of x-rays that has low initial and operating costs. The frequent service required by gas tubes, and their vacuum systems, weighs heavily against them today when the trend is toward facilities that offer push-button operation.

HOT CATHODE TUBES. In hot cathode tubes the electrons are supplied by a heated filament, which also acts as the cathode. The air pressure is much lower (of the order of 0.0001 mm Hg), which thus provides a near-vacuum in which no appreciable discharge can occur in the residual gas and ensures that all the current through the tube is carried by electrons emitted by the filament. The intensity of the beam generated by an x-ray tube depends on the availability of electrons. This is governed by the temperature of the filament in a hot cathode tube, and therefore the intensity of the beam developed by such a tube can be controlled with the current on the filament. For a given filament current, an increase in the voltage beyond the K excitation potential of the target element increases the intensity of the characteristic component of the beam only to the point at which all the electrons emitted by the filament are being directed at the target. A voltage increase beyond this level merely increases the speed of the available electrons and affects the quality of the continuous spectrum. At higher voltages the current in the tube is solely a function of the number of available electrons, which in turn is controlled by the filament current. Thus, in contrast to a gas tube, tube current and voltage are nearly independent variables in the operation of a hot cathode tube.

Most commercial tubes are of the hot cathode type and are permanently sealed off. The cost of such a tube is naturally higher than that of a gas tube manufactured in a university workshop, and a separate tube is required for each type of radiation. A commercial hot cathode tube has the advantage that it gives up to 2000 hours of trouble-free operation and provides the ultimate in convenience. One disadvantage must be mentioned. Vapor from the filament, which is usually made of tungsten wire, gradually plates on to the target and eventually contaminates the beam. Ignorance of this condition has on more than one occasion been responsible for the publication of incorrect data.

COMMERCIAL DIFFRACTION TUBES. The quality, versatility, and convenience of commercial diffraction tubes has been steadily improved in recent years. Beam intensity has been increased in a number of ways. One obvious method involves the use of lightweight window materials that offer better transmission (less absorption) of the beam. Most manufacturers now use thin-sheet beryllium (about 0.02 inch thick) or mica backed with beryllium. This is particularly important for the more easily absorbed long-wavelength radiations obtained from targets constructed of elements that are lighter than copper.

Beam intensity is also improved, in commercial tubes, by focusing the electron stream from the cathode to strike the perpendicularly situated target in a line focus, rather than a point focus. The focal spot is about 1 mm wide and 10 to 15 mm long in all-purpose diffraction tubes. The spot can be viewed either along its length or perpendicular to its length to give x-ray beams with distinctly different characteristics. When the spot is viewed end-on (Fig. 2-6*a*) at a small takeoff angle (angle between target surface and viewing direction), its long dimension is foreshortened to give a beam with small cross-sectional dimensions. Experience shows that the intensity of the beam decreases slightly as the takeoff angle is decreased to 10°, but falls off more rapidly below this point and drops sharply to zero below 3°. An angle of 6° proves to be a good compromise between the loss of intensity suffered at small takeoff angles and the gain of intensity obtained from a foreshortened view of a linear spot—at 6° the long dimension is foreshortened by a factor of 10.

The dimensions of the line focus are chosen to give a nearly square (roughly 1.0 by 1.0 mm) projected focal spot at this angle. This view of the target, often called the spot-focus view, and the symmetrical view from the opposite side of the target provide two highly concentrated x-ray beams of small, nearly equidimensional cross section. They are suitable for the majority of diffraction methods after they have been passed through pinhole systems to limit beam divergence and effective cross section in accordance with the particular needs of a project. More recently, so-called microfocus tubes with smaller focal spots (measuring

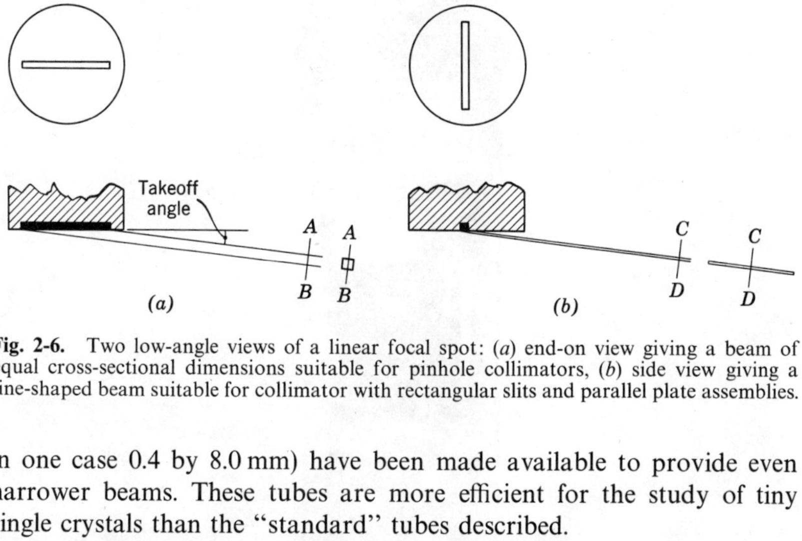

Fig. 2-6. Two low-angle views of a linear focal spot: (*a*) end-on view giving a beam of equal cross-sectional dimensions suitable for pinhole collimators, (*b*) side view giving a line-shaped beam suitable for collimator with rectangular slits and parallel plate assemblies.

in one case 0.4 by 8.0 mm) have been made available to provide even narrower beams. These tubes are more efficient for the study of tiny single crystals than the "standard" tubes described.

The short dimension of the focal spot is foreshortened when the view is perpendicular to the length of the spot. Consequently, at small takeoff angles the projection of the spot is very narrow and line-shaped. A 1.0 by 10.0 mm focal spot has a 0.1 by 10.0 mm projection when viewed at a takeoff angle of 6°. The side view is usually described as the line-focus view. Line-focus beams proved to be too weak for the small equant specimens (usually only a fraction of a millimeter in diameter) that are used in most types of diffraction studies and were largely ignored. At least two makes of commercial diffraction units did not provide access windows to these beams. The development of the North American Philips powder diffractometer, however, created an important use for the beams. The prime object of the diffractometer (described in detail in Chapter 5) is to attain the ultimate in resolution of the diffraction effects, aided by a long specimen-to-recorder distance. To this end, the instrument requires a powerful, highly collimated beam. It was found that the takeoff angle for the spot-focus beam had to be reduced to $1\frac{1}{2}$ or 2° to obtain a reasonable degree of resolving power, with a consequent considerable loss of intensity. This naturally led to the use of a large sample surface to strengthen the intensities to acceptable levels.

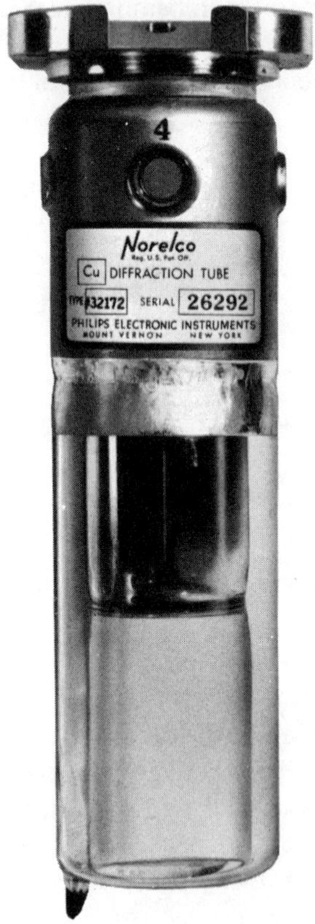

Fig. 2-7. Norelco x-ray diffraction tube. (Courtesy Philips Electronic Instruments).

It then became evident that the line-focus beam could be used to advantage with the diffractometer—the large sample surface (about 20 mm square) avoided the usual large loss of intensity with this beam, and an important gain in resolution was realized, even with the takeoff angle increased to 3° to improve the intensity. As a result of this discovery and of the increasing importance of the diffractometer, the market now offers several tubes, in place of one, with windows for viewing one or both of the line-focus beams.

X-Ray Generators

It is not difficult to construct a basic x-ray generator that will give years of satisfactory service; custom units, some equipped with demountable gas tubes, are not uncommon in university laboratories. Generally speaking, it is advisable to purchase a commercial unit if funds permit. Such a unit is the product of years of research and is normally equipped with many desirable features: convenient tube interchange without the necessity of realigning the recording instruments, automatic exposure timers, and safety devices to ensure that the tube is not operating without an adequate supply of cooling water. The unit is generally supplied with the manufacturer's offer to service the equipment. Finally, it is usually designed to take the refinements and accessories that will become a necessity as the scope of research in the laboratory broadens.

CHAPTER THREE

THE DIFFRACTION OF X-RAYS

The Geometry of Diffraction

An electron in the path of an unpolarized x-ray beam vibrates with the frequency of the incident radiation, periodically absorbing energy and emitting it as x-radiation of the same frequency. The original x-rays are unmodified in wavelength by the interaction but are radiated in all directions (Fig. 3-1). The electron has the effect of *scattering* the incident radiation and acts as a source of secondary x-rays. The nucleus, because of its high mass, makes a negligible contribution to the radiation scattered by an atom.

An atom is not a point source of x-rays. Its electrons occupy a space whose size is commensurable with the distances between adjacent atoms. To develop the geometry of diffraction, it is permissible to adopt the simple picture that rays scattered from different parts of the atom combine to give the effect of a point source. The treatment of the intensity of diffraction, however, rests on a more detailed study of atomic scattering. This shows that the intensity of the radiation scattered by an atom depends not only on the number of electrons associated with the nucleus but also on their distribution, on the angle of scattering, and on the temperature.

A crystal is a complex, although orderly, arrangement of atoms. All atoms in the path of an x-ray beam scatter x-rays simultaneously. In general, the scattered waves interfere with and destroy one another, but in certain specific directions they combine to form new wave fronts. This cooperative scattering is known as *diffraction.* The directions of possible diffraction depend only on the size and shape of the unit cell. Certain classes of diffraction are systematically extinguished by lattice centering and by the presence of space-group symmetry elements. The intensities of the diffracted waves depend on the kind and arrangement

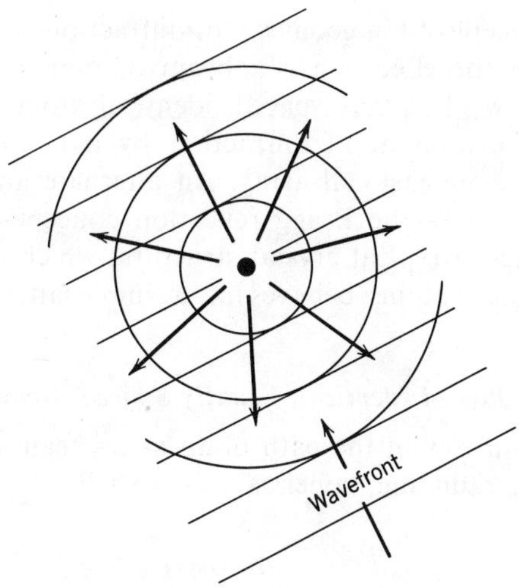

Fig. 3-1

of atoms in the crystal structure. Thus a study of the geometry of diffraction from a crystal yields the unit cell dimensions. An analysis of the missing spectra gives the symmetry (space group) of the crystal structure. The intensities are needed to work out the arrangement of atoms.

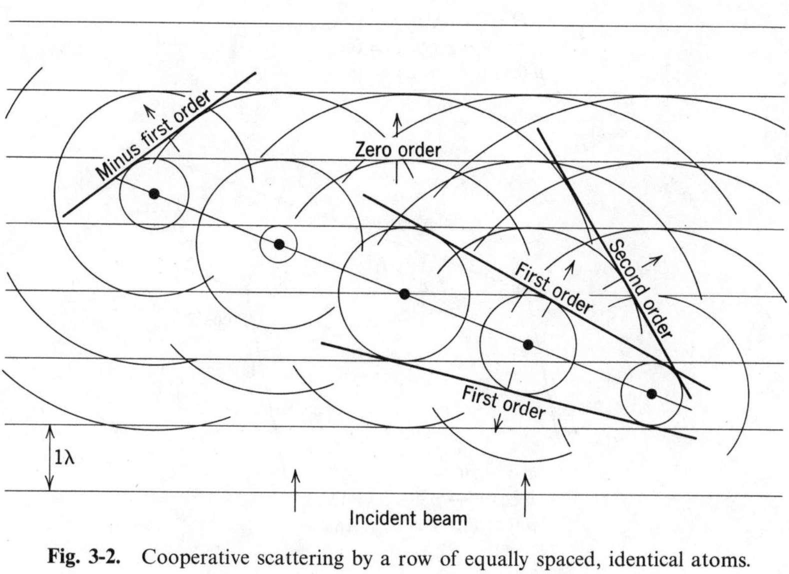

Fig. 3-2. Cooperative scattering by a row of equally spaced, identical atoms.

The Laue concept of the geometry of diffraction is developed here by considering first the effect when a beam of monochromatic x-rays is incident on a row of equally spaced, identical atoms. The treatment is extended to take account of diffraction by a two-dimensional array and then a three-dimensional array. An alternate and more workable picture of diffraction—the Bragg reflection concept—is presented and used to show that a typical crystal structure, which usually consists of two or more kinds of atoms, behaves like a simple lattice-array of identical atoms.

Diffraction by a Row of Identical, Equally Spaced Atoms

Each atom of a row in the path of an x-ray beam can be considered as the center of radiating, spherical wave shells (Fig. 3-2). In general,

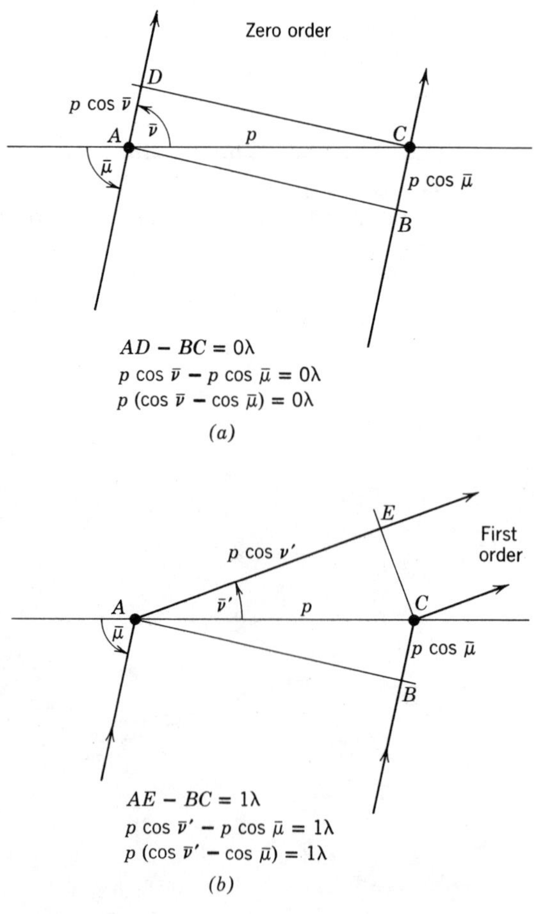

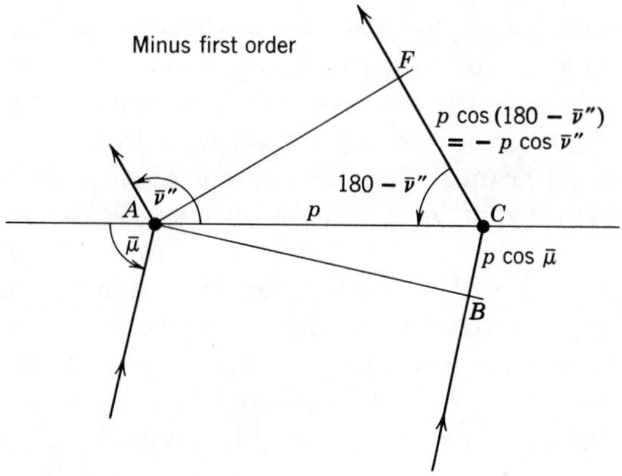

$$CF + BC = 1\lambda$$
$$-p \cos \bar{\nu}'' + p \cos \bar{\mu} = 1\lambda$$
$$p (\cos \bar{\nu}'' - \cos \bar{\mu}) = -1\lambda$$

(c)

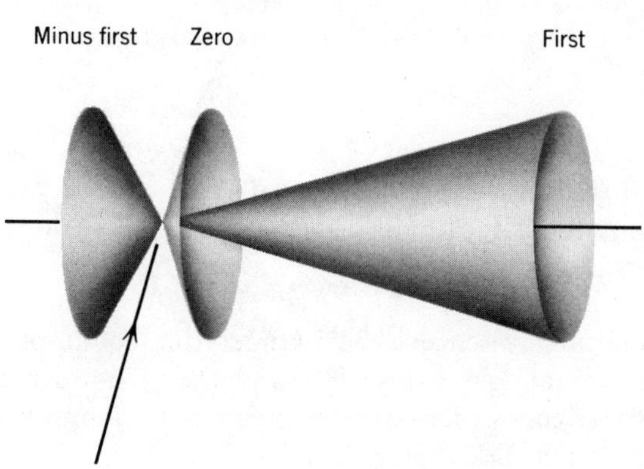

Zero and first-order diffraction cones

(d)

Fig. 3-3. Condition for diffraction from a row.

the scattered waves interfere with one another, but they combine to form wave fronts when they happen to be in phase—that is, when the difference in path length between rays scattered from adjacent atoms is zero or any whole number of wavelengths. Wave fronts are designated in accordance with the path difference of rays scattered by neighboring atoms. The path difference for a first-order wave front is one wavelength, and the diffracted beam is described as a first-order diffraction. If the diffraction is to the left of the incident beam, it is called a minus-first-order diffraction.

Figure 3-3 illustrates the condition for diffraction by a row of atoms. Two rays of an x-ray beam strike the row, of periodicity p, at an angle of incidence $\bar{\mu}$ to form zero-, first-, and minus-first-order diffractions. The angle of diffraction $\bar{\nu}$ is here measured from the positive end of the row. Clearly, the diffracted rays are in phase only if

$$p(\cos \bar{\nu} - \cos \bar{\mu}) = \pm h\lambda$$

where h in this case is 0 or 1. The condition for diffraction is evidently met not only in the directions AD, AE, and AF shown in the diagrams but in all directions that make angles of $\bar{\nu}$, $\bar{\nu}'$, and $\bar{\nu}''$ with the row. These directions outline three concentric cones, coaxial with the row and with half-apex angles equal to $\bar{\nu}$, $\bar{\nu}'$, and $\bar{\nu}''$ (Fig. 3-3*d*). A line in a cone represents a direction in which all atoms of the row are simultaneously scattering in phase (diffracting).

The expression developed in the preceding paragraph is usually called the Laue equation for diffraction by a row. Several features of this expression are worth noting. For zero-order diffractions

$$p(\cos \bar{\nu} - \cos \bar{\mu}) = 0\lambda$$

$\bar{\nu}$ is equal to $\bar{\mu}$ and independent of p. The significance of this is that the incident beam is always a line in the zero-order cone.

The Laue equations reduce to

$$p \cos \bar{\nu} = \pm h\lambda$$

when the angle of incidence is 90°. Under this condition the angle $\bar{\nu}$ for zero-order diffractions is also 90°, and the zero-order cone has the form of a disk. Higher-order cones occur in pairs, symmetrically placed about the zero-order disk (Fig. 3-4).

Diffraction by a Plane Lattice-Array of Atoms

A plane lattice-array of atoms (Fig. 3-5) may be defined by two translation periods a and b in the rows OA and OB and the angle γ. This picture of a plane provides a convenient way of describing diffraction by a plane—namely, by extending the concept of diffraction by a row to include simultaneous diffraction by two kinds of nonparallel rows.

The diffraction directions for the row OA comprise a set of concentric cones that are coaxial with OA and have half-apex angles which are characteristic for the period a in accordance with the Laue equation for diffraction by a row (Fig. 3-5*b*). The diffraction directions for the row OB comprise another set of cones (Fig. 3-5*c*). Because a line in a cone is a direction in which all atoms of a row are scattering in phase, a line of intersection of two concentric cones is a direction in which all atoms of two nonparallel rows are diffracting. In general, cones intersect along

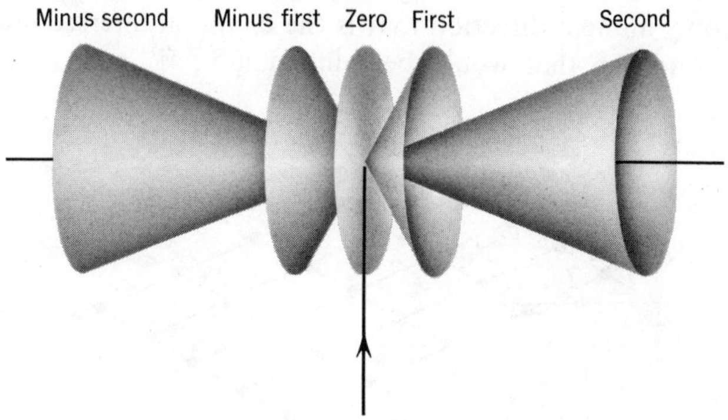

Fig. 3-4. Diffraction from a row of atoms when the angle of incidence is 90°.

two lines (such as OX and OY of Fig. 3-5*d*), and the lines are symmetrically arranged about the plane. Actually, all atoms in the plane are diffracting along OX and OY directions because each atom is the intersection of an OA and an OB type row and is therefore the site of an identical pair of intersecting cones.

The condition for diffraction by the plane may therefore be expressed in terms of the Laue equations for the rows OA and OB:

$$a(\cos \bar{\nu}_1 - \cos \bar{\mu}_1) = h\lambda$$

$$b(\cos \bar{\nu}_2 - \cos \bar{\mu}_2) = k\lambda$$

where a and b are the periods of the rows, $\bar{\mu}_1$ and $\bar{\mu}_2$ are the angles at which the beam meets the rows, h and k are the wavelength orders of the intersecting cones, and $\bar{\nu}_1$ and $\bar{\nu}_2$ are the diffraction angles referred to the rows. Diffraction occurs if the two equations are simultaneously satisfied, that is, if the angles $\bar{\nu}_1$ and $\bar{\nu}_2$ define the same direction. When applied to Fig. 3-5*d*, it means that the beam meets the plane at such an angle that the hth-order cone about OA intersects the kth-order cone about OB along OX and OY. The angle between OA and OX (and OY) is $\bar{\nu}_1$ and that between OB and OX is $\bar{\nu}_2$.

Diffraction by a Three-Dimensional Lattice-Array of Atoms

The diffraction directions for a three-dimensional array can be referred to three sets of diffraction cones coaxial with three noncoplanar reference rows. In general, when three cones intersect, each cone will make two intersections with each of the other two cones. The result is six intersections (Fig. 3-6), each of which represents a direction in which two rows are scattering in phase. Evidently, then, the three rows will scatter in phase only along a direction that is the common intersection of three cones, a condition that would be fulfilled if V, W, and Y of Fig. 3-6

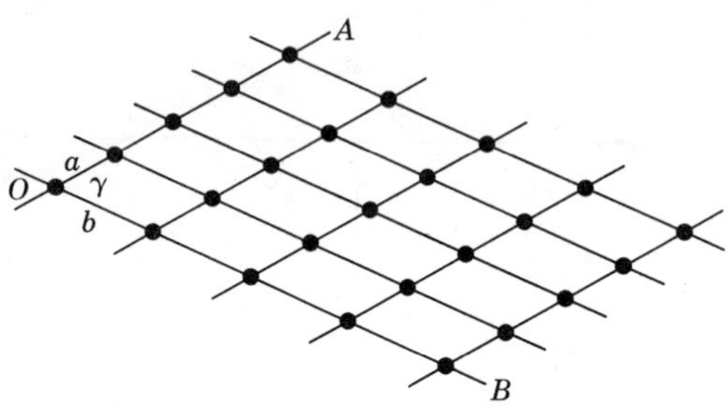

(a)

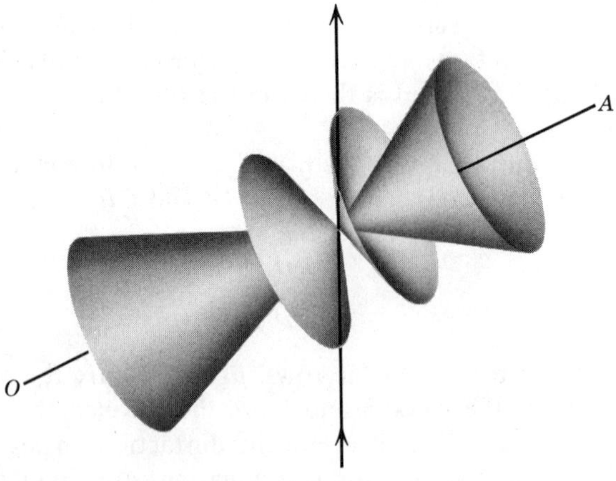

(b)

were brought to coincidence. A crystal will in general not diffract x-rays. Diffraction occurs only under special conditions, which in terms of the Laue equations

$$a(\cos \bar{\nu}_1 - \cos \bar{\mu}_1) = \pm h\lambda$$
$$b(\cos \bar{\nu}_2 - \cos \bar{\mu}_2) = \pm k\lambda$$
$$c(\cos \bar{\nu}_3 - \cos \bar{\mu}_3) = \pm l\lambda$$

requires that the three angles $\bar{\nu}_1$, $\bar{\nu}_2$, and $\bar{\nu}_3$ define a common direction. Because each diffraction direction is common to three cones which are coaxial, respectively, with the three reference rows, all directions can be referred to the cones of any one row. Thus, if the row in Fig. 3-4 is the row of a crystal, all the zero-, first-, and second-order diffractions emanating from this crystal lie on the cones shown in the figure.

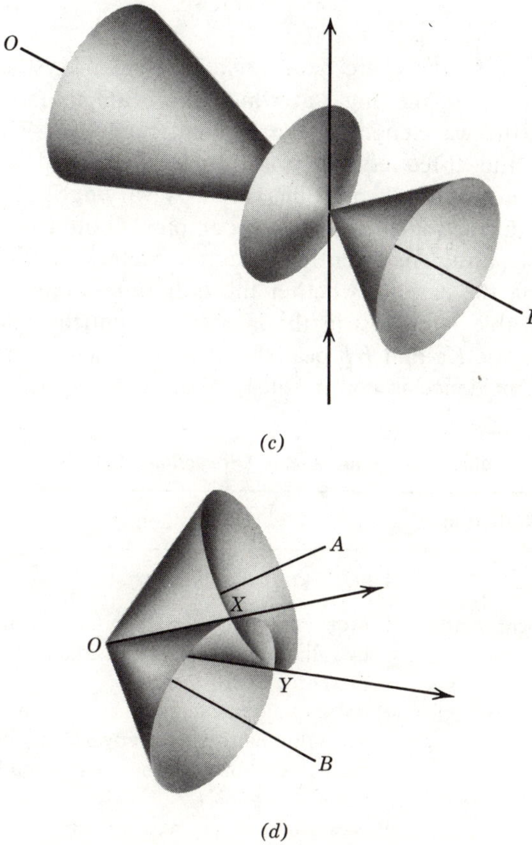

Fig. 3-5. Diffraction from a plane lattice-array of atoms.

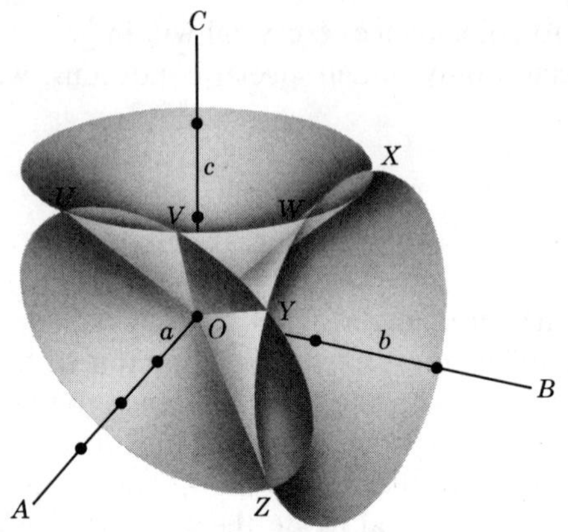

Fig. 3-6. The general case for the intersection of diffraction cones coaxial with three noncoplanar rows.

The values a, b, and c are fixed for a particular crystal. Thus the $\bar{\nu}$ values depend on $\bar{\mu}$ (the angle at which the beam meets the reference rows) and λ (the wavelength of the incident radiation). The possibility of satisfying the three equations simultaneously and achieving the condition for diffraction can be increased by varying $\bar{\mu}$ or λ during the analysis. The diffraction effects are too complex if both factors are varied simultaneously, and consequently diffraction methods fall naturally into two groups on the basis of whether the radiation is monochromatic or continuous (Table 3-1). Most methods employ monochromatic radiation (the characteristic $K\alpha$ and $K\beta$ peaks); if the specimen is a single crystal, the angle of incidence is continuously altered during the exposure by

Table 3-1. Some X-Ray Diffraction Methods

Radiation	Method
White	Laue: stationary single crystal
Monochromatic	Powder: specimen is polycrystal, and thereby all orientations are simultaneously presented to the beam
	Rotation, Weissenberg; oscillation, De Jong-Bouman: single crystal rotates or oscillates about chosen axis in path of beam
	Precession: chosen axis of single crystal precesses about beam direction

giving the crystal a motion in the path of the beam. The motion is controlled in order to ease the problem of interpreting the diffraction effects. In contrast to the fixed-wavelength methods, the Laue method uses a beam that has a range of wavelengths (the continuous spectrum), but the orientation of the crystal is fixed during the analysis.

The "Reflection" of X-Rays

A few months after the results of the original Laue experiments had been published Sir W. L. Bragg discovered that the phenomenon of diffraction could be pictured as a reflection of the incident beam from

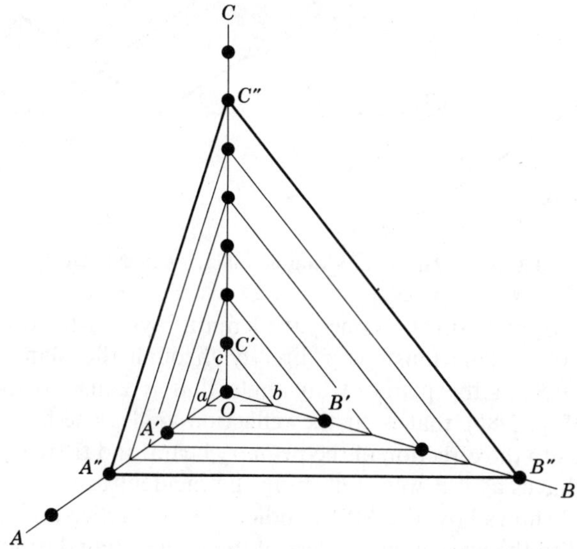

Fig. 3-7

lattice planes. He developed a mathematical equation for the condition for reflection, which is equivalent to the simultaneous solution of the three Laue equations but provides a much simpler and more workable expression of the process of diffraction.

Suppose that a beam of x-ray strikes a three-dimensional lattice-array of atoms which has been referred to rows OA, OB, and OC (Fig. 3-7), at an angle such that the third-order cone about OA, the second-order cone about OB, and the first-order cone about OC intersect in a common line to satisfy the Laue condition for diffraction. This implies that x-rays scattered in the diffraction direction by adjacent atoms on OA (such as O and A') have a path difference of three wavelengths. Similarly, adjacent atoms on OB (such as O and B') and on OC (such as O and C') scatter in the diffraction direction with path differences of two and one wavelengths, respectively. It follows, then, that rays scattered at A'', B'', and C'' are six wavelengths out-of-phase with those scattered at the origin O. These three points define a plane with intercepts $2a$, $3b$, $6c$ and Miller indices (321). Because A'', B'', and C'' scatter rays differing in phase by the same number of wavelengths from those scattered at O, they scatter rays differing by zero wavelengths from one another. This is equivalent to saying that rays scattered in the diffraction direction by A'', B'', and C'' have no path difference.

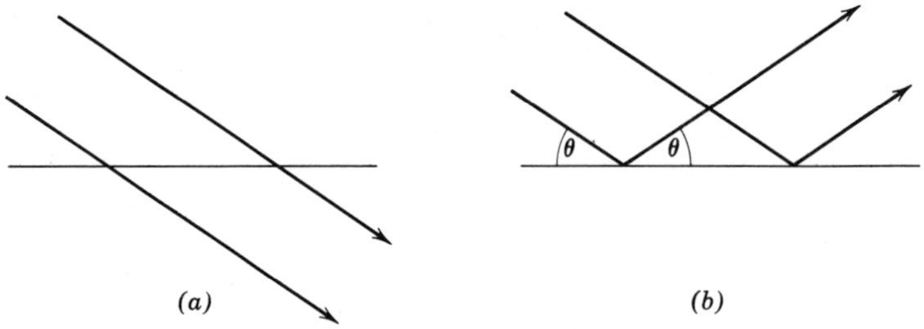

(*a*) (*b*)

Fig. 3-8. Parallel rays with no difference in path length.

In order to maintain the same path length, rays scattered in phase by the atoms of a plane must pass directly through the plane (Fig. 3-8*a*) or be deviated at the plane at an angle that is equal to the angle of incidence (Fig. 3-8*b*), that is, at the reflection angle. The first alternative corresponds to the direction of the primary beam, and therefore the plane $A''B''C''$ behaves as if it were reflecting the incident beam.

Table 3-2 shows how the Miller indices of the "reflecting" plane (321) correspond to the wavelength orders of the intersecting diffraction cones. The table implies that the reflection concept is not limited to the specific case depicted in Fig. 3-7 but applies equally well to the general case which may be stated as follows: a diffraction direction defined by the intersection of the hth-order cone about the a axis, the kth-order cone about the b axis, and the lth-order cone about the c axis is geometrically equivalent to a reflection of the incident beam from the (hkl) plane referred to these axes. One index of a reflecting plane is immediately known if the wavelength order of the diffraction cone in which it "reflects" has been identified and its axis has been labeled. For example, if the row in Fig. 3-4 is the b axis of a crystal, all planes reflecting in the second-order cone have the indices ($h2l$). In most single-crystal x-ray methods one index of each reflecting plane is predetermined to ease the problem of interpreting the diffraction record. This is accomplished, as noted in Table 3-1, by choosing one axis for special orientation in the beam and placing the film or other recording medium to intercept the cones associated with this axis, according to some scheme that identifies the wavelength order on sight.

Reflection is evidently not confined to the $A''B''C''$ plane but must occur from a stack of planes parallel to it. The reflection concept is valid only if it can be shown that reflections from successive planes cooperate—that is, that rays scattered by successive planes differ in path

Table 3-2. Relation Between the Wavelength Orders of Three Intersecting Diffraction Cones and the Miller Indices of the "Reflecting" Plane

Row	Wavelength Orders of Intersecting Cones	The "Reflecting" Plane: Intercepts of Plane on Three Rows	The "Reflecting" Plane: Miller Indices
OA	3	$2a$	
OB	2	$3b$	$\frac{1}{2}\frac{1}{3}\frac{1}{6} = (321)$
OC	1	$6c$	

length by a whole number of wavelengths. There are six (321) planes from the origin to the plane $A''B''C''$ (Fig. 3-7). Because rays scattered at A'', B'', and C'' are six wavelengths out-of-phase with those scattered at the origin, it follows that rays reflected by successive (321) planes differ by one wavelength and cooperate. The reflection may therefore be described as a first-order reflection and may be designated by the symbol 321.

It is interesting now to consider the reflection that corresponds to the diffraction direction defined by the intersection of the sixth-order cone about OA, the fourth-order cone about OB, and the second-order cone about OC. Using the procedure followed in Table 3-2, the indices of the reflecting plane are (642), which reduces to (321). Therefore, the same planes reflect as in the previous case, but now rays scattered at A'', B'', and C'' differ by twelve wavelengths from those scattered at the origin. Consequently, rays scattered by successive (321) planes differ by two wavelengths, and the reflection can be described as a second-order 321 reflection. Alternatively, and this proves more useful, the indices that correspond to the wavelength orders of the Laue cones can be retained, and the reflection can be described by the symbol 642.

THE CONDITION FOR REFLECTION. The preceding discussion furnishes the basis for deriving a mathematical expression of the condition for reflection. Figure 3-9 shows a beam of parallel x-rays penetrating a stack of planes of spacing d, at a glancing angle of incidence θ. Each plane must be pictured as reflecting a fraction of the incident beam. The reflected rays combine to form a diffracted beam if they differ in phase

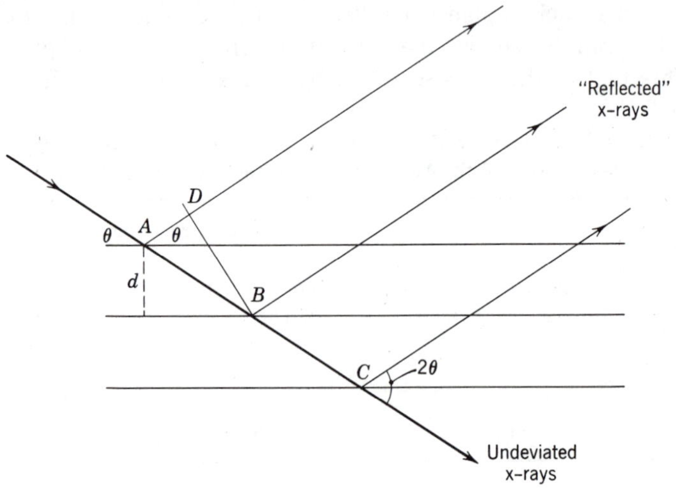

Fig. 3-9. The condition for reflection—the Bragg law.

by a whole number of wavelengths—that is, if the path difference $AB - AD = n\lambda$, where n is 1, 2, 3, etc.

$$AB = \frac{d}{\sin\theta}$$

$$AD = AB\cos 2\theta = \frac{d}{\sin\theta}(\cos 2\theta)$$

Therefore

$$n\lambda = \frac{d}{\sin\theta} - \frac{d}{\sin\theta}(\cos 2\theta)$$

$$= \frac{d}{\sin\theta}(1 - \cos 2\theta) = \frac{d}{\sin\theta}(2\sin^2\theta)$$

$$= 2d\sin\theta$$

This is the Bragg condition for reflection.

The Bragg equation gives the angle θ, at which a set of planes of spacing d cooperatively reflects x-radiation of wavelength λ in the nth order. The equation has a number of solutions for θ for particular values of d and λ

$$\theta = \sin^{-1}\left(\frac{1\lambda}{2d}\right)$$
$$= \sin^{-1}\left(\frac{2\lambda}{2d}\right)$$
$$= \sin^{-1}\left(\frac{3\lambda}{2d}\right) \quad \text{etc.}$$

Because θ cannot exceed 90° and $\sin 90° = 1$, the series of solutions corresponds to all values of n for which the expression $n\lambda/2d$ does not exceed unity. These statements imply that, for a given experimental arrangement with monochromatic radiation (fixed λ), a set of planes gives a series of reflections when rotated through positions at which the phase differences are $1, 2, 3, \ldots$ wavelengths (Fig. 3-10). Because $n\lambda/2d$ cannot exceed unity, the maximum phase difference $n\lambda$ is equal to $2d$. This condition places a limit on the number of reflections that can emanate from a crystal. A crystal has an infinite number of sets of planes, representing a large range in values of d. Because only those reflections are possible for which $n\lambda/2d \leq 1$, only those planes for which $d \geq \lambda/2$ can reflect.

The Bragg equations for reflections from a stack of (123) planes have the form

$$1\lambda = 2d_{123} \sin \theta'$$
$$2\lambda = 2d_{123} \sin \theta''$$
$$3\lambda = 2d_{123} \sin \theta'''$$

The reflections are properly described as first-, second-, and third-order (123) reflections. A more convenient designation for the reflections, alluded to in the preceding section, includes the wavelength order (value of n) in the index description. Under this scheme the three reflections are designated by the symbols 123, 246, and 369 and are regarded as first-order reflections. The Bragg equations then have the form

$$\lambda = 2d_{123} \sin \theta'$$
$$\lambda = 2d_{246} \sin \theta''$$
$$\lambda = 2d_{369} \sin \theta'''$$

The multiple-index spacing symbols d_{246} and d_{369} are equivalent, respectively, to $d_{123}/2$ and $d_{123}/3$. This notation is now standard practice in x-ray crystallography.

Reflection by a Typical Crystal Structure. The preceding description of the process of diffraction was developed for a simple lattice-array of atoms. It remains to be shown that the Bragg law holds also for a typical crystal structure, which is usually a much more complicated arrangement and normally consists of several kinds of atom (Fig. 3-11).

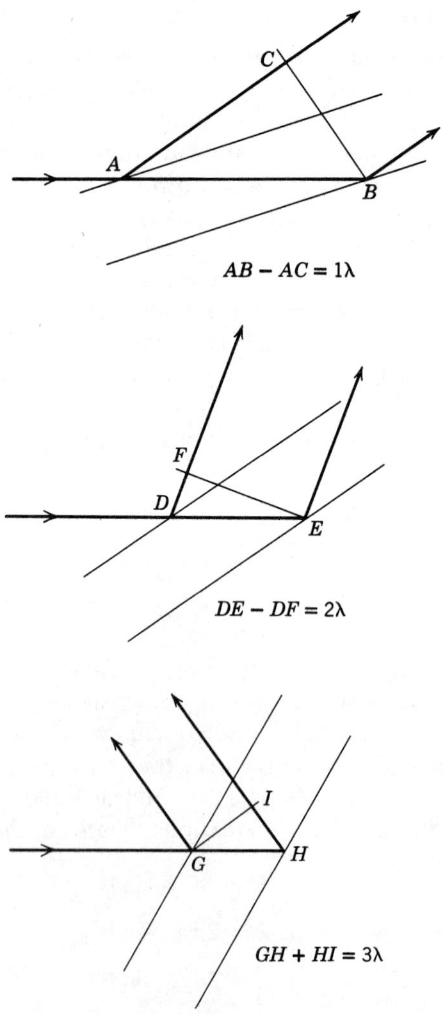

Fig. 3-10. Three orders of reflection from the same set of planes.

The space lattice blocks out units of the structure that are exactly repeated. Each lattice point can be regarded as representing a repeat unit of the structure. Figure 3-11 emphasizes a feature of all crystal structures. If the lattice framework is shifted parallel to its translations to make one of its points coincide with an atom, each lattice point will coincide with an identical atom in other parts of the structure. This

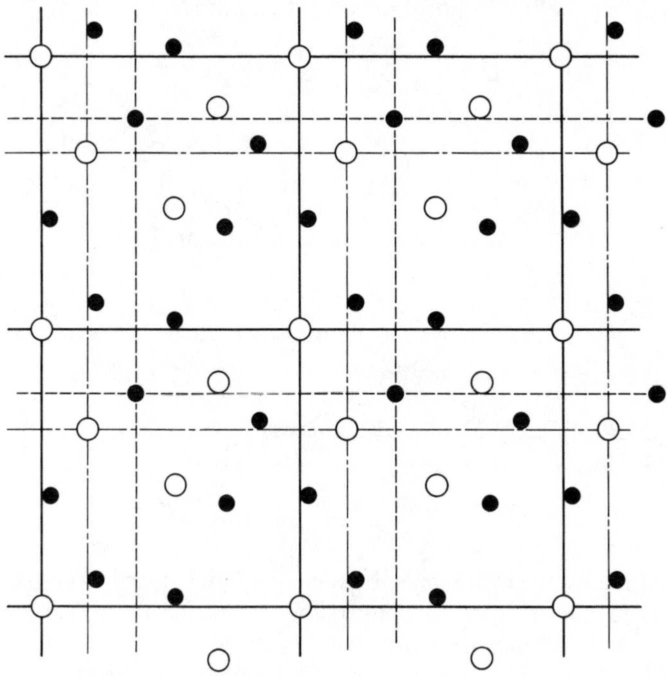

Fig. 3-11

holds for every atom in the repeat unit. Then each atom of the unit helps to form its own simple lattice-array, which is parallel to, and has the same dimensions as, the space lattice. The number of such lattice-arrays is equal to the number of atoms in the repeat unit. This allows us to picture a crystal structure as a group of simple, interpenetrating, and parallel lattice-arrays of atoms, all identical in size. Diffraction from a typical crystal structure can then be treated in terms of simultaneous reflections from this group of interpenetrating lattice-arrays.

Figure 3-12 shows part of a hypothetical structure consisting of two kinds of atom, *A* and *B*. In general, these atoms lie on separate sets of planes, but, for reasons given in the preceding paragraph, the sets have the same interplanar spacings. The Bragg law states that the angles at which x-radiation is reflected by a set of planes depends only on the spacing, if the wavelength is fixed. Consequently, the A and B planes reflect at the same angle. However, the two sets of reflections differ in phase to an extent that depends on d, λ, and s (the separation of A and B planes), and they differ in amplitude if they arise from different atoms. Hence the resultant wave from the two sets of planes differs from its constituent waves in both phase and amplitude.

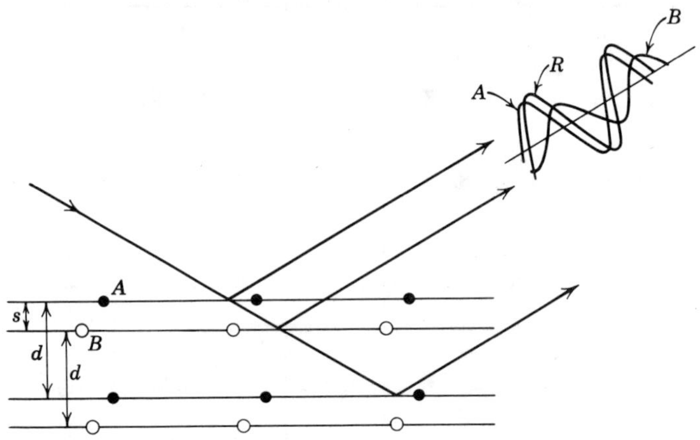

Fig. 3-12. Reflection from a crystal structure.

In summary then—angles of reflection depend only on the lattice spacings parallel to the reflecting planes, provided that the wavelength is constant. It is convenient for this reason to regard the reflections as emanating from *lattice* planes and to refer to the d_{hkl} component of the Bragg equation as a *lattice* spacing. The intensities of the various reflections are proportional to the squares of the amplitudes of the composite reflected waves and depend on the kind of atoms in the crystal structure and their distribution between the lattice planes. Thus a critical study of reflection angles leads to a determination of the shape and size of the unit cell; the intensities of the reflections are used to deduce the atomic arrangements within the cell.

The Intensity of Diffraction

Scattering by an Electron

The characteristic radiation obtained from a diffraction tube is generally considered to be unpolarized. However, the process of scattering causes the diffracted beams to be partially polarized, the amount of polarization depending on the angle of scattering. As a consequence, the intensity of scattering by an electron varies with the angle. According to the classical formula of Sir J. J. Thomson, the intensity I_e at a distance R from a free electron is

$$I_e = I_0\left(\frac{e^4}{m^2c^4R^2}\right)\sin^2\alpha$$

where I_0 is the intensity of the incident beam, e and m are the charge and mass of the electron, c is the speed of light, and α is the angle between the direction of scattering and the acceleration direction of the electron.

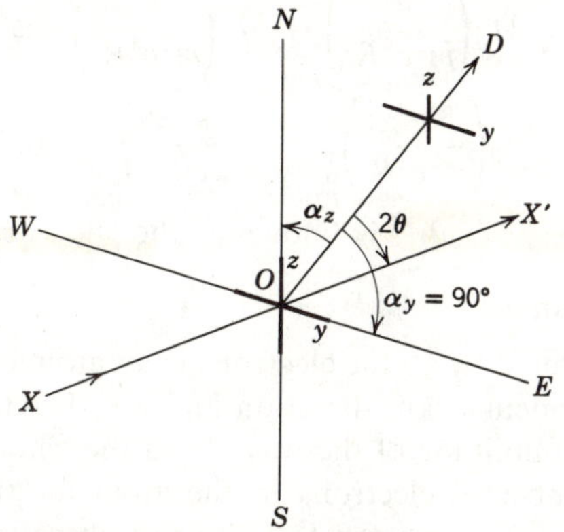

Fig. 3-13. Polarization of radiation scattered by an electron.

The unpolarized beam (XX' of Fig. 3-13) vibrates in all directions normal to the direction of travel. The vibration can be resolved into two plane-polarized components, y along WE and z along NS, with electric vectors having the relation

$$E^2 = E_y{}^2 + E_z{}^2$$

On the average $E_y = E_z$ and, because E is a measure of amplitude and intensity is proportional to amplitude squared,

$$\tfrac{1}{2}I_0 = I_{0y} = I_{0z}$$

Consider now an electron at O scattering the incident beam along a diffraction direction OD, which makes an angle 2θ with the undeviated beam direction. The y component of the beam accelerates the electron in the direction OE at an angle $\alpha_y = 90°$ to the direction of scattering.

This results in a scattering component with intensity

$$I_{ey} = \tfrac{1}{2}I_0\left(\frac{e^4}{m^2c^4R^2}\right)\sin^2 90 = \tfrac{1}{2}I_0\left(\frac{e^4}{m^2c^4R^2}\right)$$

The z component accelerates the electron in the ON direction at an angle $\alpha_z = (90 - 2\theta)$ to OD. The intensity of the z scattering component is therefore

$$I_{ez} = \tfrac{1}{2}I_0\left(\frac{e^4}{m^2c^4R^2}\right)\sin^2(90 - 2\theta) = \tfrac{1}{2}I_0\left(\frac{e^4}{m^2c^4R^2}\right)\cos^2 2\theta$$

The intensity of scattering in the OD direction is the sum of the intensity components

$$\begin{aligned} I_e &= \tfrac{1}{2}I_0\left(\frac{e^4}{m^2c^4R^2}\right) + \tfrac{1}{2}I_0\left(\frac{e^4}{m^2c^4R^2}\right)\cos^2 2\theta \\ &= I_0\left(\frac{e^4}{m^2c^4R^2}\right)\left(\frac{1 + \cos^2 2\theta}{2}\right) \end{aligned} \tag{1}$$

The quantity $(1 + \cos^2 2\theta)/2$ is known as the *polarization factor*.

Scattering by an Atom

The waves scattered by all the electrons in an atom have the same path length for the incident beam direction and are therefore in phase (Fig. 3-14). Hence the amplitude of the wave from the whole atom is proportional to the number of electrons in the atom for this direction. The electrons cannot all be expected to scatter in phase in other directions, however. Electrons are distributed in an atom over a region, the size of which is of the same magnitude as the x-ray wavelengths used in diffraction. As a consequence, the constituent waves partially interfere, as shown in Fig. 3-14, and cause a decrease in the amplitude of the wave scattered by the whole atom. The waves become more out-of-phase as the scattering angle increases and the resultant amplitude decreases rapidly. Figure 3-14 indicates that the amplitude for a particular scattering angle depends on the number of electrons, hence on the atom's state of chemical combination, on the distribution of the electrons in the atom, and on the wavelength of the incident radiation.

The efficiency of scattering by an atom in a particular direction is known as its *atomic scattering factor* (f_0) and is expressed by the ratio $f_0 = A_A/A_e$, where A_A is the amplitude of the wave from the whole atom and A_e is the amplitude of the wave from a free electron. The intensity of scattering is proportional to the square of the amplitude and therefore

$$I_A = I_e f_0^2$$

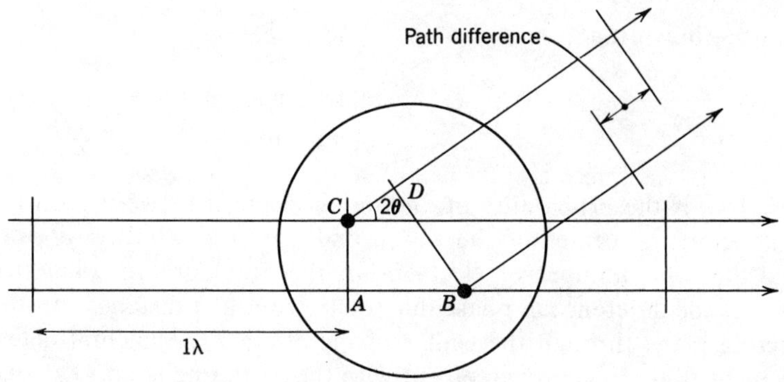

Fig. 3-14. Scattering by an atom.

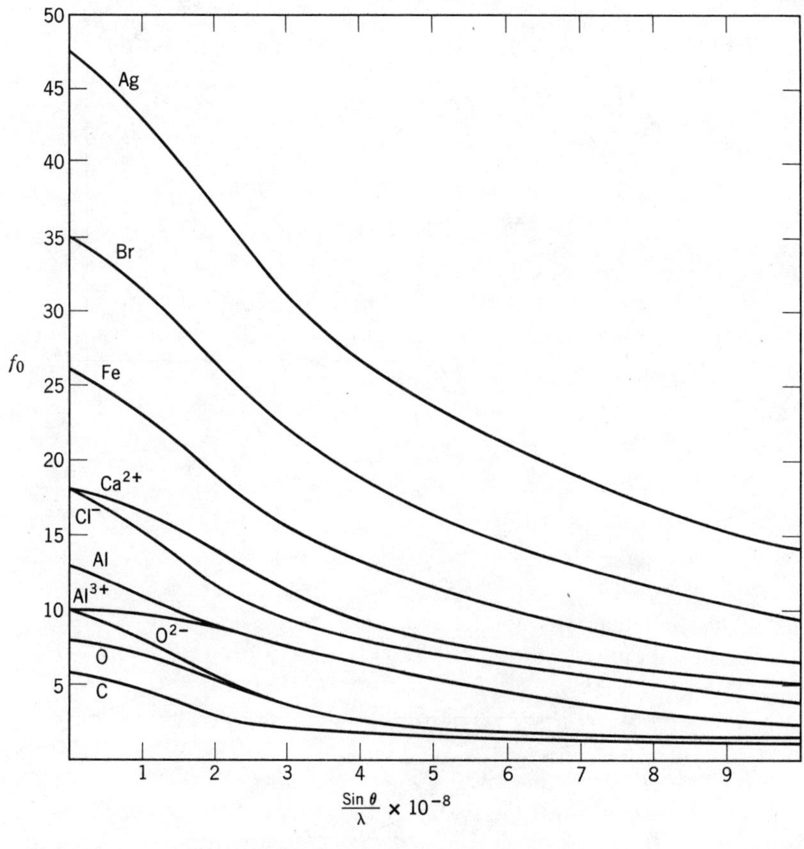

Fig. 3-15. Diffracting power of a few common atoms and ions (after C. W. Bunn, *Chemical Crystallography*, Oxford University Press, 1945).

It can be shown that

$$f_0 = \int_0^\infty U(r)\,dr \frac{\sin(4\pi r \sin\theta/\lambda)}{(4\pi r \sin\theta)/\lambda}$$

where $U(r)$ is the probability of finding an electron between radii r and $r + dr$ from the center of the atom, and $\int_0^\infty U(r)\,dr$ is therefore equal to Z, the total number of electrons in the atom or ion; $(4\pi r \sin\theta)/\lambda$ expresses the difference in phase due to electrons at a distance r from the reflecting plane through the center of the atom. Experimental determinations of f_0 are in good agreement with the scattering factors calculated by making assumptions about the distribution of electrons in the atoms.

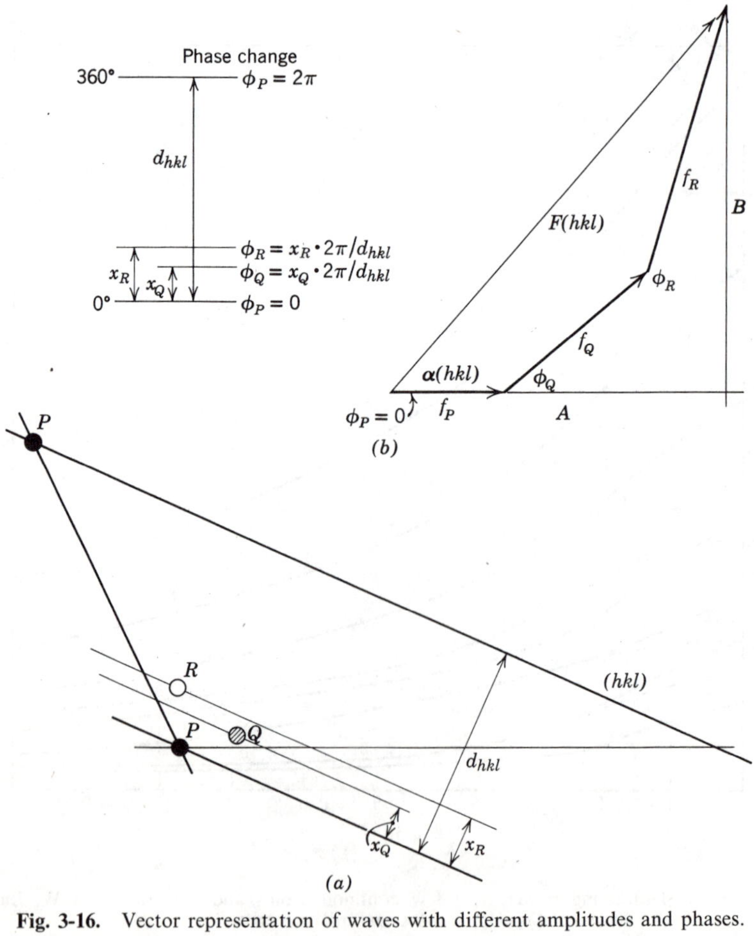

Fig. 3-16. Vector representation of waves with different amplitudes and phases.

Because the value of the factor depends primarily on $\sin\theta/\lambda$, atomic scattering data are tabulated or plotted as a function of this term. Figure 3-15 shows calculated f_0 curves for a number of atoms and ions. The curves have their maximum values when $\sin\theta/\lambda = 0$, corresponding to scattering at $\theta = 0$. At this point the values are proportional to the number of electrons in the atoms or ions. The values decrease with an increase in θ, as predicted in the preceding paragraph. A decrease in the wavelength has the same effect—the path difference (for example $AB - CD$ in Fig. 3-14) at a given scattering angle becomes a larger fraction of the wavelength and represents a greater difference in phase.

Scattering by a Unit Cell

Whereas the scattering from an atom depends on the distribution of its electrons, the scattering from a unit cell depends on the atomic arrangements. Consider a crystal structure (Fig. 3-16*a*) consisting of three kinds of atom—P (at the origin), Q, and R—reflecting in the first order from a set of (hkl) planes of spacing d_{hkl}. This means that the path difference for scattering from successive P planes is one wavelength, equivalent to a 360° or 2π radian phase change. The displacements of the Q and R planes (respectively, x_Q and x_R) are equivalent to phase changes of

$$\frac{x_Q}{d_{hkl}}2\pi = \phi_Q \quad \text{and} \quad \frac{x_R}{d_{hkl}}2\pi = \phi_R$$

The amplitudes of the hkl waves scattered by the three atoms and their phase relation can be conveniently shown on a vector diagram (Fig. 3-16*b*). Each wave is represented by a vector whose length and inclination to the horizontal are proportional, respectively, to the atom's scattering factor f_0 and the appropriate phase angle ϕ. The length of the resultant, found by adding the vectors, and its inclination give the amplitude $F(hkl)$ and the phase angle $\alpha(hkl)$ of the composite wave due to all three kinds of atom.

$F(hkl)$ is called the structure factor. X-ray intensities are proportional to amplitude squared, and therefore only the modulus of the structure factor, namely $|F(hkl)|$, can be obtained experimentally. $|F(hkl)|$ is known as the structure amplitude and is the ratio of the amplitude of the wave scattered by all atoms in the cell to the amplitude of the wave scattered by one electron. From Fig. 3-16*b*

$$\begin{aligned}|F(hkl)| &= [A^2 + B^2]^{\frac{1}{2}} \\ &= [(F\cos\alpha)^2 + (F\sin\alpha)^2]^{\frac{1}{2}}\end{aligned}$$

where

$$(F\cos\alpha)^2 = (f_P\cos\varphi_P + f_Q\cos\varphi_Q + f_R\cos\varphi_R)^2$$
$$(F\sin\alpha)^2 = (f_P\sin\varphi_P + f_Q\sin\varphi_Q + f_R\sin\varphi_R)^2$$

Then for a unit cell with r atoms

$$|F(hkl)| = [(\sum f_{0,r}\cos\varphi_r)^2 + (\sum f_{0,r}\sin\varphi_r)^2]^{\frac{1}{2}}$$

where f_r and φ_r refer, respectively, to the atomic scattering factor and

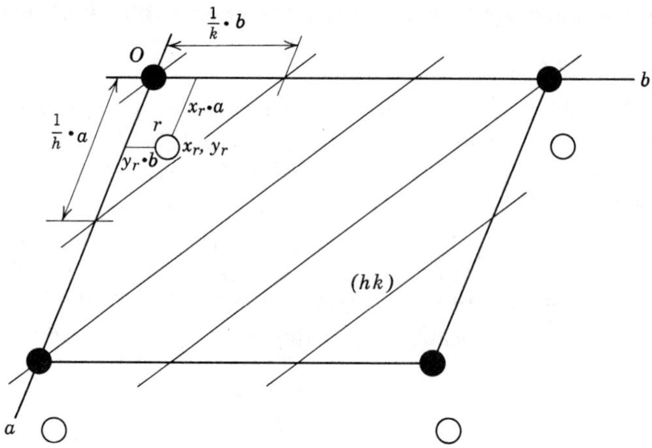

Fig. 3-17.

the phase factor of the rth atom. The phase angle for the hkl reflection from this unit cell is

$$\alpha(hkl) = \tan^{-1}\frac{B}{A} = \tan^{-1}\frac{\sum f_{0,r}\sin\varphi_r}{\sum f_{0,r}\cos\varphi_r}$$

It is useful now to express the phase factors in the structure amplitude equation in terms of the positions of the atoms in the unit cell. Consider a two-dimensional structure (Fig. 3-17) with black atoms at the corners of the cell and a white atom r with coordinates x_r and y_r (these symbols refer to fractional parts of the cell edges a and b). Assume a set of (hk) "planes" reflecting in the first order. The closest (hk) plane to the origin has intercepts $1/h \cdot a$ and $1/k \cdot b$ on the cell edges. These distances are equivalent to 2π phase changes in the a and b directions, respectively. The atom r is displaced from the origin at O by the components $x_r \cdot a$ along a and $y_r \cdot b$ along b. The distance $x_r \cdot a$ is equivalent to a phase component

$$\frac{x_r \cdot a}{a/h} 2\pi = hx_r \cdot 2\pi = \varphi_{ra}$$

along a for the r atom. Similarly, $y_r \cdot b$ is equivalent to the phase component

$$\frac{y_r \cdot b}{b/k} 2\pi = ky_r \cdot 2\pi = \varphi_{rb}$$

along b. The phase factor for the r atom is the sum of the two component phase factors and therefore

$$\begin{aligned} \varphi_r &= \varphi_{ra} + \varphi_{rb} \\ &= 2\pi(hx_r + ky_r) \end{aligned}$$

It can be shown that for a three-dimensional structure

$$\varphi_r = 2\pi(hx_r + ky_r + lz_r)$$

where the term lz_r refers to a phase component along the c axis. After inserting this form of the expression for the phase factor, the equation for the structure amplitude becomes

$$\begin{aligned} |F(hkl)| = \{&[\textstyle\sum f_{0,r} \cos 2\pi(hx_r + ky_r + lz_r)]^2 \\ &+ [\textstyle\sum f_{0,r} \sin 2\pi(hx_r + ky_r + lz_r)]^2\}^{1/2} \end{aligned} \tag{2}$$

The value of $|F|$ now clearly depends on the kinds of atom in the unit cell and on their positions in the cell. The first dependence manifests itself by the summation of the f_0 factors and the second by the summation of the two components

$$\cos 2\pi(hx_r + ky_r + lz_r) \quad \text{and} \quad \sin 2\pi(hx_r + ky_r + lz_r)$$

This latter portion of the equation is often called the geometrical structure factor, because it is solely a function of the coordinates of the atoms and therefore characteristic of the space group of the crystal. By using the new value of φ, the phase angle for the hkl reflection can be similarly expressed as a function of the coordinates of the atoms

$$\alpha(hkl) = \tan^{-1} \frac{\sum f_{0,r} \sin 2\pi(hx_r + ky_r + lz_r)}{\sum f_{0,r} \cos 2\pi(hx_r + ky_r + lz_r)} \tag{3}$$

The Temperature Factor

Atomic scattering factors are calculated for atoms at rest. At ordinary temperatures, however, atoms have thermal vibrations that increase with the temperature, so that at any instant like atoms in the cell are not separated by exactly the same distances. In terms of the Bragg reflection concept, this means that a reflecting plane is not a mathematical plane. At any instant its atoms lie above and below as well as on the plane defined by the mean positions for the atoms. As a consequence, waves reflected at the Bragg angle by successive planes will not be exactly in phase and the structure amplitude will be smaller than anticipated from the atomic scattering factors. The interference naturally

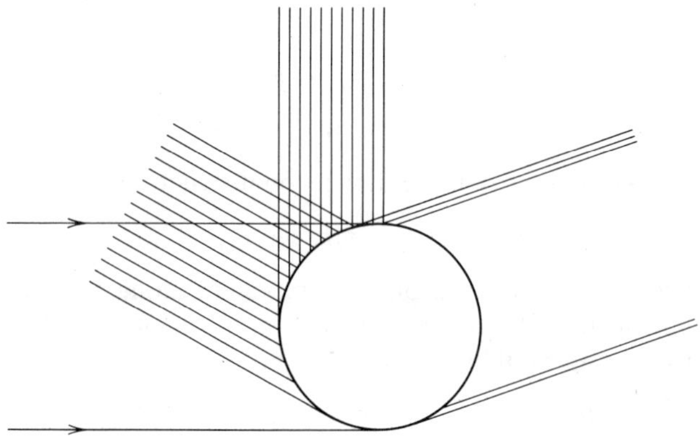

Fig. 3-18. The volume of specimen that contributes to a reflection increases with the angle θ.

increases as the displacement from the mathematical planes becomes a bigger fraction of the interplanar spacing. Hence for any crystal the effect on the structure amplitude increases with the Bragg angle [because $\theta = \sin^{-1}(n\lambda/2d)$] and with an increase in the temperature. The temperature effect is strikingly demonstrated by low-melting organic compounds. In contrast to most inorganic compounds, their diffraction intensities fall rapidly to zero with increasing θ angle.

Debye deduced on theoretical grounds in 1914 that the scattering factor of an atom at ordinary temperatures (f) is related to its scattering factor at rest (f_0) by the expression

$$f = f_0 e^{-B(\sin^2\theta)/\lambda^2}$$

where B incorporates the mean displacement of the atom from its mean position. B depends on the kind of atom, and it may also depend on the orientation of the reflecting planes in the crystal. The quantity can be estimated only approximately.

The Absorption Factor

The incident and reflected beams are partially absorbed in passing through a crystal, and consequently the intensity of reflection is less than it would be from a perfectly nonabsorbing substance. The absorption effect evidently depends not only on the absorption coefficient of the crystal but also on its cross section and, in the case of a powdered specimen, on the density of packing of the particles. As a result, the intensities of reflections vary with the Bragg angle. Figure 3-18 illustrates this effect for a highly absorbing specimen that absorbs most rays following the long-path length through the central part of the specimen. Most of the rays that contribute to a reflected beam necessarily originate at or near the surface. The figure indicates that, as θ increases, the volume of specimen contributing to a reflection also increases. Hence the absorption factor decreases with increasing Bragg angle and has the opposite effect of the temperature factor. In fact, the two factors approximately cancel each other in crystals that have moderate absorption coefficients. This naturally is not true of many organic crystals, for which the temperature effects are high and the absorption effects low.

It is possible to calculate the absorption factor for a crystal of any shape. The degree of accuracy of the correction depends on the accuracy with which the shape, and therefore the path lengths of incident and diffracted rays, can be defined. The inherent problems with a crystal of general shape may be avoided by grinding the crystal to a uniform shape, either a cylinder or a sphere, thereby simplifying the evaluation of path length and systematizing the correction factor for path direction. It is relatively easy to grind a crystal into a cylindrical shape but it is useful only if the cylinder axis is parallel to some preferred direction in the crystal structure. This may be difficult to achieve if the crystal is oddly shaped or particularly resistant to grinding. A spherical shape is the ideal form for the standard single-crystal x-ray methods, but not all crystals can be readily ground into spheres, some having a tendency towards football-like shapes. The *International Tables for X-ray Crystallography*[1] provide tables of the absorption correction factor (i.e., the number by which the apparent intensity should be multiplied to obtain the "true" intensity) for successive values of θ and μR, where μ is the linear absorption coefficient and R is the radius of the cylinder or sphere.

It is a common practice to make no correction, but to use very small specimens (of the order of 0.1 mm in size), thereby reducing the effect of absorption presumably to negligible proportions.

[1] *International Tables for X-ray Crystallography*, **2**, Kynoch Press, Birmingham (1959).

The Lorentz Factor

In most single-crystal x-ray methods the specimen is given a controlled motion in the path of the x-ray beam. Various planes in the crystal are thereby successively presented to the beam at their respective Bragg angles and are given an opportunity to reflect. The rate at which the planes pass through the Bragg condition varies with the diffraction angle, and therefore each intensity measurement must be separately corrected for the time that a particular plane spends in rotating through the Bragg angle. The form of this correction, which is known as the Lorentz factor, depends on the method of recording the reflections.

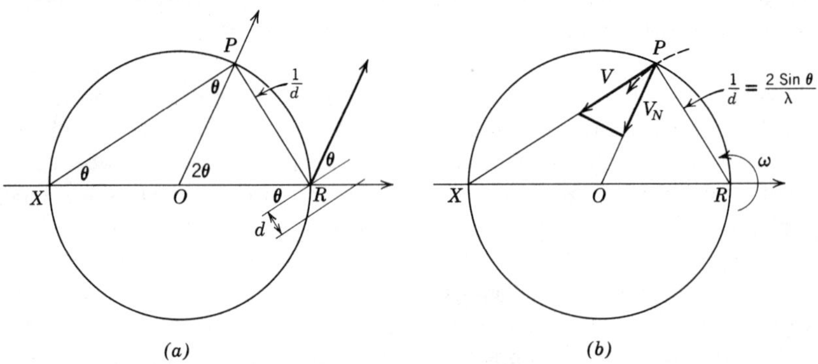

Fig. 3-19. (*a*) Geometrical representation of the Bragg condition, and (*b*) the Lorentz factor for reflection in the plane of the page.

The geometry of the Bragg condition can be conveniently illustrated by the construction given in Fig. 3-19*a*. The figure shows a set of planes of spacing *d*, in a crystal at *R* on the circumference of a circle, reflecting the incident beam *XOR*. If *PR* is constructed normal to the planes to intersect the circle, angle $PXR = \theta$. Because angle $RPX = 90°$, $PR/XR = \sin\theta$, and then *PR* and *XR* are proportional, respectively, to $1/d$ and $2/\lambda$. Also, angle $ROP = 2\theta$, and therefore *OP* is parallel to the reflection direction. Similarly, the Bragg condition for each set of planes reflecting in the plane of the page leads to the construction of a right-angle triangle resting on the same diameter, provided that the wavelength is constant.

Assume now that the crystal is rotating with a constant angular velocity ω about an axis normal to the page at *R* (Fig. 3-19*b*). The Bragg condition for the planes is satisfied, and the planes reflect during the time that it takes *P* to cross the circumference of the circle. The time involved depends on the velocity of *P* and the angle at which the point intersects the circle—namely, on the velocity component V_N

normal to the circumference of the circle. Points far from R (small spacings, large θ angles) move more rapidly than those near R, but pass through the circle at more oblique angles. The velocity V of the point is $RP \cdot \omega = (1/d)\omega = ((2 \sin \theta)/\lambda)\omega$ and at P is directed along PX. The velocity component normal to the circle has the direction PO, at an angle θ to PX. Therefore

$$V_N = V \cos \theta = \left(\frac{2 \sin \theta}{\lambda} \omega\right) \cos \theta = \left(\frac{\sin 2\theta}{\lambda}\right) \omega$$

If E is the total energy of radiation reflected by a set of planes in the time that it takes P to cross the circumference of the circle

$$E \cdot V_N = E\omega\left(\frac{\sin 2\theta}{\lambda}\right) = I_c$$

where I_c is the intensity of scattering by the crystal. The product $E\omega(\sin 2\theta/\lambda)$ is constant for a given beam intensity. If ω is halved, the planes have twice the time to reflect, and E is doubled. The relation can usefully be recast to give energy in terms of the angular velocity of the crystal

$$E\omega = I_c\left(\frac{\lambda}{\sin 2\theta}\right) \tag{4}$$

Because energy is intensity multiplied by time, the quantity $\lambda/\sin 2\theta$ expresses the time that the planes spend in reflecting energy E. Hence $1/\sin 2\theta$ expresses the *relative* time that any crystal plane spends in rotating through the small angular range of θ over which reflection occurs. This quantity is known as the Lorentz factor. The factor evidently has a different form for reflecting planes which are not parallel to the rotation axis—that is, for points which cross the circle of Fig. 3-19 at an angle to the page.

The Integrated Reflection

The expression

$$E\omega = I_c\left(\frac{\lambda}{\sin 2\theta}\right) \tag{4}$$

gives the total energy of radiation reflected by a set of (hkl) planes in the time that it takes the planes to rotate through the Bragg condition, if the crystal rotates with an angular velocity ω about an axis parallel to the planes. The expression can be expanded into a useful equation with the aid of the various factors that determine the intensity of scattering by a crystal. Remembering that intensity is proportional to amplitude squared,

$$I_c = I_e \times (|F(hkl)| \times N)^2 \times V \times T$$

where I_e is the intensity of scattering by an electron, $|F(hkl)|$ is the ratio of the amplitude of the composite wave scattered by one unit cell to the amplitude of the wave scattered by one electron, N is the number of unit cells per unit volume, V is the volume of the crystal, and T is the temperature factor. The relation

$$I_e = I_0\left(\frac{e^4}{m^2c^4R^2}\right)\left(\frac{1 + \cos^2 2\theta}{2}\right) \tag{1}$$

which was developed previously, gives I_e at a distance R from the electron. Taking R in this relation equal to $1/\lambda$ (the radius of the "reflection" circle in Fig. 3-19),

$$\begin{aligned} E\omega &= I_0\left(\frac{e^4}{m^2c^4(1/\lambda)^2}\right)\left(\frac{1 + \cos^2 2\theta}{2}\right) \times (|F(hkl)| \times N)^2 \times V \times T \times \frac{\lambda}{\sin 2\theta} \\ &= I_0\left(\frac{N^2e^4\lambda^3V}{2m^2c^4}\right)\left(\frac{1 + \cos^2 2\theta}{\sin 2\theta}\right)|F(hkl)|^2 T \end{aligned}$$

This gives the total energy received at the circumference of the "reflection" circle. The expression assumes that the primary and reflected beams are unaffected in the crystal by ordinary absorption and by the special absorption effects known as extinction, which are described in the next section. It therefore applies to the theoretical ideal case of a crystal that is infinitely small.

When recast into the form

$$\frac{E\omega}{I_0} = \left(\frac{N^2e^4\lambda^3V}{2m^2c^4}\right)\left(\frac{1 + \cos^2 2\theta}{\sin 2\theta}\right)|F(hkl)|^2 T \tag{5}$$

the relation gives the ratio of the total energy reflected by the (hkl) planes to the energy of x-radiation falling on unit area of the crystal in unit time. The ratio, usually designated by the symbol ρ, is called the integrated reflection. The measurement of the absolute integrated reflection involves a comparison of the intensity of reflection from a set of planes with the intensity of the primary beam. Measurement of the primary beam is inconvenient, experimentally, and tedious. Consequently *relative* intensities, which serve for most purposes, are generally used. The quantity $(N^2e^4\lambda^3V/2m^2c^4)$ can then be treated as a constant for a particular crystal and the relative intensities I of all reflections from planes parallel to the rotation axis are given by the relation

$$I_{hkl} \propto \left(\frac{1 + \cos^2 2\theta}{\sin 2\theta}\right) |F(hkl)|^2 T \tag{6}$$

Integrated Reflections from Real Crystals

The Effect of Extinction

The equation for the integrated reflection, developed in the preceding section, is based on the concept of a crystal which exhibits no absorption of the incident beam. The crystals used in x-ray single-crystal studies—that is, real crystals—do affect the intensity of the beam by absorption, and, when the intensity of a particular reflection is noticeably decreased, it is necessary to modify the equation by the inclusion of an absorption factor. Real crystals show two additional absorption effects—known as *primary* and *secondary extinction*—which operate at the Bragg angles while a crystal reflects. The effect of extinction varies with the degree of crystal perfection, which in turn is defined by the following terms. A crystal in which the atoms are arranged in a perfectly regular manner throughout the crystal is said to be *ideally perfect.* Certain fine-quality specimens of calcite (rhombohedral $CaCO_3$) and the diamond approach this state of perfection, but most crystals exhibit an imperfection known as *mosaic structure.* Mosaic crystals consist of more or less perfect blocks, of the order of 1000 Å (0.1 μ) in size, which deviate from parallel alignment by less than one degree. An *ideally imperfect* crystal is one in which the blocks are very tiny and in completely random orientation—a condition that is most closely realized by a fine crystalline powder.

In the process of a Bragg reflection each successive plane in a perfect region of a crystal reflects a fraction of the incident radiation, with the result that the beam becomes progressively weaker as it penetrates to lower planes. The reflected rays strike the reverse sides of the planes at the correct angle for reflection, and consequently a fraction of the once-reflected radiation is reflected back into the crystal in a direction parallel to the incident rays (Fig. 3-20). Because a $\pi/2$ phase change occurs on reflection, the twice-reflected rays have a phase difference of π with respect to the incident beam and interfere to cause an additional decrease in the intensity of the penetrating beam. The combined effect manifests itself as a pronounced increase in absorption at the Bragg angle; the increase may be as much as a hundredfold one in some ideally perfect crystals. This effect is known as primary extinction. The intensity of a reflected beam suffers through primary extinction, not only because the incident beam is weakened, but also because multiply reflected rays, traveling in the reflection direction, may interfere. A thrice-reflected ray, for example, travels in the same direction as a once-reflected ray, but is opposite in phase and interferes to reduce the intensity of the reflected beam. The effect of primary extinction is therefore a dissipation of both incident and reflected beams. The strong reflections from a crystal evidently cause the greatest reduction in the intensity of the primary beam, and consequently these reflections are relatively more affected by primary extinction than the weak reflections.

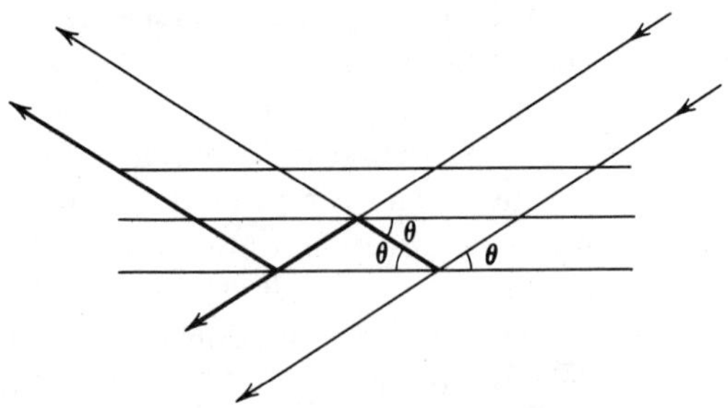

Fig. 3-20. Multiple reflections in a perfect region of a crystal.

Darwin showed that the observed integrated reflection (ρ_{obs}) can be corrected for primary extinction by the relation

$$\rho_{\text{calc}} = \frac{\rho_{\text{obs}}}{(\tanh mq)/mq}$$

where q is the amplitude of the radiation reflected by a single plane of atoms and m is the number of perfect planes in a regular succession. The quantity q equals $2Ne^2d^2F/mc^2$ (neglecting the Lorentz and polarization factors), in which N is the number of unit cells in unit volume, d is the spacing of the planes, F is the structure amplitude, and e, m, and c have their usual meaning. The product $m \cdot d$ is the thickness of the regular succession of perfect planes. Table 3-3 gives the correction factors $[(\tanh mq)/mq]$ calculated by Lonsdale[2] for $h00$ reflections from cleavage faces of NaCl of thicknesses 0.1, 1, and 10 μ. The table shows that primary extinction is important only when the thickness is in excess of 1 μ and chiefly affects the strong reflections (large values of q). For thicknesses in excess of 100 μ, the expression $(\tanh mq)/mq$ is not a good approximation of the correction factor; under these circumstances

[2] K. Lonsdale, *Mineral. Mag.*, **28**, 14 (1949).

an equation

$$\rho = \left(\frac{4Ne^2\lambda^2}{3\pi mc^2}\right)|F|\left(\frac{1 + \cos 2\theta}{\sin 2\theta}\right)$$

derived by Darwin for an ideally perfect crystal gives a better agreement between calculated and observed intensities.

Table 3-3. Values of the Expression [(tanh *mq*)/*mq*] for the *h*00 Reflections from NaCl

			(tanh mq)/mq		
Reflection	q	d (Å)	$m \cdot d = 0.1\ \mu$	$= 1\ \mu$	$= 10\ \mu$
200	2.05×10^{-4}	2.814	1	0.85	0.14
200	$0.30_7 \times 10^{-4}$	1.407	1	0.98	0.46
600	$0.08_1 \times 10^{-4}$	0.938	1	1	0.81
800	$0.02_6 \times 10^{-4}$	0.703	1	1	0.90
10·0·0	$0.00_9 \times 10^{-4}$	0.563	1	1	0.99

Primary extinction severely limits the depth of penetration of the beam in an ideally perfect crystal during a strong reflection; according to Table 3-3, the depth is of the order of 1 μ. If perfect blocks in a mosaic crystal are smaller than this, the whole of each block in the path of the beam reflects, and primary extinction is not important. However, in all crystals in which the perfect regions are larger than this, primary extinction reduces the intensity of a strong reflection below that expected from the equation for the integrated reflection.

In any mosaic crystal, whether the perfect blocks are small enough for primary extinction to be absent or not, the penetrating beam becomes progressively weaker for another reason. In such a crystal reflection takes place over an angular range because of the disorientation of the blocks. At any setting in the range a large number of blocks are precisely oriented to reflect. Each block reflects away a fraction of the intensity of the beam as it passes to greater depths; hence the intensity of the reflected beam is less than if all the blocks received the full benefit of the incident beam. This process of shielding the lower blocks by reflection from the upper blocks is known as secondary extinction. The effect, as in primary extinction, is most pronounced for the intense reflections from a crystal. With increasing disorientation, relatively fewer blocks are oriented to reflect at any position in the angular range, and therefore fewer blocks are available to shield the interior of the crystal. Secondary extinction reaches a minimum with the ideally imperfect crystal.

Because only a part of a mosaic crystal contributes to a reflection at any setting in its angular range, an ideally perfect crystal of the same crystal structure reflects more strongly at the ideal Bragg angle. However, the beam penetrates to greater depths in a mosaic crystal, and a larger volume of the crystal takes part in the over-all reflection through the angular range. Thus the curve obtained by plotting intensity against angular range shows a higher maximum for the ideally perfect crystal, but encloses a greater area in the case of the mosaic crystal. The area under the curve is a measure of the integrated reflection or total energy reflected by the crystal. An ideally imperfect crystal, which may now be defined as a crystal composed of mosaic blocks measuring less than 1 μ in size in completely random orientation, gives an integrated reflection which most closely approaches the value predicted by the integrated reflection equation developed in the preceding section. The relative integrated reflections from these crystals are therefore proportional to $|F|^2$. In contrast, the relative integrated reflections from nearly perfect crystals are roughly proportional to $|F|$ and from most real crystals to a value between these extremes.

Extinction is probably the most troublesome factor in equating measured and calculated integrated reflections from single crystals. The mosaicity can be increased, and therefore primary extinction reduced, by subjecting the crystal to a thermal shock in liquid air. Secondary extinction can be allowed for by adding a factor to the ordinary absorption coefficient, but the factor varies with the specimen and can be determined satisfactorily only if the crystal structure is known. Secondary extinction can be reduced to a minimum by grinding the crystal to a fine powder. However, this is not always a solution, because the powder method does not ensure that the reflections will be well resolved and therefore measurable; many diffraction problems can be satisfactorily treated only with single-crystal studies. Grinding to a powder also reduces primary extinction, but to attain the condition on which this factor is negligible it is necessary to grind to a particle size of 1 to 0.1 μ. Presumably this also eliminates differences in mosaicity between different specimens of the same compound, and therefore the integrated reflections for a given Bragg angle should be the same for all specimens of the same compound. Unfortunately, it is difficult and, in fact, often impossible to achieve this particle size. In general, then, the intensities of reflections from single crystals and even fine powders are affected by extinction.

The integrated reflections obtained in ordinary diffraction studies are subject to two additional intensity factors. Like the Lorentz factor, these factors vary with the method used to record the reflections. The *multiplicity* factor takes account of the number of equivalent planes that contribute to a reflection. The other factor takes account of the fact that, in cylindrical-film powder methods, the reflections at any given Bragg angle lie on the surface of a cone; the degree of blackening varies with the cone angle. These factors are best discussed after the methods have been fully described.

The Determination of Crystal Structures

The equation for the integrated reflection

$$\frac{E_\omega}{I_0} = \left(\frac{N^2 e^4 \lambda^3 V}{2m^2c^4}\right)\left(\frac{1 + \cos^2 2\theta}{\sin 2\theta}\right)|F(hkl)|^2 T \tag{5}$$

contains the factor $|F(hkl)|$ which has been expressed in terms of the positions of atoms in the unit cell

$$|F(hkl)| = \left\{\left[\sum_r f_r \cos 2\pi(hx_r + ky_r + lz_r)\right]^2 + \left[\sum_r f_r \sin 2\pi(hx_r + ky_r + lz_r)\right]^2\right\}^{\frac{1}{2}} \tag{2}$$

The equation therefore permits the calculation of the intensities of reflections when the crystal structure is known. The measured intensities, suitably corrected for absorption and extinction, and the special factors which arise from the method of recording the reflections, can be expected to confirm the calculations. It may seem that the converse also applies—namely, that the measured intensities from a crystal of unknown structure offer a straightforward method of calculating the positions of the atoms.

Atoms are positions of relatively high electron density in the crystal. Thus the electron density varies throughout the crystal, reaching maximum values at locations of atom centers, lower values at chemical bonds, and falling virtually to zero elsewhere in the structure. If the electron density could be mapped throughout the unit cell, and each atom gave a recognizable peak on the map, the structure could be easily deduced and then proved by a comparison of intensities calculated for the tentative structure with the intensities measured on the crystal. The first step in implementing such a procedure is to adopt a method of evaluating the electron density. Atoms are arranged in a periodic fashion in space, and therefore the electron density also varies periodically throughout the crystal. The density at any level x between the (100) lattice planes, for example, is a periodic function, which repeats its value whenever x is increased by an amount a equal to the cell edge. The function can be expressed by a type of trigonometric summation of an infinite series known as a Fourier series

$$f(x) = A_0 + A_1 \cos(2\pi \cdot x - \alpha_1) + A_2 \cos(2\pi \cdot 2x - \alpha_2) + \\ + A_3 \cos(2\pi \cdot 3x - \alpha_3) + \cdots$$

$$= \sum_0^\infty A_h \cos(2\pi \cdot hx - \alpha_h)$$

The expression is the summation of a number of cosine terms, of which $A_0, A_1, A_2, A_3, \ldots$ are the amplitudes and $\alpha_1, \alpha_2, \alpha_3, \ldots,$ their phases. Now, a diffracted beam has been shown to be geometrically equivalent to a reflection of the incident beam from a set of lattice planes; the intensities of the various orders of reflections from the planes depend on the distribution of atoms—that is, on the distribution of electron density— between the planes. Hence the absolute electron density $\rho(x)$ at any level x between (100) planes of spacing d_{100} can be expressed by a Fourier synthesis of all the $h00$ reflections

$$\rho(x) = \frac{1}{d_{100}}\{F(000) + F(100)\cos[2\pi \cdot x - \alpha(100)] \\ + F(200)\cos[2\pi \cdot 2x - \alpha(200)] + \cdots\}$$

$$= \frac{1}{d_{100}} \sum_0^\infty |F(h00)| \cos[2\pi \cdot hx - \alpha(h00)]$$

where $F(000)$, the structure amplitude for the zero-order diffraction, is the number of electrons in the unit cell, the other F's are the structure amplitudes of successive $h00$ diffractions, and the α's are their phases.

Similarly, the density of scattering matter at any point xyz in the unit cell can be expressed by a three-dimensional Fourier series in which there is one term for each reflection, its coefficient being the corresponding value of F

$$\rho(xyz) = \frac{1}{V_c}\left\{\sum h \sum_{-\infty}^{\infty} k \sum l |F(hkl)| \cos[2\pi(hx + ky + lz) - \alpha(hkl)]\right\}$$

where V_c is the volume of the unit cell. The evaluation and representation of a three-dimensional synthesis is tedious. A simplification, which is adequate for the preliminary investigation of crystal structures, consists of projecting the electron density on, say, three faces in turn. The data are obtained by three two-dimensional Fourier syntheses, one for each of three zones of reflection. For example, the calculation of the electron density at a large number of xy points from $hk0$ reflections with the aid of the summation

$$\rho(xy) = \frac{1}{A} \sum h \sum_{-\infty}^{\infty} k |F(hk0)| \cos[2\pi(hx + ky) - \alpha(hk0)]$$

gives a picture of the structure as viewed along the c axis. A in the equation is the area of the unit cell projected along the c axis. When the

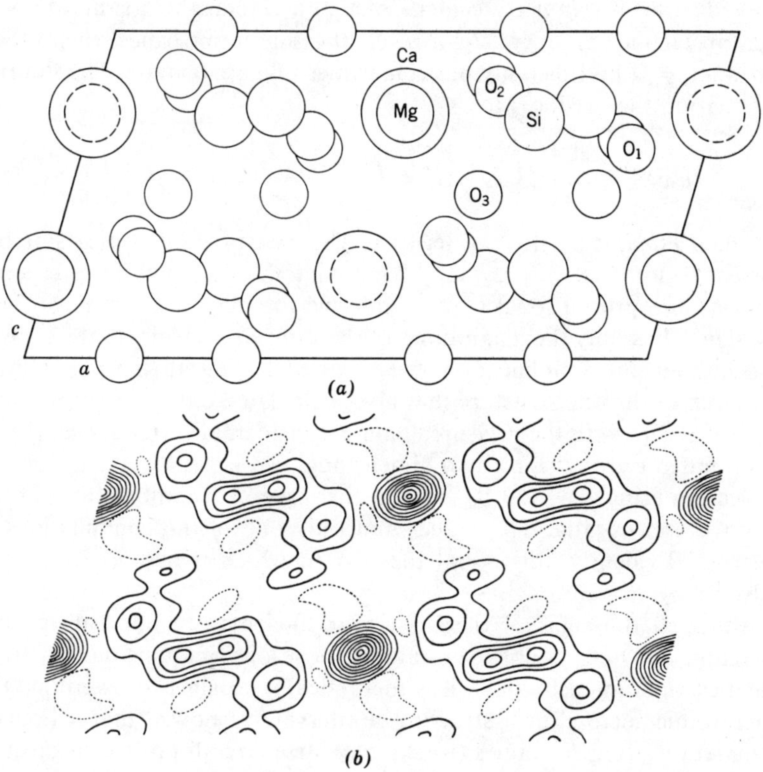

Fig. 3-21. Atomic positions (*a*) and summation of Fourier series (*b*) of diopside ($CaMg(SiO_3)_2$ projected on (010). (W. L. Bragg, *Z. Krist.*, **70**, 488, 1929).

values of $\rho(xy)$ are plotted and contoured (Fig. 3-21), the result is a projection of the electron density onto the c face.

It seems, then, that a Fourier series provides an elegant mathematical way of using experimental x-ray data (the measured intensities, which are proportional to amplitude squared) to solve structures. A fundamental difficulty to the direct determination of crystal structures by means of

Fourier analysis is the presence of the phase component $\alpha(hkl)$ in the summation. This component, which was shown to be equal to

$$\alpha(hkl) = \tan^{-1}\left[\frac{\Sigma f \sin 2\pi(hx + ky + lz)}{\Sigma f \cos 2\pi(hx + ky + lz)}\right]$$

is not provided by the experimental data. A considerable simplification does occur if the crystal is centrosymmetrical and the origin is taken to coincide with a symmetry center. In such a crystal there is an atom at $\overline{xyz}$ for each atom at xyz. As a result the sine terms cancel each other, $\tan \alpha(hkl) = 0$, and the only possible values of α are 0 and π. The Fourier summation then reduces to

$$\rho(xyz) = \frac{1}{V_c}\left\{\sum h \sum_{0}^{\infty} k \sum l \pm F(hkl) \cos 2\pi(hx + ky + lz)\right.$$

in which $F(hkl)$ is positive or negative. The measured intensities, suitably corrected for the various modifying factors described in preceding sections, are proportional to $|F|^2$ and give the magnitude of F, but not the sign. Thus only the magnitudes of the Fourier coefficients are known, and, despite the simplification, the labor of testing all the possibilities in search of the one structure that gives calculated data that are in satisfactory accord with the measured data is considerable. McLachlan (1957) has pointed out that for a two-dimensional synthesis, "if one is doing a projection using, say, 150 F_{hk0} values, there are 2^{150} combinations of the positive and negative F_{hk0} values and only one of the combinations is correct." Evidently, this is not the way in which structures have been solved.

Although the signs of the Fourier coefficients are not known, and therefore it is not possible to locate the positions of atoms directly from the measured intensity data, it is theoretically possible to determine the interatomic vectors for a structure. Patterson[3,4] showed that a Fourier summation using F^2 values (which, of course, are all positive regardless of the F signs)

$$P(xyz) = \frac{1}{V_c}\sum h \sum_{-\infty}^{\infty} k \sum l |F(hkl)|^2 \cos 2\pi(hx + ky + lz)$$

gives a vectorial pattern of the distances between the atoms in the structure. To illustrate, a plot of the P function for the three atoms in Fig. 3-22 should consist, in addition to the peak at the origin, of six

[3] A. L. Patterson, *Phys. Rev.*, **46**, 372 (1934).
[4] A. L. Patterson, *Z. Krist.*, **90**, 517 (1935).

peaks marking the interatomic vectors *AB*, *BA*, *AC*, *CA*, *CB*, *BC*, the densities of the peaks being proportional to the product of the electron densities of pairs of atoms. The origin of a Patterson map marks the interatomic vector of each atom with itself; the density of this peak is correspondingly high. Ordinarily, relative intensities are used in a Patterson synthesis, although absolute intensities could add a useful measure of significance between the densities of peaks and the pairs of atoms they represent.

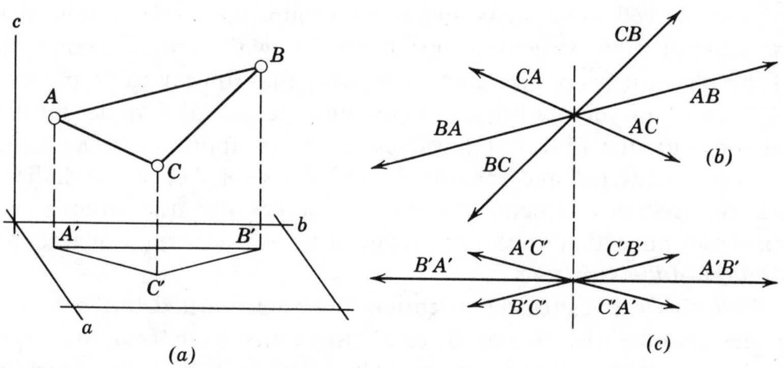

Fig. 3-22. Three atoms (*a*) and their Patterson interatomic vectors in space (*b*) and projected on the *ab* plane (*c*).

The value of a Patterson synthesis is limited by a characteristic common to all vector maps—the vectors have a common origin. The more complicated the structure, the more difficult it is to identify vectors with atoms and their actual positions in the structure. It is very helpful if the structure includes a few heavy atoms which give outstandingly dense peaks and therefore easily identifiable vectors. Despite the inherent difficulties, the Patterson synthesis has proved so useful that it has attained the status of a routine procedure for complex structures. Usually this takes the form of a two-dimensional synthesis. A synthesis of the *hk*0 intensities, for example, using the summation

$$P(xy) = \sum h \sum k |F(hk0)|^2 \cos 2\pi(hx + ky)$$

gives the Patterson peaks projected onto (001). If the Patterson synthesis gives indications of the positions of some of the atoms, it may be possible to decide on the signs of some of the *F*'s. This information can then be used to carry out a preliminary Fourier synthesis, leading to a first electron density map and a rough approximation of the crystal structure, presumably the part that accounts for most of the scattering power. The map may suggest adjustments to more probable positions for these atoms. The coordinates of the new positions, together with any new data deduced or surmised from the first synthesis about the positions of the weakly scattering atoms, may be included in a second synthesis to give a more accurate electron density map. This procedure can be repeated in a series of steps to refine a structure successively.

Many structures have been solved by a trial-and-error method. A structure is assumed on the basis of the chemical composition, a knowledge of its cell dimensions and space group, the configuration of the molecule or other structural unit, if it is known, a critical examination of the physical properties, and other data that appear to be pertinent. The assumed structure is used to compute theoretical F values for comparison with the observed F values. Even an approximate agreement between calculated and measured data is useful, because it indicates that the assumed structure at least resembles the true structure and offers the hope that successive refinements of the data will lead to a solution of the structure.

Since the war, increasing attention has been directed to the problem of determining the phases directly. Relations have been discovered between amplitudes and phases, which indicate that the amplitude alone may serve for certain crystal-structure analyses. The methods have limits of practicability. All structures cannot be solved by known methods of analysis, and research on the phase problem continues.

Aids are desirable for computing structure factors and electron densities in two-dimensional analyses and are essential in three-dimensional work. These aids take a number of forms, the most advanced of which are high-speed electronic digital computers. Such machines can be programmed to carry out differential Fourier series for a set of atom coordinates known to be approximately correct, to evaluate new coordinates, and to calculate new structure factors in cycles until the final set of required coordinates is obtained.

Diffraction or Laue Symmetry

The arrangement and stacking of atoms is identical parallel to symmetry-related planes in a crystal. The reflections that arise from these planes are therefore also identical. In general, the arrangement and stacking is different for planes unrelated by symmetry, and these planes might be expected to produce unidentical reflections. It would seem then that the intensities of reflections should reveal the point group symmetry of a crystal. Actually, ordinary intensity studies have only a limited value for deducing crystal symmetry because, in general, the intensities of reflections with the same indices, but opposite in sign (hkl and $\overline{hkl}$), are identical whether or not the crystal has a center of symmetry. The reason for this can be demonstrated with the crystal structure of cubic ZnS (Fig. 3-23*a*). The point-group symmetry of this compound is $\bar{4}3m$, and therefore the (111) and $(\bar{1}\bar{1}\bar{1})$ faces of the crystal are not related by symmetry. The planes of atoms, which are parallel to and bounded by these faces, have the sequence Zn · · S · · · · Zn · · S

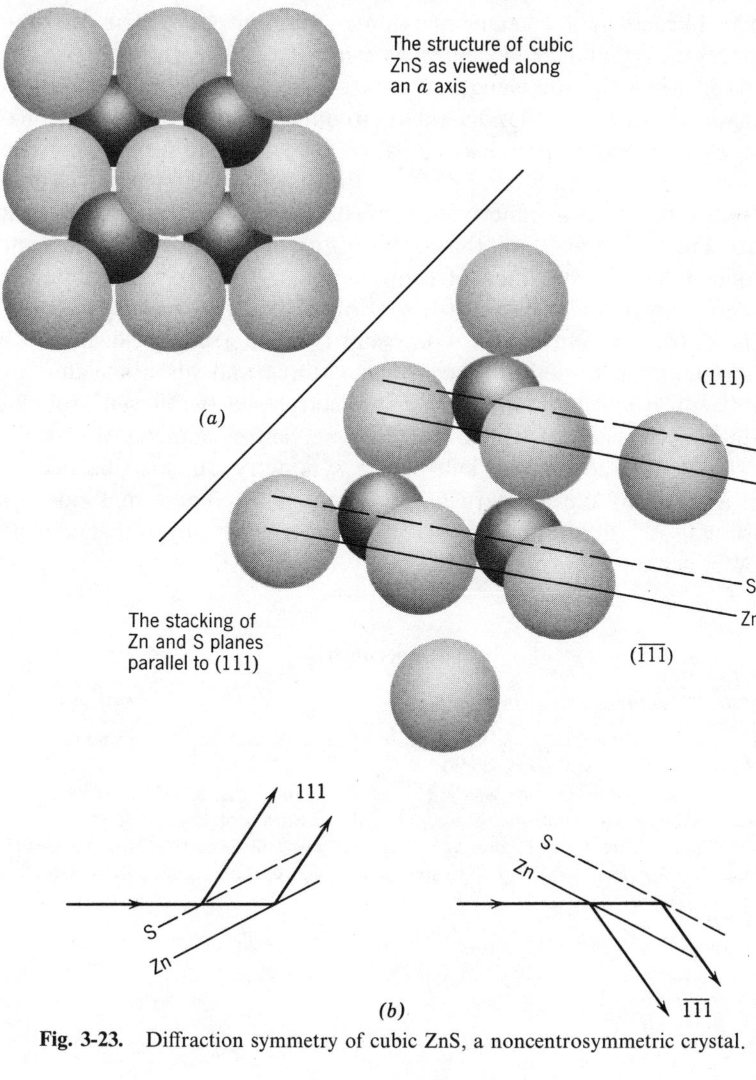

Fig. 3-23. Diffraction symmetry of cubic ZnS, a noncentrosymmetric crystal.

in one direction and S · · Zn · · · · S · · Zn in the opposite direction, portraying the absence of a symmetry center. A reflection from the top of the stack corresponds to the 111 reflection, and from the bottom of the stack, to the $\overline{1}\overline{1}\overline{1}$ reflection. The reflections are composite waves from the same two components. Although the phase of the Zn component is ahead in one reflection and behind in the other (Fig. 3-23*b*), the amplitudes of the composite waves are the same. Hence the intensities of the reflections are also the same. In fact, the intensities of all pairs of reflections, differing only by being opposite in sign, are the same, and the crystal diffracts as if a center of symmetry were present. This is true of most crystals that lack a center of symmetry. Exceptions[5] have been observed when the wavelength of x-rays is near the absorption edge of atoms in the structure. Under these circumstances an anomalous phase change may be induced in one of a pair of reflections. In general, however, ordinary x-ray methods cannot detect the absence of a symmetry center, and the diffraction symmetry of a crystal is its point-group symmetry plus a center of symmetry—that is, the diffraction symmetry corresponds to one of the eleven kinds of centrosymmetry (column 3, Table 1-3). This effect, which was first noted by Friedel, is called Friedel's law.

The diffraction symmetry of a crystal is most strikingly displayed by the original x-ray method used by von Laue and his associates. For this reason it is often called Laue symmetry, and the eleven groups of crystal classes, each of which includes one centrosymmetrical class, are described as Laue classes. Diffraction symmetry can also be deduced from a series of Weissenberg or precession films. These methods have additional advantages, which have brought them to prominence in the study of crystal symmetry.

References

Crystal Structure Analysis

Buerger, M. J. *Vector Space and its Application in Crystal-Structure Investigation*, John Wiley and Sons, New York (1959).

Buerger, M. J. *Crystal-Structure Analysis*, John Wiley and Sons, New York (1960).

Lonsdale, Kathleen. *Crystals and X-rays*, G. Bell and Sons, London (1948).

McLachlan, Jr., Dan. *X-ray Crystal Structure*, McGraw-Hill Book Co., New York (1957).

Nyburg, S. C. *X-ray Analysis of Organic Structures*, Academic Press, New York and London (1961).

[5] D. Coster, K. S. Knol, and J. Prins, *Z. Physik*, **63**, 345 (1930).

CHAPTER FOUR

SYSTEMATICALLY ABSENT X-RAY REFLECTIONS AND THE DETERMINATION OF SPACE GROUP

The record of the indices of reflections from any crystal invariably shows omissions, which indicate that the reflections from certain planes are either too weak to be detected or are totally absent, having been extinguished by the phase relations. These phase relations may be peculiar to the particular atomic structure or to lattice centering and/or the presence of glide planes and screw axes. Structural absences vary from compound to compound and do not lend themselves to a systematic classification, but absences due to lattice centering and the presence of space-group symmetry elements are systematic and readily predicted and can be classified for distinguishing between space groups with the aid of indexed x-ray patterns.

Systematic Absences due to Lattice Centering

One Face-Centered Lattice (*C*, *B*, *A*)

Figure 4-1 shows three selected sets of planes—(100), (110) and (111)—superimposed on a *C*-centered lattice. The lattice points at the face centers fall on (110) and (111) planes outlined by the primitive-lattice points but lie midway between (100) planes. The effect of *C*-centering on these three sets of planes is to double the density of points on the (110) and (111) planes and to create new planes, identical in density with and midway between the (100) planes. The spacing of the (100) planes has thereby been halved.

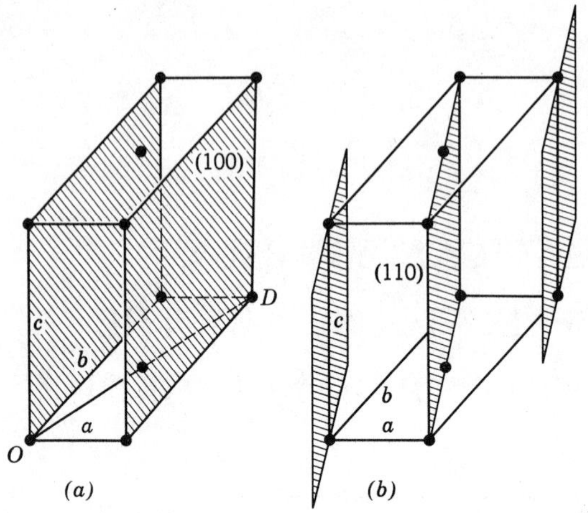

Fig. 4-1. Planes in a *C*-centered lattice.

The effect of a *C*-centering on the spacings of other planes is readily predicted. A set of (320) planes, for example, divides a in a primitive lattice into $h = 3$ parts and b into $k = 2$ parts (Fig. 4-2). A line in the lattice, such as OL, can be expressed by a vectorial equation

$$OL = 2a + 1b$$

where a and b are the unit vectors along the crystallographic axes. The number of parts into which this line is divided by the planes is given by the sum of the parts on $2a$ and $1b$, namely,

$$(2 \times h) + (1 \times k) = (2 \times 3) + (1 \times 2) = 8$$

In the general case any lattice line expressed by the vector sum

$$OL = ua + vb + wc$$

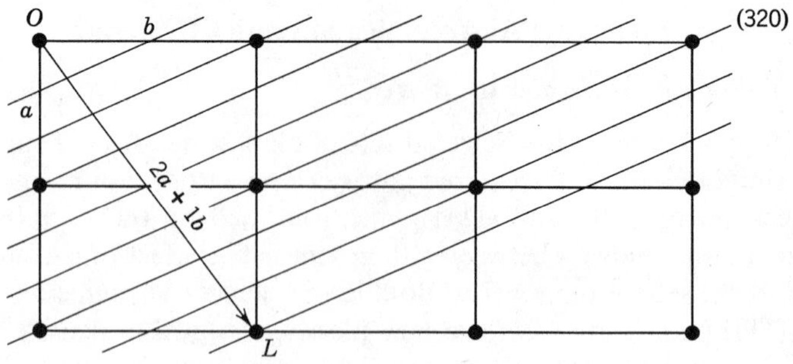

Fig. 4-2. A set of (320) planes divides OL into $2h + 1k = 8$ parts.

is divided into

$$uh + vk + wl \text{ parts}$$

by a set of (hkl) planes.

The face diagonal OD of Fig. 4-1 includes the centering lattice point and is defined by the vector sum

$$OD = 1a + 1b$$

The diagonal is therefore divided into $1h + 1k$ parts by a set of (hkl) planes. If the sum $h + k$ is odd, as in (100), the centering points fall midway between the planes formed by the primitive-lattice points, and the spacing is halved. The 100 reflection develops from a primitive lattice

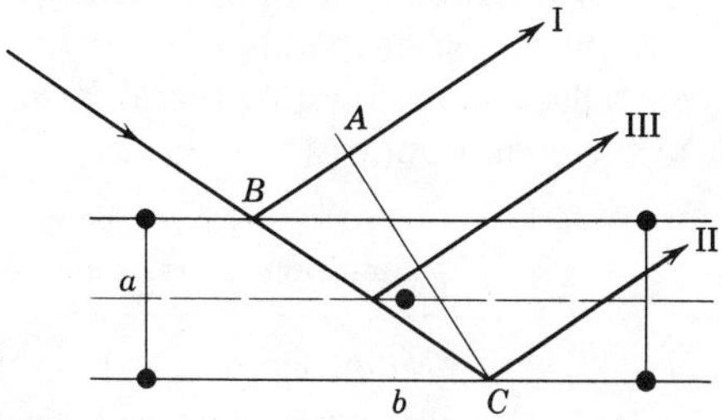

Fig. 4-3. Reflection from (100) planes in a C-centered lattice.

when the path difference $BC - BA$ (Fig. 4-3) between rays I and II, reflected from successive (100) planes, is one wavelength. If the lattice is C-centered, as shown in Fig. 4-3, the path difference between ray III from the new plane and rays I and II is exactly one half wavelength. The waves from successive planes in the centered lattice are therefore opposite in phase, and complete cancellation occurs. However, waves from successive planes do cooperate, if the path difference between rays I and II is equal to an even number of wavelengths. This is equivalent to stating that only the even-order reflections—200, 400, 600, etc.—can occur. This example serves for all reflections whose spacings are halved by centering the base; only the even orders of reflections can survive; hence the reflections have indices with $h+k$ even.

If the planes have indices with $h + k$ even, the centering lattice points fall on the planes outlined by the corner points. The spacings of these planes are unaffected by C-centering, and consequently all orders of reflections are possible. Since the indices of these reflections also have $h + k$ even, the indices of all reflections from a C lattice conform to the condition

$$hkl \text{ present only with } h + k = 2n$$

where n is any integer, including zero.

The absence of reflections with indices $h + k$ odd may be explained in another way. The (100) planes, like all planes with $h+k$ odd in a base-centered lattice, are characterized by a halved spacing. In the terminology of x-ray crystallography, it is proper to refer to this spacing as d_{200}, hence to the planes by the indices (200). In effect the (100) planes are missing in this lattice, and no 100 reflection is possible. The reflections must then be regarded as arising from (200) planes. The successive orders of reflections from these planes have indices 200, 400, 600, etc. The general case may be stated as follows: C-lattice centering eliminates all planes with $h + k$ odd. The reflections then arise from planes with $h+k$ even, and the indices of these reflections must have $h + k = 2n$.

The conditions for reflection imposed by A and B centered lattices bear a simple relation to the condition for the C lattice.

Lattice	Condition for Possible Reflection
C	hkl present only with $h + k = 2n$
B	hkl present only with $h + l = 2n$
A	hkl present only with $k + l = 2n$

All-Face-Centered Lattice (F)

The reflection condition for the all-face-centered lattice is obtained by combining the conditions for A, B, and C lattices. Thus hkl is present only with $h + k = 2n$, $k + l = 2n$, and $h + l = 2n$. The condition is equivalent to requiring that h, k, and l be all odd or all even. Reflections such as 010, 110, and 123 are extinguished in an all-face-centered lattice; 020, 220, 111, 222, 246 are possible.

Body-Centered Lattice (I)

The centering point in a body-centered lattice lies midway on the body diagonal defined by the vectorial equation $OD = (1a + 1b + 1c)$, where a, b, and c are the unit vectors on the axes of the cell (Fig. 4-4). A set of (hkl) planes therefore divides the diagonal into $1h + 1k + 1l$ parts. If the sum $h + k + l$ is odd, as shown in Fig. 4-4 for the (111) planes, the centering point is midway between the planes, and the spacing is halved. In effect, then, all planes with $h+k+l$ odd are destroyed by body-centering the lattice. The indices of reflections from such a lattice must

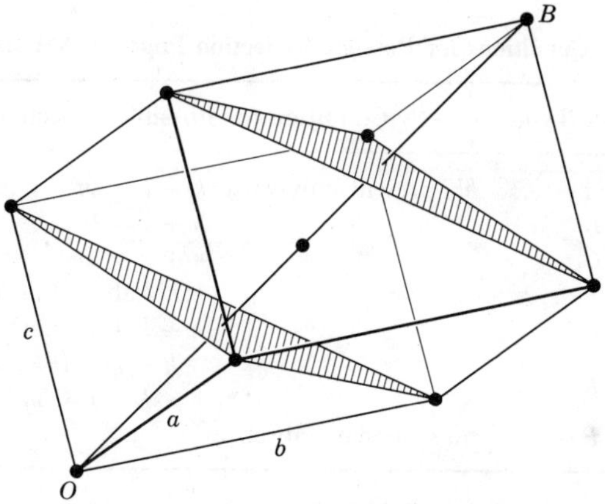

Fig. 4-4. (111) planes divide OB into $1h + 1k + 1l = 3$ parts.

therefore conform to the condition

$$hkl \text{ present only with } h + k + l = 2n$$

The Rhombohedral Lattice (*R*)

The rhombohedral lattice differs from the primitive hexagonal lattice by the addition of lattice points situated at $\frac{1}{3}$ and $\frac{2}{3}$ on a long body diagonal. This leads to two types of cell (Figs. 4-5*a* and *b*), depending on how the *a* axes are identified. The two body diagonals are defined by the

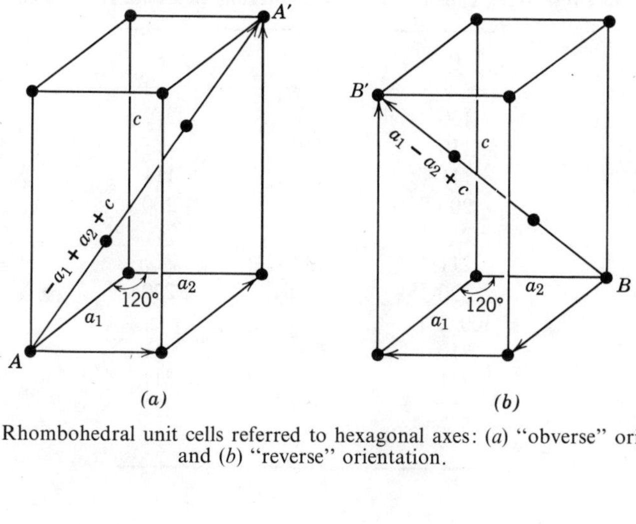

Fig. 4-5. Rhombohedral unit cells referred to hexagonal axes: (*a*) "obverse" orientation and (*b*) "reverse" orientation.

Table 4-1. Conditions for Possible Reflection Imposed by Lattice Centering

Lattice Type	Condition for Possible Reflection
A	hkl present only with: $k + l = 2n$[a]
B	$h + l = 2n$
C	$h + k = 2n$
F	h, k, l all odd or all even
I	$h + k + l = 2n$
R	$-h + k + l = 3n$ (obverse) $h - k + l = 3n$ (reverse)
P	no systematic absences

[a] n is any integer, including zero.

vectorial equations

$$AA' = -1a_1 + 1a_2 + 1c$$

$$BB' = 1a_1 - 1a_2 + 1c$$

where a_1, a_2, and c are the unit vectors in the cells. It follows from these equations that ($hkil$) planes divide the diagonals into $-h + k + l$ and $h - k + l$ parts, respectively. The planes evidently include the two centering points only if the sums are equal to $3n$. By analogy with the preceding descriptions, it follows that the conditions for reflection from

Table 4-2. Effect of Lattice Centering on Cubic Reflections

P	I	F
100	...	...
110	110	...
111	...	111
200	200	200
210	...	...
211	211	...
220	220	220
300, 221	...	...
310	310	...
311	...	311
222	222	222
320	...	...

the two cells are, respectively

$$-h + k + l = 3n$$
$$h - k + l = 3n$$

Representative possible reflections for the first type of cell are 0003, $30\bar{3}0$, $33\bar{6}3$, $32\bar{5}1$; extinguished reflections include 0001, 0002, $10\bar{1}0$, $20\bar{2}0$, $11\bar{2}1$, and $22\bar{4}2$.

Table 4-1 summarizes the conditions for possible reflection imposed by lattice centering. It is evident from the table that lattice centering drastically reduces the number of reflections to be expected from a crystal. This is further illustrated by Table 4-2, which gives, on the left, the first twelve possible reflections from a primitive cubic lattice in decreasing order of spacing (decreasing value of $d_{hkl} = a/\sqrt{h^2 + k^2 + l^2}$). The possibilities are reduced to six for a body-centered lattice and to five for an all-face-centered lattice.

Systematic Absences Due to Space-Group Symmetry Elements

Glide Planes

The operation defined as a glide plane combines a reflection across a plane with a translation of $\frac{1}{2}$ or $\frac{1}{4}$ of an identity period in the lattice

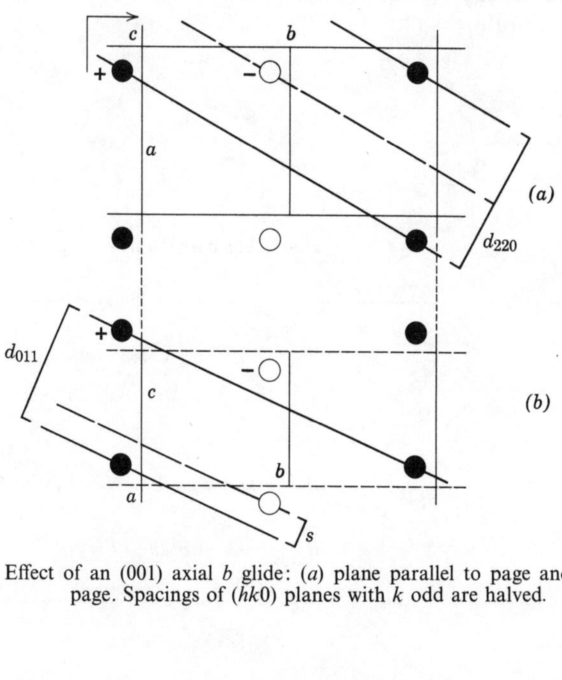

Fig. 4-6. Effect of an (001) axial *b* glide: (*a*) plane parallel to page and (*b*) normal to page. Spacings of (*hk*0) planes with *k* odd are halved.

Table 4-3. Systematic Absences Due to Glide Planes

	Orientation	Translation Component	Symbol	Condition for Possible Reflection
Axial	(100)	$\frac{b}{2}$	b	$0kl$ with $k = 2n$
	(100)	$\frac{c}{2}$	c	$0kl$ with $l = 2n$
	(010)	$\frac{a}{2}$	a	$h0l$ with $h = 2n$
	(010)	$\frac{c}{2}$	c	$h0l$ with $l = 2n$
	(001)	$\frac{a}{2}$	a	$hk0$ with $h = 2n$
	(001)	$\frac{b}{2}$	b	$hk0$ with $k = 2n$
	$(1\bar{1}0)$	$\frac{c}{2}$	c	hhl with $l = 2n$
	$(1\bar{1}00)$	$\frac{c}{2}$	c	$hh2\bar{h}l$ with $l = 2n$
	$(11\bar{2}0)$	$\frac{c}{2}$	c	$h\bar{h}0l$ with $l = 2n$
Diagonal	(100)	$\frac{b}{2} + \frac{c}{2}$	n	$0kl$ with $k + l = 2n$
	(010)	$\frac{a}{2} + \frac{c}{2}$	n	$h0l$ with $h + l = 2n$
	(001)	$\frac{a}{2} + \frac{b}{2}$	n	$hk0$ with $h + k = 2n$
	$(1\bar{1}0)$	$\frac{a}{2} + \frac{b}{2} + \frac{c}{2}$	n	hhl with $2h + l = 2n$ (equals $l = 2n$)
Diamond	(100)	$\frac{b}{4} + \frac{c}{4}$	d	$0kl$ with $k + l = 4n$ and $k, l = 2n$
	(010)	$\frac{a}{4} + \frac{c}{4}$	d	$h0l$ with $h + l = 4n$ and $h, l = 2n$
	(001)	$\frac{a}{4} + \frac{b}{4}$	d	$hk0$ with $h + k = 4n$ and $h, k = 2n$
	$(1\bar{1}0)$	$\frac{a}{4} + \frac{b}{4} + \frac{c}{4}$	d	hhl with $2h + l = 4n$

parallel to the plane. New positions created by the operation lie on planes which coincide with some of the original planes and interleave others (Fig. 4-6). By reason of the magnitude of the translation component, the interleaving has the effect of halving or quartering the spacings of certain planes normal to the glide plane (Fig. 4-6a). For all other planes the interleaving bears no simple relation, except by chance, to the interplanar spacing (Fig. 4-6b). It follows then that reflections from the zone of planes normal to a glide plane can be expected to show systematic absences.

AXIAL GLIDES (a, b, c). Figure 4-6 depicts an (001) axial glide with translation component $b/2$. The spacings of planes normal to the glide —the ($hk0$) planes— are halved for those planes that have k odd. Hence

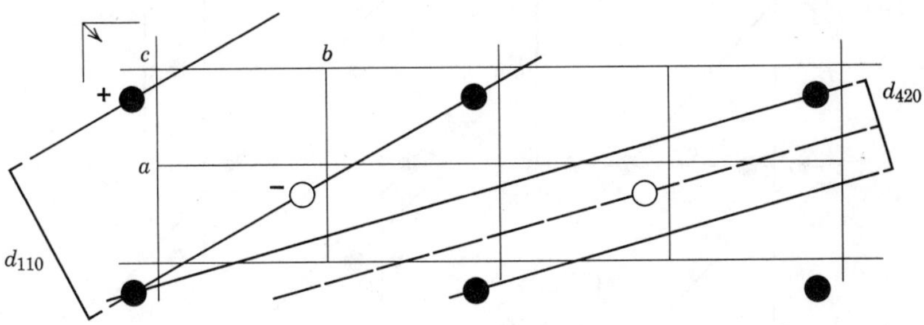

Fig. 4-7. Effect of a diagonal glide with translation component $a/2 + b/2$. Spacings of ($hk0$) planes with $h + k \neq 2n$ are halved.

the first-order reflections from such planes as (010), (110), and (210) cannot survive. Possible reflections include the even-order reflections from these planes—020, 040, 220, 420, etc.—and all orders of reflections from planes with k even—100, 200, 120, 240, etc. The condition for possible reflection clearly requires that $hk0$ reflections have $k = 2n$, where n is any integer.

The extinction conditions for other axial glides (summarized in Table 4-3) are readily predicted from this example. The orientation of the glide plane determines the class of reflection affected, whereas the translation component fixes the nature of the condition.

DIAGONAL GLIDES (n). Figure 4-7 shows a diagonal glide parallel to (001) with translation component $a/2 + b/2$. The spacings of all ($hk0$) planes with $h + k$ odd are halved by the glide operation; hence the indices of possible reflections from the ($hk0$) zone of planes must have $h+k$ even. It is apparent from this example that the condition for reflection in the presence of a diagonal glide involves two indices of the reflection symbol. The conditions for possible reflection in the presence of diagonal glides in other orientations are summarized in Table 4-3.

DIAMOND GLIDES (*d*). The condition for reflection in the presence of a diamond glide is readily deduced from the preceding argument. The glide component ($a/2 + b/2$) of the diagonal glide in Fig. 4-7 is equivalent to a translation of $\frac{1}{2}$ the cell dimension parallel to the diagonal of the glide plane. By comparison, the translation in a (001) diamond glide is $a/4 + b/4$ and equivalent to $\frac{1}{4}$ of the same cell dimension. This has the effect of halving the spacings of all ($hk0$) planes with $h + k$ odd. The

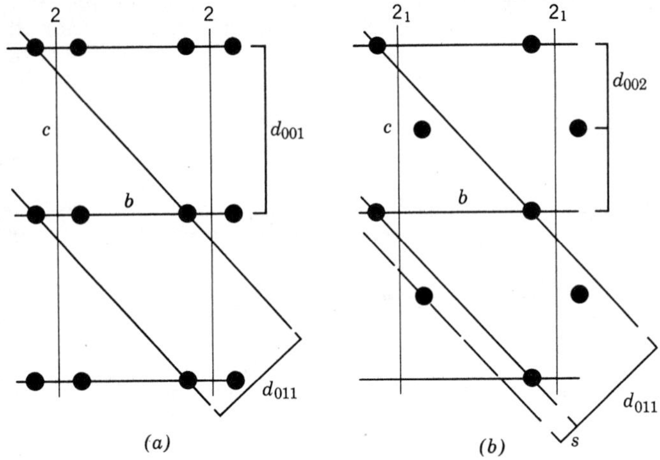

Fig. 4-8. Effect of (*a*) twofold rotation axis and (*b*) 2_1 screw axis.

net result is that ($hk0$) planes such as (110), (310), and (130) reflect only in the even orders: 220, 440, 620, 260, etc.; only every fourth order of reflection is possible from planes such as (100), (010), (210), and (120) with $h + k$ odd: 400, 800, 040, 840, 480, etc. The condition for possible reflection is therefore

$$hk0 \text{ present only with } h + k = 4n \text{ and } h \text{ and } k \text{ both even}$$

With this as a basis, the conditions for diamond glides in other orientations are easily deduced (Table 4-3).

Screw Axes

A screw axis implies the existence of atoms or groups of atoms related by translations, which amount to $\frac{1}{6}$, $\frac{1}{4}$, $\frac{1}{3}$, or $\frac{1}{2}$ of the unit cell dimension along the axis. The axes are regularly spaced in accordance with the symmetry of the structure, and consequently the atoms are systematically distributed throughout the crystal.

The effect of screw axes on the spacings of planes is depicted in Fig. 4-8. The figure shows orthorhombic unit cells with 2_1 screw axes situated along the vertical edges and, for comparison, the same cells with simple twofold rotation axes. The significant difference between the two structures lies in the spacing of planes normal to the symmetry axes. In the structure with screw axes, the spacing of the basal planes has been

Table 4-4. Systematic Absences Due to Screw Axes

Orientation	Translation Component	Symbol	Condition for Possible Reflection
[100]	$\frac{a}{2}$	$2_1, 4_2$	$h00$ with $h = 2n$
[100]	$\frac{a}{4}$	$4_1, 4_3$	$h00$ with $h = 4n$
[010]	$\frac{b}{2}$	$2_1, 4_2$	$0k0$ with $k = 2n$
[010]	$\frac{b}{4}$	$4_1, 4_3$	$0k0$ with $k = 4n$
[001]	$\frac{c}{2}$	$2_1, 4_2$	$00l$ with $l = 2n$
[001]	$\frac{c}{4}$	$4_1, 4_3$	$00l$ with $l = 4n$
[110]	$\frac{a}{2} + \frac{b}{2}$	2_1	$hh0$ with $h = 2n$
[0001]	$\frac{c}{2}$	6_3	$000l$ with $l = 2n$
[0001]	$\frac{c}{3}$	$3_1, 3_2, 6_2, 6_4$	$000l$ with $l = 3n$
[0001]	$\frac{c}{6}$	$6_1, 6_5$	$000l$ with $l = 6n$

halved; hence only the even-order 00*l* reflections are possible. Except by chance, the interleaving of planes caused by the screw axis bears no simple relation to the spacings of other planes [for example to the (011) planes] and, as in the structure with twofold rotation axes, all orders of reflections are possible.

The spacings of planes normal to screw axes are evidently always simple submultiples ($\frac{1}{6}$, $\frac{1}{4}$, $\frac{1}{3}$, $\frac{1}{2}$) of the unit-cell dimension along the axis. The submultiplication of the spacing, hence the condition for possible reflection, is readily predicted from the translation component of the axis. Table 4-4 lists the reflection conditions for all types of screw axes.

The Determination of Space Group

The preceding sections have shown that centered lattices are characterized by systematic absences in the general (*hkl*) class of reflections. Space-group symmetry elements, on the other hand, are characterized by absences in special classes of reflections—glide planes have a systematic influence on the reflections from a zone of planes normal to the glide; screw axes affect only the reflections from planes normal to the axis. The extinction data, assembled and classified in Tables 4-1, 4-3, and 4-4, indicate that systematic absences observed in the reflections from a crystal can be used to deduce the lattice type and determine the presence and orientation of space-group symmetry elements in the structure. This information in turn leads to the space group of the crystal or, at least, to a choice from a limited number of space groups.

To illustrate the use of extinction data, suppose that by some means the reflections from an orthorhombic crystal have been identified as follows:

Class of Reflection	Condition for Reflection	Cause of Condition
hkl	No systematic absences	*P* lattice
0*kl*	$k = 2n$	*b* glide parallel (100)
*h*0*l*	$h = 2n$	*a* glide parallel (010)
*hk*0	$h + k = 2n$	*n* glide parallel (001)

The absence of systematic extinctions in the *hkl* class of reflections indicates that the lattice is primitive. Systematic extinctions in the three zones of reflections (0*kl*, *h*0*l*, and *hk*0) implies the presence of glide planes parallel to (100), (010), and (001). The first two glides are axial glides because only one index is involved in each extinction condition;

the third is a diagonal glide because two indices are involved and their sum is divisible by two rather than four. Reference to Table 4-3 shows that the glides have the symbols *b*, *a*, and *n*, respectively. The space group is therefore *Pban*.

It is worth noting that, although systematic extinctions imply the presence of space-group symmetry elements, the absence of systematic extinctions does not imply the presence of alternate symmetry elements without translation components. Suppose that an orthorhombic crystal gives the following reflection data:

Class of Reflection	Condition for Reflection	Cause of Condition
hkl	No systematic absences	*P* lattice
0*kl*	$k = 2n$	*b* glide parallel (100)
*h*0*l*	$h = 2n$	*a* glide parallel (010)
*hk*0	No systematic absences	

These data identify the lattice type, establish the presence and type of glide planes parallel to (100) and (010), and prove the absence of a glide plane parallel to (001). The data do not, however, imply the presence of a mirror plane parallel to (001). The reflections are compatible with two of the three orthorhombic point groups, *mmm* and *mm*2, and consequently the space group is either *Pbam* or *Pba*2.

Space groups can be determined conveniently with the aid of tables that list the extinction conditions for the 230 space groups in some systematic order. Such tables are found in Volume One of the *Internationale Tabellen* for 1935. Similar tables have been published by Donnay and Harker.[1] Appendix 3 of this book contains tables which combine features of both the *Internationale* and the Donnay–Harker tables. The tables take account of the two monoclinic and the six orthorhombic settings of the space groups, as given in Appendices 1 and 2. The procedure for identifying a space group is best explained by referring to Appendix 3. The appendix consists of six tables for the six crystal systems. Each table is divided vertically into as many columns as there are point groups in the system and as many groups of columns as there are Laue classes in the system. All the space groups in a column of a table are isomorphous with one point group, and all of the space groups in a group of columns show the same diffraction symmetry. The tables are divided horizontally into rows of space groups that show the same

[1] J. D. H. Donnay and D. Harker, "Nouvelles tables d'extinctions pour les 230 groups de recouvrements cristallographiques," *Naturaliste Can.*, **67**, 33 (1940).

systematic absences—that is, into rows of space groups that are said by Donnay and Harker to have the same "aspect."

Assume for an illustration that a crystal is known to be tetragonal and gives the following systematic absences:

hkl present only with $h + k + l = 2n$

$hk0$ present only with $h + k = 2n$

hhl present only with $l = 2n$

$0kl$ present only with $k + l = 2n$

$00l$ present only with $l = 2n$

According to Table 4 of Appendix 3, eight tetragonal space groups

$I4/mmm$ $I422$ $I4mm$ $I\bar{4}2m$ / $I\bar{4}m2$ $I4/m$ $I4$ $I\bar{4}$

are compatible with these conditions and indistinguishable on the basis of the systematic absences. The space groups contain no symmetry elements with translation components—the systematic absences in the $hk0$, hhl, $0kl$, and $00l$ classes of reflections are imposed by the body-centered lattice. However, if the Laue symmetry is known through use of one of the single-crystal x-ray methods that display the diffraction symmetry of the crystal, the possibilities are narrowed to five or three, depending on whether the symmetry is $4/mmm$ or $4/m$. Ordinary x-ray methods cannot improve on this.

When the Laue symmetry is known, the 230 space groups can be arranged into 120 groups (not counting alternate settings in the monoclinic system and the "nonstandard" orthothombic permutations) on the basis of differences in the systematic absences. Of these, 50[2] contain only a single space group (and therefore lead to a unique determination), nine[3] contain only a single enantiomorphous pair of space groups, and one[4] contains two pairs of (trigonal) enantiomorphous space groups. This accounts for 72 space groups—the remaining 158 cannot be uniquely determined by ordinary x-ray methods. There are, however, other methods, by which the correct space group can sometimes be established or, failing this, the choice of space groups narrowed. A study of crystal form may fix the point group if the crystal is morphologically well-developed. If the point group is known, the number of space groups which cannot be uniquely determined is reduced to 22, comprising eleven

[2] They consist of monoclinic 1, orthorhombic 15, tetragonal 19, hexagonal 1, cubic 14.

enantiomorphous pairs,[3,4] plus four,[5] comprising two special pairs. Eighteen[6] of the remaining 204 occur as pairs belonging to the same point group and are distinguishable if the orientation of the point group relative to the lattice is known. Tests for pyro- or piezoelectricity, if positive, establish the absence of centrosymmetry. The demonstration of rotatory polarization indicates that the point group is holoaxial. The configuration of a molecule or other repeat unit of the structure, when known, may help to rule out one or more alternatives, because its symmetry must be compatible with the space group.

In recent years x-ray methods have been devised that give more symmetry information, in favorable cases, than is predicted by the Friedel law (this law implies that x-ray methods generally cannot detect the presence or absence of centrosymmetry in a crystal because the *hkl* and $\overline{hkl}$ intensities are identical). Only two methods are mentioned here:

(1) Coster, Knol, and Prins[7] showed that the Friedel law fails in certain circumstances. They found that *hkl* and $\overline{hkl}$ intensities may differ if the crystal is noncentrosymmetric and an atom of the structure anomalously absorbs x-rays. Thus by employing a wavelength that is near the resonance level of an atom in the crystal the method can be used to detect the absence of a symmetry center. The method can also be used to distinguish between enantiomorphous pairs (determine the "hand"), provided that the structure is known. The *hkl* and $\overline{hkl}$ intensities are calculated on the assumption that the structure is, say, right-handed; if the observed intensities are the reverse, the structure is left-handed.

The method is particularly valuable if the physical evidence (such as pyro- and piezoelectricity) for noncentrosymmetry is very weak. Unfortunately, relatively few wavelengths are available for diffraction studies, and therefore the variety of chemical compositions to which the method can be applied is limited.

(2) A second method due mainly to Wilson[8] is based on a statistical study of the distribution of the reflection intensities. Certain zones or rows may be found to have average intensities which are a small-integral multiple of the average of the whole array. These anomalous zones or rows indicate the presence, type, and orientation of all non-translatory symmetry elements, except centrosymmetry, in the point group. When

[3] ($P4_1$–$P4_3$) ($P4_122$–$P4_322$) ($P4_12_12$–$P4_32_12$) ($P3_1$–$P3_2$) ($P6_1$–$P6_5$) ($P6_2$–$P6_4$) ($P6_122$–$P6_522$) ($P6_222$–$P6_422$) ($P4_132$–$P4_332$).

[4] ($P3_121$–$P3_221$) and ($P3_112$–$P3_212$).

[5] ($I222$–$I2_12_12_1$) and ($I23$–$I2_13$).

[6] ($Pmc2_1$–$Pma2$) ($Cmm2$–$Amm2$) ($Cmc2_1$–$Ama2$) ($P\bar{4}2m$–$P\bar{4}m2$) ($I\bar{4}m2$–$I\bar{4}2m$) ($P312$–$P321$) ($P3m1$–$P31m$) ($P\bar{3}1m$–$P\bar{3}m1$) ($P\bar{6}m2$–$P\bar{6}2m$).

[7] D. Coster, K. S. Knol, and J. Prins, *Z. Physik*, **63**, 345 (1930).

[8] A. J. C. Wilson, *Acta Cryst.*, **2**, 318 (1949).

this information is combined with a knowledge of the Laue symmetry and the systematically absent reflection data, most of the space groups can be uniquely determined. The exceptions are the individuals of enantiomorphous pairs, the pairs ($I222$–$I2_12_12_1$) and ($I23$–$I2_13$), and no distinction can be made between $P1$ and $P\bar{1}$, $P3$ and $P\bar{3}$, and $R3$ and $R\bar{3}$.

CHAPTER FIVE

POLYCRYSTAL OR POWDER METHODS

The first powder patterns were prepared by Friedrich shortly after the original Laue diffraction experiments had been performed. Friedrich did not arrive at a correct interpretation of his powder diffraction effects, and it remained for Debye and Scherrer in Germany and Hull in the United States to develop the very useful powder method. Research on the method was carried on independently in the two countries during World War I, and the results were published almost simultaneously in 1917.

The powder method derives its name from the use of a polycrystal sample which may take many physical forms but is usually a powder. Ideally, all possible orientations of all possible lattice planes are present in a powder sample, and when a beam of characteristic radiation is directed at the sample the Bragg condition for reflection is simultaneously satisfied for all types of lattice plane.

The principle of the powder method is illustrated by the diagrams of Fig. 5-1. Figure 5-1*a* shows a particular set of lattice planes in one crystallite oriented at the appropriate Bragg angle θ to the incident beam. The reflected beam makes an angle 2θ (the diffraction angle) with the undeviated beam. The figure depicts only one of an infinite number of ways in which this, or an identical set of planes in another crystallite, can be oriented at the same angle θ to the beam. The reflected beams which correspond to these orientations outline a cone, which is coaxial with the incident beam and has a half-apex angle 2θ (Fig. 5-1*b*). Simultaneously, other lattice planes with different spacings satisfy the Bragg condition and generate reflection cones, which are also coaxial with the beam but have different half-apex angles (Fig. 5-1*c*). To ensure the existence of a sufficient number of the required orientations to give dense reflection cones, it is usually necessary to rotate, oscillate, or otherwise move very small samples continuously during the exposure. Reflections developed at 2θ angles of less than 90° are described as front or forward reflections and at angles greater than 90° as back reflections. Reflections are recorded on film or are detected by a radiation counter which will actuate a pen on a strip-chart to produce a graphical record. This apparatus is known as an x-ray diffractometer.

The characteristic features of an x-ray powder spectrum are the angles at which reflections occur and the relative intensities of the reflections.

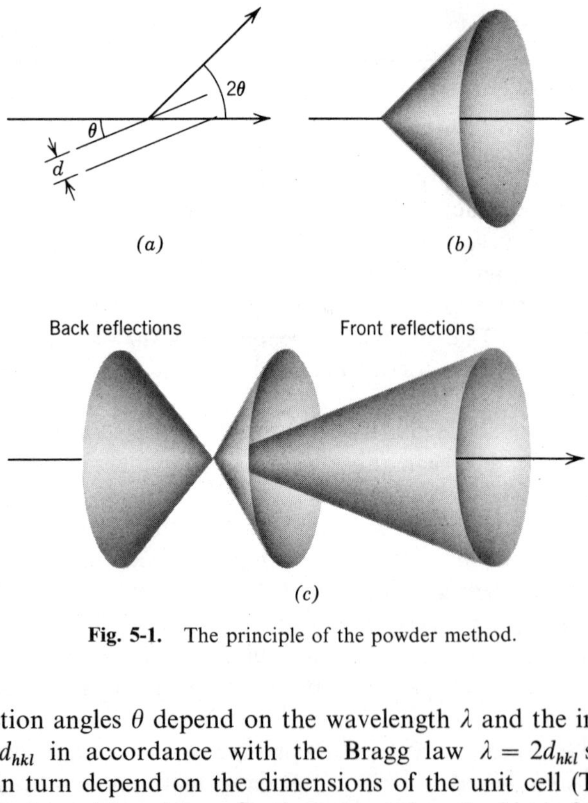

Fig. 5-1. The principle of the powder method.

The reflection angles θ depend on the wavelength λ and the interplanar spacings d_{hkl} in accordance with the Bragg law $\lambda = 2d_{hkl} \sin \theta$. The spacings in turn depend on the dimensions of the unit cell (Table 1-2). The relative intensities of the reflections depend on the kind of atoms and their arrangement between the lattice planes. A powder spectrum is therefore a reasonably complete display of the diffraction effects from a compound.

The procedure in analyzing the powder spectrum of an unknown compound consists of measuring the diffraction angles, calculating the spacings of the reflecting planes, and then deducing the dimensions of the unit cell. The unit cells of cubic, tetragonal, and hexagonal compounds can be determined with graphical methods and generally present no serious difficulties. The cells of compounds with lower symmetry are characterized by three, four, or six lattice constants. This necessitates the use of laborious analytical methods, which lead to results that are not always without ambiguity. The single-crystal x-ray methods described in later chapters are preferable for low-symmetry compounds, but can be used only if suitable crystals are available.

The powder method has a variety of special applications, a few of which are described on the following pages. Perhaps the most important application is as a tool for the identification of unknown compounds. For this purpose it outranks all other x-ray methods because of three main advantages:

(1) It is the only method that is applicable to all crystalline substances.
(2) The entire spectrum can be recorded on a short strip of film to give a comprehensive diffraction picture or x-ray fingerprint of the compound.
(3) The preparation of the sample and its presentation to the beam is very convenient compared with other methods which require the selection and orientation of a single crystal.

Part 1. Film Techniques

The Debye–Scherrer Method

A number of methods are employed to record powder x-ray reflections on film. Each is designed to exploit particular applications. The so-called Debye–Scherrer technique has proved to be the most useful for the routine identification of unknown compounds. Furthermore, it is capable of yielding lattice spacings of high precision. The apparatus is relatively inexpensive, and the technique is easily mastered in a short time. For these reasons one or more Debye–Scherrer cameras are the basic recording instruments in most diffraction laboratories.

The Debye–Scherrer method makes use of a cylindrical film, whose axis coincides with the rotation axis of the sample and is normal to the beam. In the simplest instrumental arrangement (Fig. 5-2), which has been called the regular or Bradley–Jay film position, the beam enters between the ends of the film and passes out of the camera through a hole punched in the center of the film. Each reflection cone intersects the film in a pair of arcs, and all the cones generated by the specimen are represented on the film. The front reflection arcs are curved about the center of the film, the curvature decreasing to zero at $2\theta = 90°$.

The back-reflection arcs are curved in the opposite direction, the curvature increasing with increasing 2θ angle. The distance $2s$ (Fig. 5-2*b*,*c*) between a pair of arcs on the film bears a simple relation to the apex

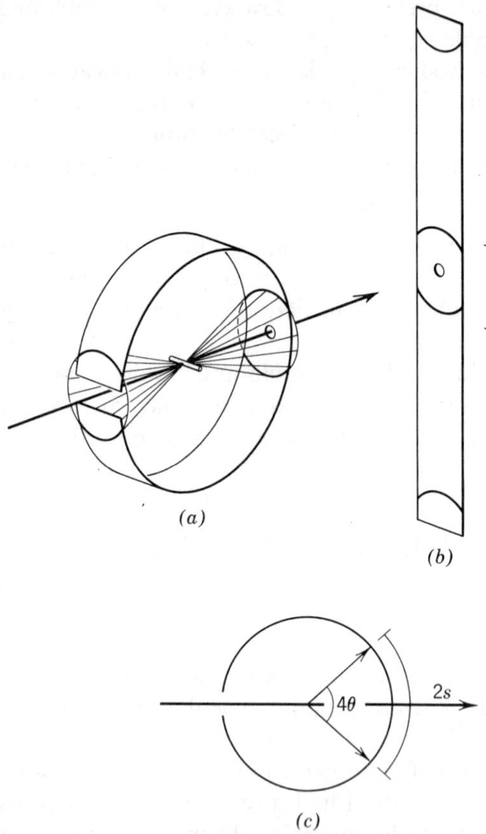

Fig. 5-2. The regular or Bradley-Jay film position in a Debye–Scherrer camera (after L. V. Azaroff and M. J. Buerger, *The Powder Method*, McGraw-Hill, New York, 1958).

angle 4θ of the corresponding reflection cone

$$\frac{2s}{4\theta} = \frac{\text{circumference of film}}{360^\circ} = \frac{2r\pi}{360}$$

where r is the radius of the film. Consequently

$$\frac{\theta}{s} = \frac{360}{4\pi r}$$

Camera Sizes

The sizes of most commercial North American Debye–Scherrer cameras are designed to facilitate the conversion of film measurements s to θ angles. The two common sizes have radii of $360/4\pi = 28.65$ mm and $360/2\pi = 57.3$ mm, respectively. The relation derived reduces to $\theta/s = 1$ for the first size and $\theta/s = \frac{1}{2}$, for the second. In other words, $1° \theta$ is represented by 1 mm s in the smaller camera and 2 mm s in the larger one.

The larger camera is evidently more efficient at resolving closely spaced reflections and has an advantage when a high degree of accuracy in measurement is required. The intensity of blackening on the film varies inversely as the square of the radius, and consequently the advantage in resolution and accuracy is gained at the cost of about four times the exposure. The resolution of the small camera is adequate for routine identification work, and, because exposures are short, this camera is the basic tool in many diffraction laboratories.

Convenience in converting film measurements to θ angles is of less concern in some research laboratories. Instead, importance is attached to the advantages that can be gained by the use of the special camera sizes. Thus extra large sizes are used to achieve the ultimate in resolution and accuracy of measurement, and intermediate sizes have been developed for a better balance between resolution and length of exposure than is afforded by the so-called standard sizes.

The Relative Intensities of the Reflections

The Lorentz factor expresses the relative time that any crystal plane spends in rotating through the small angular range of θ, over which the Bragg condition is satisfied. In Chapter 3 the constant $1/\sin 2\theta$ was derived in connection with Fig. 3-20 as the form of the factor for a single crystal rotating about an axis parallel to the reflecting planes. Ideally, the particles in a powder sample are in completely random orientation; hence the complete range of orientations, corresponding to the angular range of θ, is simultaneously presented to the beam. The effect along any line in a powder reflection cone is therefore equivalent to the effect of rotating a single crystal, oriented to reflect along the line, through its Bragg condition. Thus the form $1/\sin 2\theta$ of the factor applies also to the powder method.

The intensities of Debye–Scherrer reflections are subject to several additional geometrical factors. Figure 5-3 represents a set of crystal planes at the center of a reference sphere of radius r, reflecting to R on the surface of the sphere. The normal ON to the reflecting plane intersects the sphere at N. The normals to identical planes in other orientations that satisfy the Bragg condition for reflection intersect the sphere in a circle NN' of radius $r \cos \theta$. If the particles in the sample are in completely random orientation, the size of this circle is a measure of the number of planes that contribute to a reflection, and the intensity is proportional to $r \cos \theta$. The reflection directions that correspond to all the normals through the circle NN' outline a second circle RR', of radius $r \sin 2\theta$. The smaller the circle, the greater the intensity of the

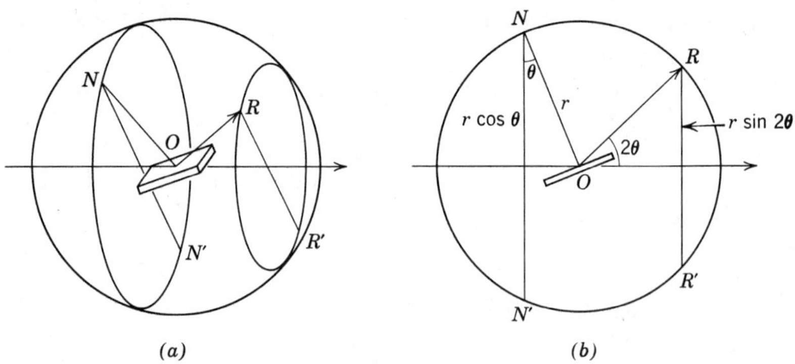

Fig. 5-3. The derivation of the factor $\cos \theta / \sin 2\theta$; (*b*) is a section of the sphere in (*a*) taken parallel to the plane defined by the incident and reflected beams.

blackening on a film arranged to record the circle. Hence the intensity of blackening caused by the cone angle is proportional to $1/r \sin 2\theta$. After combining the effect produced by the number of planes that contribute to a reflection with the effect of the cone angle, the factor has the form

$$\frac{r \cos \theta}{r \sin 2\theta} = \frac{\cos \theta}{\sin 2\theta}$$

Another factor—the *multiplicity* factor—takes account of the number of equivalent planes that contribute to a reflection. In powder photographs all planes with the same spacings reflect to the same arc. A cubic crystal has six equivalent $\{100\}$ planes—(100), ($\bar{1}$00), (010), (0$\bar{1}$0), (001), and (00$\bar{1}$)—and eight equivalent $\{111\}$ planes—(111), ($\bar{1}$11), ($\bar{1}\bar{1}$1), (1$\bar{1}$1), (11$\bar{1}$), ($\bar{1}$1$\bar{1}$), ($\bar{1}\bar{1}\bar{1}$), and (1$\bar{1}\bar{1}$). The probability that a cubic crystallite in a powder sample will satisfy the Bragg condition for a 100 reflection is six compared with eight for a 111 reflection. The probability depends not only on the indices of the reflecting planes but also on the symmetry of the crystallite. A tetragonal crystal has four equivalent $\{100\}$ planes—

(100), ($\bar{1}$00), (010), and (0$\bar{1}$0); an orthorhombic crystal has two—(100) and ($\bar{1}$00). Thus the probability for a 100 reflection is reduced from six for a cubic crystal to four and two, respectively, for tetragonal and orthorhombic crystals. Values for the multiplicity factor are given in Table 5-1.

Table 5-1. Multiplicity Factors for the Powder Method

Laue Class	*h*00	*hhh*	*hh*0	*hk*0	*hhl*	*hkl*	
*m*3*m*	6	8	12	24	24	48	
*m*3	6	8	12	2(12)[a]	24	2(24)	
	00*l*	*h*00	*hh*0	*hk*0	*h*0*l*	*hhl*	*hkl*
6/*mmm*	2	6	6	12	12	12	24
6/*m*	2	6	6	2(6)	12	12	2(12)
$\bar{3}$*m*	2	6	6	12	2(6)	12	2(12)
$\bar{3}$	2	6	6	2(6)	2(6)	2(6)	4(6)
	00*l*	*h*00	*hh*0	*hk*0	*h*0*l*	*hhh*	*hkl*
4/*mmm*	2	4	4	8	8	8	16
4/*m*	2	4	4	2(4)	8	8	2(8)
	*h*00	0*k*0	00*l*	0*kl*	*h*0*l*	*hk*0	*hkl*
mmm	2	2	2	4	4	4	8
	*h*00	0*k*0	00*l*	*h*0*l*	0*kl*	*hk*0	*hkl*
2/*m*	2	2	2	2	4	4	4
	all types						
$\bar{1}$	2						

[a] A multiplicity given as 2(12) indicates two sets of reflections at the same angle but with different intensities.

When the equation derived in Chapter 3 for the relative integrated reflection from a single crystal is modified by the inclusion of the factors described, the new expression accounts for the relative integrated reflections of Debye–Scherrer powder arcs,

$$I \propto \left(\frac{1+\cos^2 2\theta}{\sin 2\theta}\right)\left(\frac{\cos\theta}{\sin 2\theta}\right)F^2 p$$

where p is the multiplicity factor. The expression neglects the effect of absorption and temperature. Because $\sin 2\theta = 2\sin\theta\cdot\cos\theta$, this can be recast into the more convenient form,

$$I \propto \left(\frac{1+\cos^2 2\theta}{\sin^2\theta\cdot\cos\theta}\right)F^2 p$$

When x-radiation falls on a film, silver halide crystals in the emulsion are energized. The silver is freed from these crystals during the developing process to form an opaque area. The area has maximum opacity if all the silver in this region of the film has been freed. In film used for diffraction studies the optical density below this level varies linearly to low values with the amount of radiation received by the area. It follows then that the densities of powder arcs can be used to determine the relative intensities of the reflections.

The total amount of radiation that is reflected from an x-ray sample can be controlled by varying the length of the exposure and the intensity of the incident x-ray beam. According to the reciprocity law of optics, a reduction in the length of the exposure or the intensity of the beam by a certain factor reduces the density of the blackened area by the same factor. If the purpose of the film is to grade the intensities of the reflections, the exposure should be controlled to record the weakest reflection and to allow none of the strong reflections to produce maximum blackening. This is not possible for most compounds because of the great difference in intensity between the very weak and the very strong reflections. One procedure under these circumstances is to prepare one film to record the weak reflections and a second, with a shorter exposure, to differentiate the strong reflections. According to the reciprocity law, the densities measured on the two films are related by a factor, which is readily determined by comparing the densities of arcs that appear on both films. The application of this factor permits the two sets of measurements to be combined to grade the intensities of the whole spectrum. In another method the camera is loaded with two or more strips of film. Each layer of film reduces the densities of the powder arcs by a factor, which is again determined by a consideration of arcs that appear on two or more films. An alternative method, when the camera design does not allow this procedure, covers half the film with thin metal foil.

The relative intensities of powder reflections are now usually expressed on a 0 to 100 scale, on which the strongest reflection is assigned a value of 100. The values are given in steps of ten, if the measurements are based on rough visual estimates, and more exactly if the measurements are obtained with a densitometer or by a visual comparison with a scale of standard densities. One method of preparing such a scale consists of exposing successive parts of a strip of film for increasing periods of time, with the object of producing a series of densities in a range from very faint to maximum blackening. According to the reciprocity law, the densities of the exposed areas are directly proportional to the lengths of the exposures. The film is shielded by a sheet of metal in which a narrow aperture has been cut. The film is placed at a considerable distance from the x-ray source to ensure uniform irradiation of the aperture. At the end of an exposure the x-ray source is screened off, the film is shifted to a new position opposite the aperture, and the exposure is repeated for the next chosen period of time.

The Sample

Ordinarily, the finely ground compound is mixed with a binder and fashioned into a rod. Collodion is a suitable agent because it does not appreciably absorb x-rays, does not give a diffraction pattern, and dries rapidly. Deliquescent powders and compounds soluble in binders are packed into lead-free glass capillary tubes. Metals may be machined into the shape of a wire or reduced to a fine state by filing and are then rolled into a collodion mount. Working a metal distorts the structure, and these strains must be removed by annealing.

The optimum thickness of a Debye–Scherrer specimen depends on the absorptive characteristics of the compound and on the density with which it is packed into the collodion mount or capillary tube. Ideally, the most intense part of a reflected beam should arise from the center of the specimen. Consequently, the specimen should be sufficiently dilute or thin to pass x-rays through its center, yet dense enough to give good intensities in a reasonable exposure time. In general, a diameter of about $\frac{1}{2}$ mm proves satisfactory.

It is essential that the reflections be represented by smooth arcs on the film, if the powder photograph is to furnish accurate θ angles and intensities. The presence of "spotty" arcs in a pattern obtained with a specimen of average size is an indication that the crystallites are too coarse to provide the required number of orientations to give dense reflection cones. The maximum crystallite size for a small powder sample depends partly on the symmetry of the compound. When based on the number of equivalent planes, the chance that a cubic crystallite will satisfy the Bragg condition for reflection from {100} planes, for example, is three times as good as that of a triclinic crystallite; for the {111} planes it is four times as good. Hence triclinic compounds must be ground to a finer state to provide the requisite number of crystallites in a given amount of sample. In general, cubic compounds ground to pass through a 325-mesh screen (a particle size of less than 44 microns) give satisfactory arcs. The comparable size for triclinic compounds is below 400 mesh. In practice, powder samples are usually not sized. The ground material is examined under a binocular microscope, and the crystallographer estimates visually when an adequate reduction of the grains has been achieved. With some experience, this usually yields satisfactory patterns.

One convenient method for mounting most samples is illustrated in Fig. 5-4. Wax is stuck to one end of a brass rod of a size to fit into the sample holder of the camera. A fiber drawn from a solid glass rod is

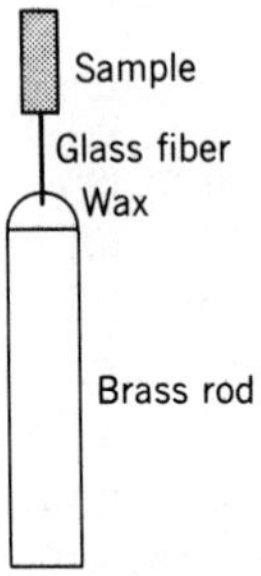

Fig. 5-4. An x-ray powder mount.

inserted into the wax. Under a medium-power binocular microscope, the ground sample is mixed into a drop of collodion with a needle until a paste is formed. When this is sufficiently tacky to cling to the needle, it is brought into contact with the end of the fiber and stretched into a rod measuring 0.5 mm or less in diameter. Alternatively, the paste can be rolled into a rod between two ground-glass slides; when dry enough to maintain its shape, the rod is attached to the end of the glass fiber with collodion.

In the course of research it is often necessary to work under conditions which preclude the fashioning of the "ideal" Debye–Scherrer specimen. The mineralogist, for example, may have occasion to identify a single microscopic grain that he has observed embedded in a polished or thin section. The specimen takes the form of a few fragments extracted

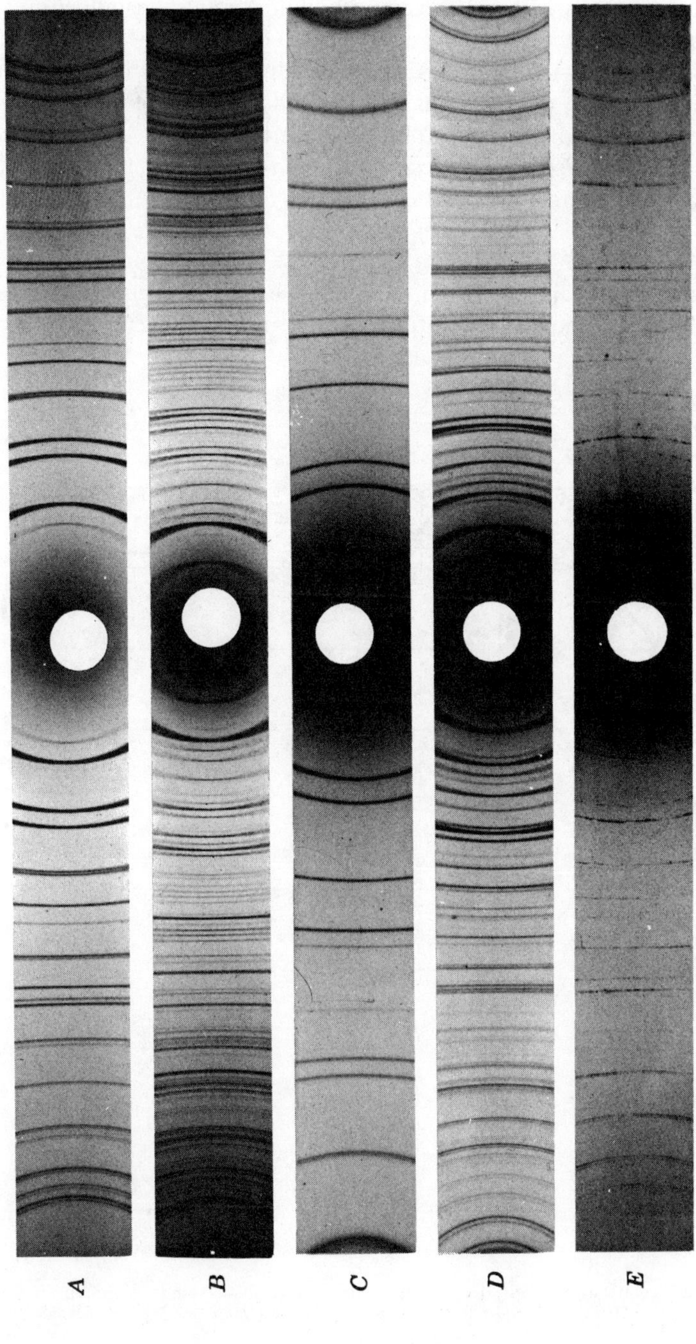

Fig. 5-5. X-ray powder patterns. Camera radius 28.65 mm—actual-size contact prints. *A*—niccolite (NiAs), *B*—scheelite ($CaWO_4$), *C*—gold, *D*—zircon ($ZrSiO_4$), *E*–*A* "spotty" pattern due to insufficient sample grinding.

with a sharp needle, which are mixed with collodion and fashioned into a rough sphere. At other times he may wish to obtain a powder pattern of a few tiny crystals too precious to crush. Under these circumstances the resulting powder photographs are likely to consist of spotty arcs (Fig. 5-5). Although the photographs may not be suitable for precision measurements, they are nevertheless useful for rough measurements and for comparison with reference patterns.

The Choice of Radiation

It was noted in Chapter 2 that undesirable effects occur when the wavelength of the incident radiation is close to but shorter than the

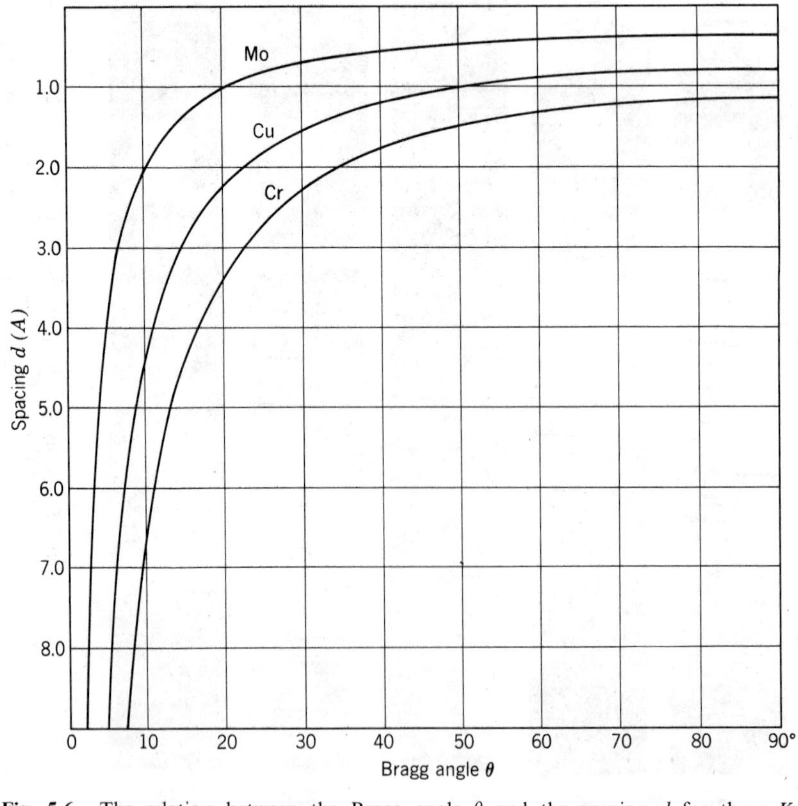

Fig. 5-6. The relation between the Bragg angle θ and the spacing d for three $K\alpha$, radiations: $\lambda(\text{Mo}) = 0.71$, $\lambda(\text{Cu}) = 1.54$, and $\lambda(\text{Cr}) = 2.29$.

absorption limit of an element in the sample—the radiation is strongly absorbed by the element, with a consequent weakening of the pattern, and fluorescent x-rays are generated and scattered to fog the film. If the fluorescing element is a major constituent of the sample, the film may be so badly fogged that essential details of the pattern are obscured. Table 2-2 lists the elements that fluoresce strongly with the kinds of radiation used in diffraction studies and serves as a guide to the satisfactory choice of a radiation when the chemical composition is known.

The Bragg equation, when rewritten in the form

$$\theta = \sin^{-1}\left(\frac{\lambda}{2d}\right)$$

shows that the reflection angle is proportional to the wavelength and inversely proportional to the spacing of the reflecting plane. The relation between θ and d is graphically illustrated in Fig. 5-6 for three $K\alpha_1$ radiations—Cr ($\lambda K\alpha_1 = 2.29$), Cu ($\lambda K\alpha_1 = 1.54$), and Mo ($\lambda K\alpha_1 = 0.71$). The graph indicates that the use of a long wavelength radiation, such as Cr$K\alpha_1$, has the effect of dispersing the diffraction effects to higher 2θ angles; a short wavelength radiation tends to concentrate the effects into the front-reflection region. Thus Mo radiation has the undesirable effect with large unit cells of crowding the simple-index (large spacings—small 2θ angles) reflections into the low-angle region of the film. A better resolution of these important reflections is obtained with the longer Cu and Cr radiations. However, dispersion of the low-angle reflections is not the only consideration in the choice of radiation. The limiting high-angle (smallest-spacing) reflection corresponds to a spacing of

$$d = \frac{\lambda}{2 \sin 90^\circ} = \frac{\lambda}{2}$$

and is equal to 2.29/2 for Cr$K\alpha_1$ compared with 1.54/2 for Cu$K\alpha_1$ and 0.71/2 for Mo$K\alpha_1$. Thus the longer the wavelength, the more limited the range of reflections generated by the sample. In general, the radiation is selected to achieve a satisfactory dispersion of the low-angle reflections while retaining a suitable range of reflections. It follows from the foregoing that the ultimate choice of wavelength rests on the size of the unit cell, unless the generation of fluorescent radiation is an issue. Cu radiation is satisfactory for most inorganic compounds; Co and Fe are useful alternatives if the sample contains cobalt, iron, or manganese—elements that fluoresce strongly in the radiation from a Cu target. Cr radiation is a frequent choice for the large cells of organic and complex inorganic compounds.

The β component of the beam produces its own diffraction pattern superimposed on the α pattern. This adds to the difficulty of interpreting an unknown pattern, and consequently in preliminary studies the β component is removed by filtering (Table 2-3). However, the precision measurement of known patterns is often aided by the presence of β arcs. The α component is a closely spaced wavelength doublet that is not

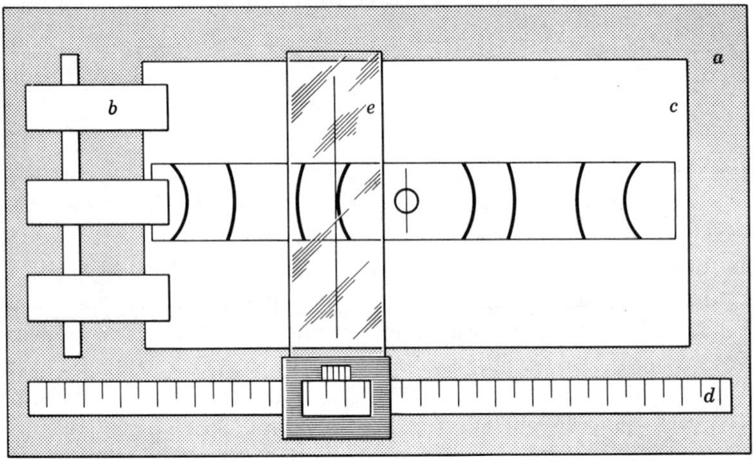

Fig. 5-7. A device for measuring powder photographs; (*a*) metal frame, (*b*) clamp, (*c*) glass plate, (*d*) millimeter scale, and (*e*) cursor with vernier riding on scale.

resolved at low 2θ angles. The α arcs are relatively broad in this region of the film, and measurements are correspondingly inaccurate. In contrast, the β component is virtually monochromatic radiation, and the resulting arcs are sharp and offer the opportunity of precise measurements.

Measurement of the Film

A device designed especially for measuring powder films is an essential instrument in a diffraction laboratory. Figure 5-7 shows a typical commercial instrument that has proved convenient for most routine measurements. The film is held in place with clamps and is viewed by placing the instrument on a light box. The scale in such an instrument is generally graduated in millimeters and is equipped with a vernier that permits the setting of the hairline to be estimated to a twentieth (0.05) of a millimeter.

One method of making the measurements and assembling the data is illustrated by Table 5-2. The data were obtained from the powder pattern of a specimen of chemically pure gold (Fig. 5-5*a*). The fourth and fifth columns of the table give the measurements of the positions of the arcs. The difference between the readings for a pair of arcs (column 6) is the distance $2s$. The camera used in this experiment has a radius of $360/4\pi =$ 28.64 mm, plus an appropriate allowance for the thickness of the film. In this instance it was assumed that the length of the film did not change as a result of developing, fixing, washing, and drying and the diffraction

Table 5-2. X-Ray Powder Data for Gold; Nickel-Filtered Copper Radiation; Film Radius $360/4\pi = 28.64$ mm (1 mm $s = 1°\theta$)

Reflection Number	I	λ	Right	Left	$(R - L) = 2s = 2\theta$	d (meas)	
1	100	α	101.40	63.00	38.40°	2.34 Å	
2	70	α	104.40	59.90	44.50	2.04	
3	50	α	114.65	49.80	64.85	1.438	
4	60	α	121.20	43.30	77.90	1.226	
5	20	α	123.20	41.35	81.85	1.177	
6	10	α	131.40	33.00	98.40	1.018	
7	40	α	137.85	26.65	111.20	0.9343	
8	40	α_1	139.95	24.45	115.50	0.9108	
9	40	α_1	149.90	14.45	135.45	0.8324	0.8322_5
	20	α_2	150.40	14.15	136.25	0.8321	
10	50	α_1	161.15	3.30	157.85	0.7849	0.7848_5
	20	α_2	161.95	2.50	159.45	0.7848	

Wavelengths: $CuK\alpha_1 = 1.54050$, $\alpha_2 = 1.54434$, $\alpha = (2\alpha_1 + \alpha_2)/3 = 1.5418$.

angles 2θ were taken as numerically equal to the $2s$ values (column 6). In precision work it is necessary to determine the effective or actual radius of the film at the time it is measured and to apply a suitable correction. The methods for accomplishing this objective are discussed in some detail in the section on the attainment of accuracy.

The final column in Table 5-2 gives the solution of the Bragg equation ($d = \lambda/2 \sin \theta$) for d in terms of the measured 2θ angles. A number of publications, listed at the end of this chapter, provide this solution in convenient intervals of 2θ or θ for various wavelengths. The first seven arcs in the powder pattern represent unresolved $K\alpha$ reflections; the corresponding d values were obtained from a set of tables which give d in terms of the weighted average of the wavelengths of the unresolved radiation components: $(2\alpha_1 + \alpha_2)/3$. The eighth arc is sufficiently well resolved to permit a measurement of the α_1 reflection; the appropriate tables for this reflection give d in terms of the $K\alpha_1$ wavelength value. The radiation reflected by the remaining two planes is resolved into two well-defined $\alpha_1\alpha_2$ doublets. Each doublet allows two independent determinations of the spacing of the reflecting plane and gives a degree of confidence in the average value for the spacing that is not afforded by the single determination from a $K\alpha$ reflection.

The Interpretation of X-Ray Powder Data

Indexing the Reflections and Deriving the Dimensions of the Unit Cell

One procedure for indexing the reflections of a powder pattern when the crystal system and unit-cell dimensions are known is to calculate all d values for the cell with the aid of the appropriate spacing formula (Table 1-2). A comparison of measured and calculated spacings gives the indices of the reflections.

When only the crystal system is known, the process of indexing is essentially one of deriving a unit cell, the calculated spacings of which are in agreement with the measured spacings. If the cell does not prove to be the conventional one, the indices must be transformed to the proper setting. A number of graphical methods are entirely satisfactory for deducing the indices of reflections for crystals of cubic, tetragonal, and hexagonal symmetry. However, for orthorhombic crystals, whose unit cells are defined by three constants, analytical methods are preferable, and for crystals of monoclinic and triclinic symmetry, analytical methods are essential. The Ito method—which consists of indexing the reflections on the basis of an arbitrary primitive cell and then deducing the conventional cell and indices—has gained wide acceptance for these three crystal systems.

The task of indexing is increased if the pattern is completely unknown. A reasonable plan of attack is to test the pattern for cubic, tetragonal, and hexagonal symmetry with the appropriate graphical methods and, if this fails, to proceed directly with the Ito method.

It is the intention here to present a brief account of some of the graphical methods. The reader who wishes to pursue this subject and to become familiar with the standard analytical methods should consult books devoted primarily to the powder method. A number of excellent references are listed at the end of this chapter.

INDEXING A CUBIC PATTERN. The spacing of planes in a cubic lattice are related to the corresponding Miller indices hkl and the cell edge a as

$$d_{hkl} = \frac{a}{\sqrt{h^2 + k^2 + l^2}}$$

The spacings are therefore directly proportional to $(h^2 + k^2 + l^2)^{-1/2}$ and differ from one crystal to the next only in scale, which is determined by the value of a. Thus the relation between the spacing of a given plane and the cell edge is linear and can be expressed in a simple graphical form, such as is shown in Fig. 5-8. Each line on the graph represents a solution of the spacing formula for a particular lattice plane. As a preliminary to constructing the graph, a list of cubic planes in order of decreasing spacing is prepared by systematically increasing the numerical value of $(h^2 + k^2 + l^2)$ in the expression $(h^2 + k^2 + l^2)^{-1/2}$. Table 5-3 gives these data for the graph of Fig. 5-8. No combination of integers gives a total of 7, 15, 23, or 28 for the expression $(h^2 + k^2 + l^2)$. In a number of cases two combinations of integers have identical values for the expression. For example, 300 and 221 sum to nine, indicating that the planes represented by these indices have identical spacings and will give rise to a single reflection. Because θ is inversely proportional to spacing, the order of listing the planes is also the order of increasing θ angle.

The graph of Fig. 5-8 can be used as a "cubic crystal analyzer" to identify the Miller indices of reflecting planes on a powder photograph and to provide a graphical determination of the cell edge of the compound. The spacings, as derived by measurement of the film, are plotted

Table 5-3. The Relative Spacings of Cubic Planes

$(h^2 + k^2 + l^2)$	Miller Indices	$\frac{d}{a} = (h^2 + k^2 + l^2)^{-1/2}$	$(h^2 + k^2 + l^2)$	Miller Indices	$\frac{d}{a} = (h^2 + k^2 + l^2)^{-1/2}$
1	(100)	1.00	16	(400)	0.250
2	(110)	0.707	17	(410), (322)	0.242
3	(111)	0.577	18	(411), (330)	0.236
4	(200)	0.500	19	(331)	0.229
5	(210)	0.447	20	(420)	0.224
6	(211)	0.408	21	(421)	0.218
7	...	...	22	(332)	0.213
8	(220)	0.354	23	...	...
9	(300), (221)	0.333	24	(422)	0.204
10	(310)	0.316	25	(500), (430)	0.200
11	(311)	0.302	26	(510), (431)	0.196
12	(222)	0.289	27	(511), (333)	0.192
13	(320)	0.277	28	...	...
14	(321)	0.267	29	(520), (432)	0.186
15	...	...	30	(521)	0.182

on a strip of paper to the scale of the graph. The strip is then moved up the graph in search of a match between the plotted spacings on the strip and the lines on the graph. Usually a number of positions of agreement can be found. The lowest position corresponds to the simplest cell. The

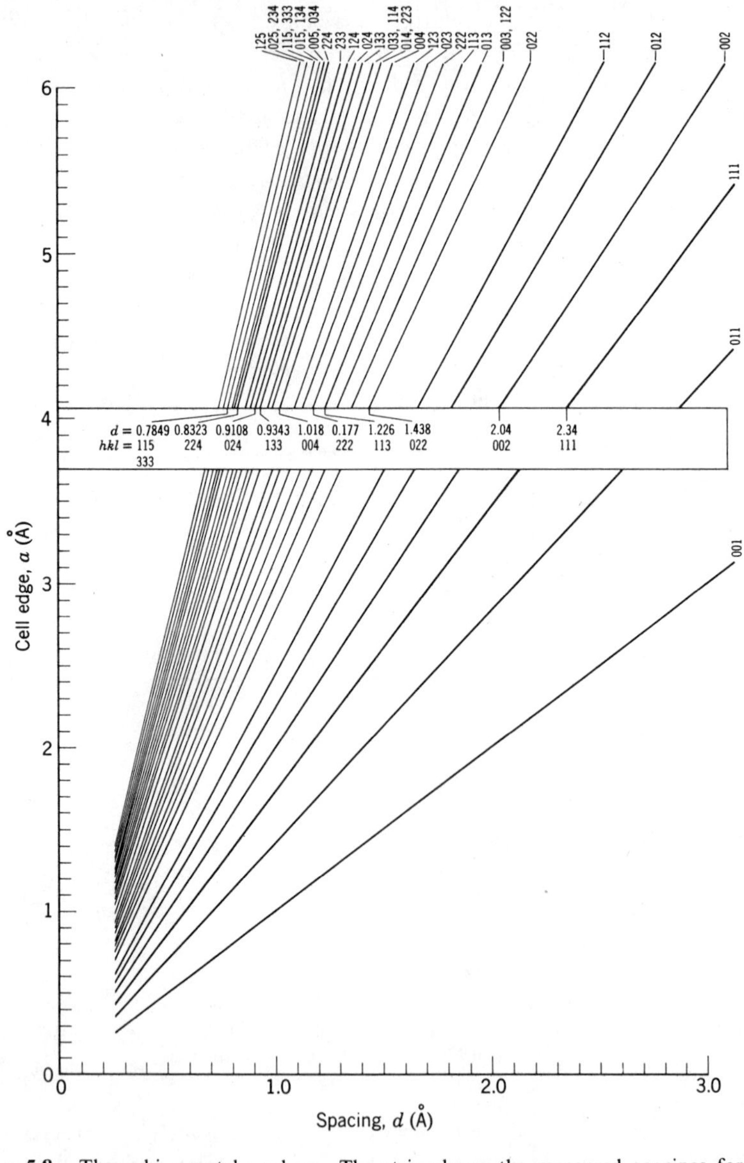

Fig. 5-8. The cubic crystal analyzer. The strip shows the measured spacings for gold (Table 5-2) and the indices of the reflecting planes as derived from the analyzer.

lines on the graph identify the Miller indices of the reflecting planes; the intersection of the edge of the strip with the left edge of the graph gives the cell edge.

The spacing data plotted on the strip in Fig. 5-8 were taken from Table 5-2. The figure indicates that when the strip is positioned to correspond to a cell edge of 4.07 Å, a satisfactory match is realized for each reflection. This value of the cell edge compares with the precision value of 4.0781 Å listed in Donnay and Nowacki.[1]

INDEXING A TETRAGONAL PATTERN. The spacings of tetragonal lattice planes are given by

$$d_{hkl} = \left(\sqrt{\frac{h^2 + k^2}{a^2} + \frac{l^2}{c^2}}\right)^{-1/2}$$

The effect of two lattice dimensions a and c, in place of one in cubic cells, is to reduce the number of equivalent planes represented by most sets of Miller indices and thereby to increase the number of possible reflections. For example, the spacings of the cubic planes (100), (010), and (001) are identical and can develop only one reflection. The corresponding spacing relationship in tetragonal crystals is given by $d_{100} = d_{010} \neq d_{001}$. Consequently, two reflections may be developed. Table 5-4 lists on the left the indices of planes that contribute to the first eight possible cubic reflections. On the right the same indices are arranged into groups of tetragonal planes with identical spacings. The table shows that eighteen tetragonal reflections may be generated in place of the eight cubic reflections. In general, then, tetragonal powder patterns are more complex than cubic pattern and proportionately more difficult to index.

The tetragonal spacing formula reduces to

$$d_{hk0} = \frac{a}{\sqrt{h^2 + k^2}}$$

for planes which do not intersect the c axis—that is, for planes identified by the general indices ($hk0$). This is identical with the spacing relation for cubic ($hk0$) planes, and thus the ratios of the spacings of these planes are identical in the two crystal systems: $d_{100}:d_{110}:d_{200}:d_{210}:d_{220}:d_{300}:d_{310}:d_{320}:\text{etc.} = 1.00:0.707:0.500:0.447:0.354:0.333:0.316:0.277$, etc. (Table 5-3). It is possible to index a tetragonal pattern by first identifying the $hk0$ reflections on the basis of these ratios and then assigning likely indices to the remaining reflections by trial and error. This is obviously a laborious procedure, and the results may not be without ambiguity.

[1] J. D. H. Donnay and W. Nowacki, "Crystal Data," *Geol. Soc. Am. Mem.*, **60** (1954).

The spacing formula, when recast in the form

$$d_{hkl}^2 = \frac{a^2}{(h^2 + k^2) + l^2/(c^2/a^2)}$$

indicates that the spacings of tetragonal crystals with the same c/a ratios are identical except for scale; the value of the a cell constant determines the scale or absolute value of the spacings. If the scale factor is eliminated by taking $a = 1$ in the numerator of the equation and values of d_{hkl} are then plotted for a range of c/a ratios, the result is a chart of the variation of *relative* spacings with c/a ratio. The reflections of a tetragonal powder pattern can be indexed on such a chart by matching the measured spacings against the spacings plotted on the chart, provided that the measured absolute spacings are first reduced to relative spacings. This can be accomplished by preparing a chart in which the log values of d_{hkl} are plotted against c/a and the log values of the measured spacings are then matched to the chart.

One form of a logarithmic chart is shown in Fig. 5-9. The logarithmic

Table 5-4. A Comparison of Possible Debye–Scherrer Reflections in Cubic and Tetragonal Crystals

Possible Reflections	Equivalent Cubic Planes	Equivalent Tetragonal Planes	Possible Reflections
1	(100) (010) (001)	{(100) (010) (001)}	2
1	(110) (101) (011)	{(110) (101) (011)}	2
1	(111)	(111)	1
1	(200) (020) (002)	{(200) (020) (002)}	2
1	(120) (210) (201) (021) (102) (012)	{(120) (210) (201) (021) (102) (012)}	3
1	(121) (211) (112)	{(121) (211) (112)}	2
1	(220) (202) (022)	{(220) (202) (022)}	2
1	(300) (030) (003) (221) (122) (212)	{(300) (030) (003) (221) (122) (212)}	4
8			18

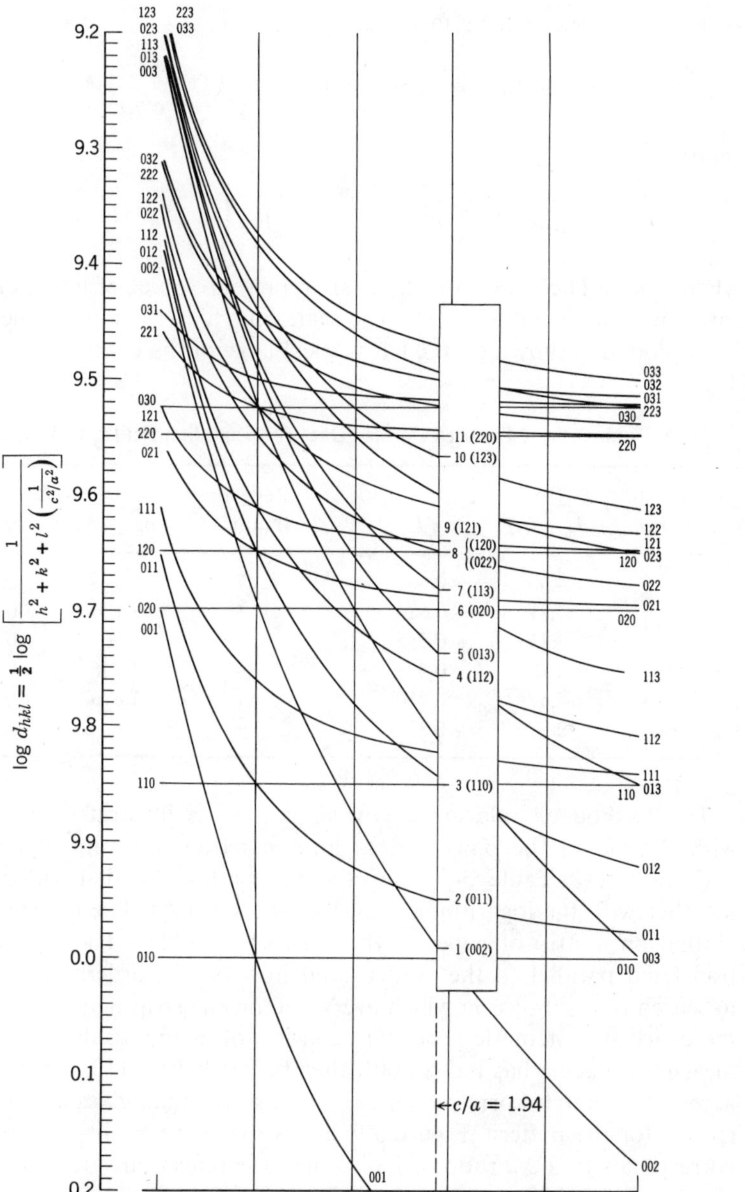

Fig. 5-9. A tetragonal crystal analyzer. The strip shows the best match for the log values of the measured spacings for the mineral tapiolite (Table 5-5).

equivalent of the spacing formula

$$2 \log d_{hkl} = 2 \log a - \log h^2 + k^2 + \frac{l^2}{c^2/a^2}$$

reduces to

$$\log d_{hkl} = -\frac{1}{2}\log h^2 + k^2 + \frac{l^2}{c^2/a^2} = \frac{1}{2}\log \left(h^2 + k^2 + \frac{l^2}{c^2/a^2}\right)^{-1}$$

when $a = 1$. The curves on the chart represent values of $\log d_{hkl}$ for c/a ratios ranging from 0.5 to 3.0. The scale on which the $\log d$ values have been plotted is shown on the left vertical edge of the chart.

Table 5-5. Measured Powder Data for Tapiolite, $Fe(Ta, Cb)_2O_6$

Reflection Number	d_{meas} (Å)	$\log d_{meas}$	Reflection Number	d_{meas} (Å)	$\log d_{meas}$
1	4.66	0.668	7	2.27	0.356
2	4.21	0.624	8	2.12	0.326
3	3.35	0.525	9	2.06	0.314
4	2.70	0.431	10	1.746	0.242
5	2.57	0.410	11	1.676	0.224
6	2.37	0.375			

The method of indexing a powder pattern is illustrated in Fig. 5-9 with the aid of the powder data for the tetragonal mineral tapiolite, $Fe(Ta, Cb)_2O_6$. Table 5-5 gives these data for the first eleven arcs, together with the logarithmic values of the spacings. The log values are plotted on a strip of paper to the scale of the chart. The strip is then translated parallel to the vertical and horizontal borders of the chart in search of positions on which every line on the strip matches a line on the chart. A systematic procedure consists of assuming that the largest measured spacing has indices 001, then 002, 010, 011, 110, etc. The most acceptable match corresponds to the position that gives the simplest indices for the pattern. Figure 5-9 shows that for tapiolite this position corresponds to a c/a ratio of 1.94. When the reflections are indexed, the absolute value of a (or the value of c if the $00l$ reflections appear more reliable) is calculated from $0k0$ reflections. For the eleventh reflection $hkl = 220$ and $d_{220} = 1.676$ (measured). Hence $a = d_{hk0}\sqrt{h^2 + k^2} = d_{220}\sqrt{4 + 4} = 1.676 \times \sqrt{8} = 4.74$ Å. Because $c/a = 1.94$ and $a = 4.74$, $c = 1.94 \times 4.74 = 9.20$ Å.

Large-sized logarithmic charts of a somewhat different design have been prepared by C. W. Bunn for c/a ranges of 0.224 to 1 and 1 to 5 and are commercially available.

INDEXING A HEXAGONAL PATTERN. The spacing relation for hexagonal crystals

$$d_{hkl}^2 = \frac{1}{\frac{4}{3}(h^2 + hk + k^2)(1/a^2) + l^2(1/c^2)} = \frac{a^2}{\frac{4}{3}(h^2 + hk + k^2) + l^2/(c/a)^2}$$

is very similar to the relation for tetragonal crystals. The logarithmic equivalent

$$2 \log d_{hkl} = 2 \log a - \log \frac{4}{3}(h^2 + hk + k^2) + \frac{l^2}{c^2/a^2}$$

reduces to

$$\log d_{hkl} = \frac{1}{2} \log \left[\frac{4}{3}(h^2 + hk + k^2) + \frac{l^2}{c^2/a^2}\right]^{-1}$$

when a in the numerator is taken equal to 1. Thus it is possible to construct similar logarithmic charts giving the variation of relative spacings with c/a ratio. Bunn charts for the hexagonal system are also available.

Calculation of the Unit Cell Contents

The unit cell of a compound weighs

$$V(\text{cm}^3) \times \rho\,(\text{g/cm}^3)\ \text{g}$$

(where V is the volume of the cell and ρ is the density) or

$$V(\text{Å}^3) \times 10^{-24} \times \rho\,(\text{g/cm}^3)\ \text{g}$$

if the volume is given in cubic Angstrom units. The weight of one molecule is

$$\frac{V(\text{Å}^3) \times 10^{-24} \times \rho\,(\text{g/cm}^3)}{Z}\ \text{g}$$

if the cell contains Z molecules.

The molecular weight of a compound is the weight of 6.02×10^{23} molecules of that compound. Hence the weight of one molecule can also be expressed as

$$\frac{MW}{6.02 \times 10^{23}} = MW \times 1.66 \times 10^{-24}\ \text{g}$$

and therefore

$$\frac{V(\text{Å}^3) \times 10^{-24} \times \rho(\text{g/cm}^3)}{Z} = MW \times 1.66 \times 10^{-24}$$

This can be reduced and rearranged to express $MW \times Z$, the weight of the cell referred to $0 = 16$, in terms of V (in Angstrom units) and ρ (in g/cm^3)

$$MW \times Z = \frac{V(\text{Å}^3) \times \rho(\text{g/cm}^3)}{1.66}$$

The relation has a number of important uses. If the molecular weight of a compound is known, say by chemical analysis, and the cell dimensions and density have been measured, the value of Z, which is of fundamental importance in determining the atomic arrangements within the cell, can be calculated. Magnetic iron oxide, Fe_3O_4, for example, has a cubic cell with $a = 8.374$ Å and $\rho = 5.18$ by measurement; $MW = 231.55$. Then

$$Z = \frac{(8.374)^3 \times 5.18}{231.55 \times 1.66} = 7.9$$

Of the two experimental values, a can generally be more accurately determined than ρ, and consequently the departure of Z from the ideal value of eight is probably largely the result of an error in measuring the density.

The relation $MW \times Z = V \times \rho/1.66$ provides a powerful nonchemical test of the accuracy of chemical analysis. It has served in the field of mineralogy to improve the chemical formulas assigned to many minerals on the basis of analyses performed, often of necessity, on impure samples. The mineral fülöppite is an interesting example. In the original description[2] of this species the chemical composition $Pb_2Sb_6S_{11}$ was proposed on the basis of the analysis listed in column 1 of Table 5-6. The sample used for the analysis consisted of crystals picked from an intimate association of fülöppite with at least one other mineral of a similar composition. The density was measured as 5.23. Later, x-ray studies[3] showed that the monoclinic cell has dimensions $a = 13.39$, $b = 11.69$, $c = 16.91$ Å, $\beta = 94° 41'$; measurement of the density gave 5.22. The value of $MW \times Z$ is therefore

$$\frac{V \times \rho}{1.66} = \frac{13.39 \times 16.91 \times \sin 85° 19' \times 11.69 \times 5.22}{1.66} = 8295$$

[2] I. de Finály and S. Koch, *Mineral Mag.*, **22**, 179 (1929).
[3] E. W. Nuffield, *Univ. Toronto Studies, Geol. Ser.*, **50**, 49 (1946).

With the aid of this value and the chemical analysis, the content of the unit cell was calculated in three steps as shown in Table 5-6:

(1) The analysis was recalculated to sum 100 per cent after subtracting impurities.
(2) The weight of each element in the cell was then found from the relation:

$$\text{weight of element in cell} = \text{weight fraction of element} \times 8295$$

This gave the values listed in column 3.
(3) Finally, the weights were converted to numbers of atoms by dividing each weight by the corresponding atomic weight (column 4).

Table 5-6. Cell Content of the Mineral Fülöppite

$$\text{weight of cell} = \frac{V \times \rho}{1.66} = 8295$$

	1[a]	2	3	4	5	6
Pb	28.29	28.32	2349	11.3	12	29.94
Sb	47.50	47.55	3944	32.4	32	46.90
S	24.10	24.13	2002	62.4	60	23.16
	100.08[b]	100.00	8295			100.00

[a] 1. Weight per cent by chemical analysis; analyst de Finály (in de Finály and Koch, *Mineral. Mag.*, **22**, 179, 1929). 2. Analysis recalculated to sum 100 per cent after subtracting SiO_2. 3. Weight of each element in cell = (weight fraction × 8295). 4. Atoms in the unit cell = weight of each element/atomic weight. 5. Probable ideal cell content. 6. Percentage composition calculated for $Pb_3Sb_8S_{15}$.
[b] Including SiO_2 0.19.

The three steps lead to the calculated cell content $Pb_{11.3}Sb_{32.4}S_{62.4}$. If the formula $Pb_2Sb_6S_{11}$ derived by the analyst is correct, the ideal cell content must be one of

$$Pb_{10}Sb_{30}S_{55} = 5(Pb_2Sb_6S_{11})$$

$$Pb_{11}Sb_{33}S_{60\frac{1}{2}} = 5\tfrac{1}{2}(Pb_2Sb_6S_{11})$$

$$Pb_{12}Sb_{36}S_{66} = 6(Pb_2Sb_6S_{11})$$

Because Z is an integer in the properly chosen cell, the second possibility is immediately ruled out. The remaining two possibilities differ greatly from the calculated cell content. Furthermore, the ideal densities calculated from the relation

$$\rho = \frac{Z \times MW_{Pb_2Sb_6S_{11}} \times 1.66}{V}$$

are 4.71 and 5.65 for Z values of five and six respectively, as compared to the independently measured values of 5.22 and 5.23. Thus the physical measurements are not compatible with the composition deduced by the analyst, and it must be concluded that the formula $Pb_2Sb_6S_{11}$ does not represent the chemical composition of fülöppite.

Two other ideal cell contents, $Pb_{11}Sb_{32}S_{59}$ and $Pb_{12}Sb_{32}S_{60} = 4(Pb_3Sb_8S_{15})$, are in reasonable accord with the calculated cell content. Indeed, the Pb/Sb ratio of the first possibility is very close to that prescribed by the formula adopted in the original description of the mineral. However, the calculated density for this cell content is 5.08, in wide disagreement with the measured values, and consequently this evaluation of the cell content is in error. The density calculated for the second possibility is 5.23, in exact agreement with one of the measured values. It seems, then, that the chemical and physical measurements are in best accord with a cell that contains four formula weights (value of $Z = 4$) of $Pb_3Sb_8S_{15}$.

In summary—the physical data have refuted the formula proposed for fülöppite solely on the basis of one analysis and have provided an alternate estimate of the chemical constitution. A comparison of columns 2 and 6 in Table 5-6 shows that the percentage composition calculated for the new formula does not differ markedly from the analysis; the small differences may be attributed to a contamination of the analytical sample by admixed minerals of similar composition.

The Determination of Lattice Type and Space Group

It was noted in Chapter 4 that lattice centering and the presence of screw axes and glide planes have the effect of systematically extinguishing certain reflections. The effects are summarized in Tables 4-1, 4-3, and 4-4 and are arranged in space-group determination tables in Appendix 3.

The indexed powder data for gold (Fig. 5-8) and the tetragonal mineral tapiolite (Fig. 5-9) serve as useful examples to illustrate the value of powder patterns for deducing lattice-type and space-group symmetry. The cubic space groups are found in Table A3-6. The *hkl* reflections of the gold pattern conform to the condition *hkl* present only with $h + k = 2n$ and $k + l = 2n$, indicating that the lattice is all-face centered. The special classes of reflections show no conditions other than the general condition imposed by the lattice centering:

hhl present only with $h + l = 2n$

$0kl$ present only with $k + l = 2n$

$00l$ present only with $l = 2n$

Consequently, no symmetry elements involving a translation operation are present. As noted previously, powder films, unlike Laue, Weissenberg, and precession single crystal films, do not display diffraction symmetry. Thus it is not possible to decide, on the basis of indexed powder reflections, whether gold belongs to the $m3m$ or the $m3$ Laue class; that is, whether the space group is one of $Fm3m$, $F\bar{4}3m$, and $F432$, or one of $Fm3$ and $F23$.

The powder data given for tapiolite in Fig. 5–9 may be classified as follows:

hkl all orders

$hk0$ all orders

hhl all orders

$0kl$ present only with $k + l = 2n$

According to Table A3-4, the space group is $P4_2/mnm$ (or $P4_2nm$ or $P\bar{4}n2$). The absence of an $hk0$ condition is based on the assumption that the (120) plane contributes to the eighth reflection. Actually, the measured d value for this reflection matches the chart positions for (120) and (022) equally well. The powder pattern offers no evidence to indicate whether one or both of these planes are responsible for the reflection. These planes do, however, reflect to distinct positions on single-crystal x-ray films, enabling one to reach a reliable decision concerning the reflection condition for the $hk0$ planes.

In general, indexed powder patterns lead to a positive identification of the lattice type. However, the usefulness of the powder method for deducing the space group is limited by two factors—planes with the same spacings reflect to the same position on the film, and the films do not display diffraction symmetry. In favorable cases, when no superposition of reflections occurs, indexed patterns lead to a reliable determination of space-group symmetry elements, but, if the crystal system includes more than one Laue class, single-crystal films are necessary to distinguish the diffraction symmetry.

The Identification of Crystalline Substances

The principle of the identification of substances by x-ray powder diffraction was first described by Hull[4] in 1919 as follows: "Every crystalline substance gives a pattern;—the same substance always gives the same pattern; and—in a mixture of substances each produces its pattern independently of the others, so that the photograph obtained with a mixture is the superimposed sum of the photographs that would be obtained by exposing each of the components separately for the same length of time."

To illustrate the value of the method, Hull described a simple analysis that gave information not readily obtained with the conventional methods of that time. He compared the x-ray patterns of "certified-pure" NaF and specially prepared, very pure NaF and found that the former contained a number of extra reflections, indicating the presence of an impurity. He then prepared the patterns of probable impurities, such as $NaCO_3$, NaCl, and $NaHF_2$. The pattern of $NaHF_2$ accounted for all the extra reflections, and he concluded that the "C.P." NaF actually consisted of a mixture of NaF and $NaHF_2$.

Clearly, an extensive library of reference films or of powder data obtained from standard substances was needed to develop the method as a tool for routine identification. Despite the obvious advantages inherent in the method, no concerted efforts were made to compile the necessary data, and for the next twenty years powder techniques were used on only a limited scale. Finally in 1938, Hanawalt, Rinn, and Frevel[5] of the Dow Chemical Company published powder data for 1000 chemical substances and a simple scheme for classifying and using the data to identify an unknown. Meanwhile, Harcourt[6] was accumulating powder data on the ore minerals and in 1942 published tables for their identification. General use of the method now became practical, and a Joint Committee of the American Society for Testing Materials and the American Society of X-ray and Electron Diffraction was formed to assemble and publish the Hanawalt data and other available data as a card file. These data, covering about 1300 compounds, were published in 1942 as the first edition of the ASTM X-ray Powder Data Diffraction file. The use of powder identification, expanding rapidly, stressed the need for new and revised data. The first set of cards was soon followed by additional sets as data were collected from industry, from the literature, and from research laboratories. With the aid of funds from the

[4] A. W. Hull, *J. Am. Chem. Soc.*, **41**, 1168 (1919).
[5] J. D. Hanawalt, H. Rinn, and L. K. Frevel, *Ind. Eng. Chem., Anal. Ed.*, **10**, 457 (1938).
[6] G. A. Harcourt, *Am. Mineralogist*, **27**, 63 (1942).

sale of cards, research associates were employed to review conflicting data in the card file, to prepare new patterns of standard materials, and to perform other duties to improve the value of the file. By 1964 the Joint Committee, under the auspices of the American Crystallographic Association, the British Institute of Physics, the National Association of Corrosion Engineers, and the American Society for Testing Materials, had published the fourteenth set of ASTM x-ray diffraction cards bringing the total to more than 11,000 cards.

THE ASTM X-RAY POWDER DATA FILE. An ASTM data card, typical of the more recent sets, is reproduced in Fig. 5-10. The number

9-423

d	3.03	1.85	1.59	3.03	$CuFeS_2$
I/I_1	100	80	60	100	COPPER IRON SULFIDE CHALCOPYRITE

Rad. FeKα λ 1.9373 Filter MnO Dia. 57.3MM
Cut off I/I_1 VISUAL
Ref. R.B. FERGUSON AND M.A. PEACOCK, UNIV. TORONTO, CANADA

Sys. TETRAGONAL S.G. D_{2D}^{12} - $I\bar{4}2D$
a_0 5.280 b_0 c_0 10.409 A C 1.9714
α β γ Z 4 Dx 4.283
Ref. BURDICK AND ELLIS, J. AM. CHEM. SOC. 39 2518 (1917)

εα nωβ εγ Sign
2V D 4.2 mp Color BRASS-YELLOW
Ref. D_7

SAMPLE FROM CUMBERLAND, ENGLAND.

REPLACES 1-0842, 2-0624

d Å	I/I_1	hkl	d Å	I/I_1	hkl
3.03	100	112			
2.63	5	020,004			
1.865	40	220			
1.854	80	024			
1.591	60	132			
1.573	20	116,033			
1.518	5	224			
1.323	10	040			
1.303	5	008			
1.214	10	332			
1.205	30	136,143			
1.077	60	244			
1.069	30	228			
1.018	20	152			
1.014	10	336			
1.005	5	1.1.10			

Fig. 5-10. An ASTM x-ray data card.

in the upper left corner of the card (9-423) indicates, first, the set in which it occurs and, next, its numerical position in the set. Below this are given the spacings and intensities corresponding to the three most intense reflections and of the innermost reflection (which, in this instance, is identical with the strongest reflection). The data for the three strongest reflections are used to classify the pattern in the Index as described subsequently. Below these data are listed experimental details and crystallographic, optical and physical information about the sample. The chemical formula and name of the compound are found on the upper right-hand side of the card. This is followed by a complete listing of the spacings, relative intensities, and Miller indices of the reflections. Cards which list data that are believed to be of a high reliability are marked with a star in the upper right corner.

THE INDEX TO THE X-RAY POWDER DATA DIFFRACTION FILE. The powder diffraction file is designed to be used with a book index divided into two sections. The numerical (or Hanawalt) section of the book enables one to locate a card in the file and thus identify a compound when its spacings and relative intensities are known. The alphabetical (or Davey) section of the Index lists alphabetically the names of substances represented in the X-ray Powder Data file. It is divided into four parts, namely inorganic, organic, organic formula, and mineral. The object of the alphabetical section is to enable one to find the three strongest lines and card number for any compound listed in the file.

Figure 5-11 shows part of page 143 in section one of the Index which covers the first nine sets of cards. Each entry on the page consists of the three strongest lines (x-ray diffraction term for the spacings corresponding to the three most intense reflections), their relative intensities, usually

3.55	2.18	1.86	90	100	90	$CuInTe_2$	Copper indium telluride	6-0605
3.54	2.17	3.07	100	74	34	AlSb	Aluminum antimony	6-0233
3.54	2.17	3.07	100	41	32	SnI_4	Tin IV iodide	6-0232
3.54	2.17	1.85	100	90	80	Li_5TiAs_3	Lithium titanium arsenide	7-91
3.50	2.17	1.85	52	100	90	α-CuI	Alpha copper iodide	6-0623
3.54	2.16	3.19	40	100	83	$BaCrO_4$	Barium chromate	1-1221
3.53	2.16	2.74	44	100	65	ZrB_2	Zirconium boride	6-0610
3.54	2.16	1.86	100	100	90	Li–$MnAs_4$	Lithium manganese (V) arsenide	14-43
3.53	2.16	1.85	80	100	80	α–Cu_2HgI_4 (at 90°C)	Alpha copper iodomercurate (II)	3-0967
3.52	2.16	1.84	100	67	44	GaSb	Gallium antimony	7-215
3.51	2.16	1.84	100	100	70	Cu_2Te (high temp.)	Copper tellurium	7-106
3.52	2.16	1.83	100	100	90	Li_3MnAs_2	Lithium manganese (III) arsenide	14-42
3.51	2.15	1.84	90	100	90	Li_5GeAs_3	Lithium germanium arsenide	7-100
3.51	2.15	1.84	100	51	32	HgSe	Mercury (II) selenide	8-469
3.51	2.15	1.24	60	100	100D	$AuGa_2$	Gallium gold	3-0969
3.51	2.14	3.05	100	50	45	$Rb_7U_6F_{31}$	Rubidium uranium fluoride	10-137
3.50	2.14	1.83	100	80	70	Li_5SiAs_3	Lithium silicon arsenide	7-148
3.50	2.14	1.83	100	67	54	ZnTe	Zinc telluride	1-0582
3.52	2.11	2.86	100	100	70	$Cd_{1.25}Bi_{1.5}O_2Cl_3$	Cadmium bismuth oxychloride	12-119

Fig. 5-11. Part of a page from the index to the x-ray powder diffraction file.

the chemical formula and name of the substance, and the set and card number.

The entries are arranged in the so-called "Hanawalt" order, which sorts them into 87 groups according to the value of the first line. Figure 5-11, for example, is a page from the 3.59–3.50 Hanawalt group. The preceding group in the Index contains those entries for which the first line lies in the range 3.69–3.60; the group that follows represents the range 3.49–3.45. Within each group entries are arranged in decreasing order of the second line. If two or more entries have the same second line, the value of the third line determines the sequence. The Index is made up of three entries for each data card in the sets. Of the three entries, one shows the three strongest lines in the same sequence as on the data cards, whereas the second and third entries show these lines rearranged in the sequence: second strongest, strongest, third strongest and third strongest, strongest, second strongest. The purpose of the multiple listing is to make it possible to find an entry despite minor variations in observed spacings and intensities.

Use of the Powder Data Diffraction File and the Index. To identify a compound from its diffraction data with the aid of the Index and the file

(1) locate the correct Hanawalt group for the strongest line; if the value of the strongest line is close to the limit between two Hanawalt groups, it may be necessary to consult both groups;
(2) read down the second column to find the closest match or matches with the second line;
(3) select the entry or entries which also show a match for the third line;
(4) extract the appropriate card or cards from the file and compare the remainder of the observed data with the data on the cards.

To illustrate the procedure, let us assume that an unknown compound yields the following data for the three strongest lines:

3.53 2.16 3.07 100 60 40

A search in the appropriate Hanawalt group (3.59–3.50) for a match with the second line leads to the page of the Index reproduced in Fig. 5-11. When the third line is taken into consideration, it is found that two sets of entries have values that are reasonably close to the observed values:

3.54 2.17 3.07	100 74 34	AlSb	aluminum antimony	6–0233
3.54 2.17 3.07	100 41 32	SnI_4	tin IV iodide	6–0232

Provided that it is listed in the file, the identification of the unknown has now been narrowed to two possibilities. Despite the similarity in the Index data for the three strongest lines of AlSb and SnI_4, the corresponding data cards show that the two compounds are readily distinguished when entire patterns are taken into account (Table 5-7). The unknown can be assumed to have been identified only if there is satisfactory agreement between all of the observed data and the data given for one of the two compounds. Otherwise it must be assumed that the unknown is not listed in the file.

Table 5-7. A.S.T.M. Powder Data for SnI_4 and AlSb

SnI_4		AlSb	
d (meas.)	I/I_1	d (meas.)	I/I_1
7.09	3		
5.49	10		
5.012	15		
3.543	100	3.54	100
3.404	5		
3.280	5		
3.070	32	3.067	34
2.816	3		
2.678	6		
2.617	5		
2.279	5		
2.241	4		
2.1701	41	2.169	74
etc.		etc.	

THE IDENTIFICATION OF MIXTURES. If the three strongest lines of a pattern do not lead to an identification, the sample may be a mixture of compounds. The procedure then is to attempt the identification of one of the components on the basis of two of the strong lines. The data of Table 5-8 will serve as an illustration. The three strong lines

2.97 3.43 3.12 100 90 80

cannot be matched with an entry in the Index for 1964. Ten entries in the 2.99–2.95 Hanawalt group of the Index have second lines that are reasonably close (within 0.03 Å) to 3.43. The spacing value 2.10 (or 2.11), another strong line in the unknown pattern, is present in three of the ten entries:

2.99 3.45 2.11 100 64 63 NaBr

2.95 3.44 2.10 100 80 80 EuS

2.97 3.43 2.10 100 84 57 PbS Galena

An examination of the data cards for the three compounds reveals that the agreement with lines in Table 5–8 is only approximate for NaBr and EuS. On the other hand, the PbS lines are in good agreement, and all but the weakest lines are found in the table. The three strongest lines of the

Table 5-8. X-Ray Powder Data for a Mixture of PbS and ZnS

d (meas.)	*I*		*d* (meas.)	*I*		*d* (meas.)	*I*	
3.43	90	PbS	1.361	30	PbS	0.9892	20	PbS
3.12	80		1.352	20		0.9556	10	
2.97	100	PbS	1.325	40	PbS	0.9386	20	PbS
2.70	20		1.240	20		0.9139	20	
2.10	60	PbS	1.213	30	PbS	0.9052	10	PbS
1.910	50		1.143	20	PbS	0.8952	20	PbS
1.790	50	PbS	1.102	20		0.8547	10	
1.715	40	PbS	1.050	10	PbS	0.8311	10	PbS
1.632	40		1.039	20		0.8232	10	PbS
1.483	30	PbS	1.003	20	PbS			

data that remain in the table, when the PbS lines are removed from consideration, have the values

3.12 1.910 1.632 80 50 40

The Index data for thirteen entries in the 3.14–3.10 Hanawalt group are in substantial agreement with these values. But again only one card, representing the compound ZnS, satisfactorily accounts for all of the remaining lines. The sample is therefore a mixture of PbS and ZnS.

The example just described illustrates a fairly simple problem. The identification of a mixture is more difficult if one of the components is present in small amounts. The minor constituent is usually represented by a few lines, and the certainty of the identification is correspondingly low. This situation generally applies to at least one of the components of a mixture of three or more compounds. Line superposition is another cause of difficulty in working with mixtures. This causes an apparent increase of intensity in lines that are superimposed.

The presence of a mixture can sometimes be recognized from the appearance of the film. A difference in the texture of the lines usually indicates that the sample contains two or more components that differ in degree of crystallinity, in grain size, or in grain shape. The effect can be accentuated by coarse-grinding the sample or by keeping it stationary during the exposure.

THE FINK INDEX TO THE X-RAY POWDER DIFFRACTION FILE. The Hanawalt Index is not particularly effective for identifying electron diffraction patterns, or x-ray diffraction patterns of mixtures and of specimens having a preferred orientation of the constituent grains, because the relative intensities of the lines may differ considerably from the data given in the file. In the Hanawalt Index the spacing of the strongest line determines the Hanawalt group in which the pattern is listed, or the location of the pattern within a group. Failure to recognize the strongest line of a pattern and to utilize it in the search of the Index can hamper the identification.

The Fink Index, first published in 1963, is designed to depend less critically on the strongest lines. It uses the eight strongest lines, from 10.0 Å down, to give eight entries for each pattern. The first entry lists the lines in decreasing order of d values; in the other entries the arrangement is by cyclic permutation of the values. The entries are sorted into about 100 groups on the basis of the first line. The entries in each group are sorted on the basis of the second line.

This Index is presently available only for the inorganic portion of the file.

THE MATTHEWS COORDINATE INDEX TO THE X-RAY POWDER DIFFRACTION FILE. This Index is a punched-card coordinate indexing system, named for Dr. F. W. Matthews who was mainly responsible for the development of procedures for applying the system to powder diffraction data.

Each card has 10,000 hole locations arranged in 100 equally spaced rows and columns. Each inorganic material in the X-ray Powder Diffraction File has the same location on each card. A conversion book relates the coordinates of each location to the name of the material and its file card and set number. Each card represents a chemical element, interplanar spacing, or other property, and a hole is drilled if the material identified by the location has the property represented by the card.

To retrieve data from the Index, cards representing the properties recognized in the unknown are selected and superimposed on a readout device. Negative cards, representing the absence of chemical elements, interplanar spacings, and other characteristics are available and, when added to the stack of selected cards, enable the searcher to eliminate materials that are known to have properties not present in the unknown.

When holes in the cards coincide, points of light are visible. The locations of the lights determine the identity of all materials in the file that have the known properties of the material under investigation.

The Matthews Index covers the inorganic portion of the first fourteen sets of the file.

A COMPARISON OF THE THREE INDEXES. The Fink and Matthews Indexes cannot be properly evaluated until they have been tested over a period of years. However, it is clear that each of the three indexes now available has certain advantages as well as disadvantages. Unquestionably, a user's evaluation of each Index for routine identifications will reflect his personal preference for, or success with, one of the three kinds of searching schemes.

The Quantitative Analysis of Mixtures

In his description of the principle of identification of substances by x-ray powder diffraction Hull stated that, "in a mixture of substances each produces its pattern independently of the others, so that the photograph obtained with a mixture is the superimposed sum of the photographs that would be obtained by exposing each of the components separately for the same length of time. This law applies quantitatively to the intensities of the lines, as well as to their positions, so that the method is capable of development as a quantitative analysis."

Although the intensity of each component pattern does depend on the concentration of the component in the mixture, the relation between intensity and concentration is not so simple as implied by Hull's statement. In general, the relation is not linear because of the effect of absorption—the intensities of reflections due to strongly absorbing materials in a mixture of strong and weak absorbers are enhanced, whereas those due to weakly absorbing components are attenuated.

It has been shown that the total intensity I_c of x-rays reflected by some selected plane of a component c of a mixture is given by

$$I_c = K_c \frac{V_c}{\mu_M}$$

where V_c is the volume fraction of the component and μ_M is the linear absorption coefficient of the mixture. K_c depends on a number of factors which include the nature of the component (K_c therefore includes the expression for the integrated reflection), the degree of crystal perfection (the influence of primary and secondary extinction), grain shape (the effect of preferred orientation), and such instrumental arrangements as the packing density of grains in the sample, sample size, beam intensity, and length of exposure. The first factor is a constant for a particular plane of a crystal. The crystal perfection, however, may vary markedly from specimen to specimen of the same compound, but it can be reduced to a common factor by very fine grinding. Fine grinding, as well as careful sample preparation, also serves to reduce the effect of preferred orientation. The instrumental arrangements can probably never be exactly duplicated from one day to the next.

The contribution of each component to the linear absorption coefficient of the mixture is the product of its volume fraction and linear absorption coefficient—namely, $V_c\mu_c$. Consequently, for a mixture of n components, $\mu_M = \Sigma_1^n V_n\mu_n$ and

$$I_c = K_c \frac{V_c}{\sum_1^n V_n\mu_n}$$

Because volume × density equals weight, the intensity can alternatively be expressed as

$$I_c = K_c \frac{W_c/\rho_c}{\sum_1^n (W_n/\rho_n)\mu_n}$$

where W_c and ρ_c are, respectively, the weight fraction and density of the c component. The linear absorption coefficient is dependent on the state of the absorber. It is usually more convenient to use the mass absorption coefficient, μ^*, which is independent of the physical and chemical state of the absorber. The two coefficients are related as $\mu^* = \mu/\rho$. Hence

$$I_c = K_c \frac{W_c/\rho_c}{\sum_1^n W_n\mu_n^*} \tag{1}$$

ANALYSIS OF A TWO-COMPONENT MIXTURE. The analysis of a two-component mixture is often treated in terms of the ratio of the intensity of a selected reflection of one component of the mixture to the intensity of the same reflection in the pure component. According to Eq. 1

$$I_1 = K_1 \frac{W_1/\rho_1}{W_1\mu_1^* + (1 - W_1)\mu_2^*} = K_1 \frac{W_1}{\rho_1[W_1(\mu_1^* - \mu_2^*) + \mu_2^*]}$$

where W_1, ρ_1, and μ_1^* are the weight fraction, density, and mass absorption coefficient of component 1 and $(1 - W_1)$, ρ_2, and μ_2^* are the same constants for the second component. The intensity of the pure component is

$$I_{1(\text{pure})} = K_1 \frac{W_1/\rho_1}{W_1\mu_1{}^*} = K_1 \frac{1}{\rho_1\mu_1{}^*}$$

The ratio of the intensities is therefore

$$\frac{I_1}{I_{1(\text{pure})}} = \frac{W_1\mu_1{}^*}{W_1(\mu_1{}^* - \mu_2{}^*) + \mu_2{}^*} \tag{2}$$

The equation confirms the statement that the relation between intensity and concentration is not linear. An exception is the special case in which

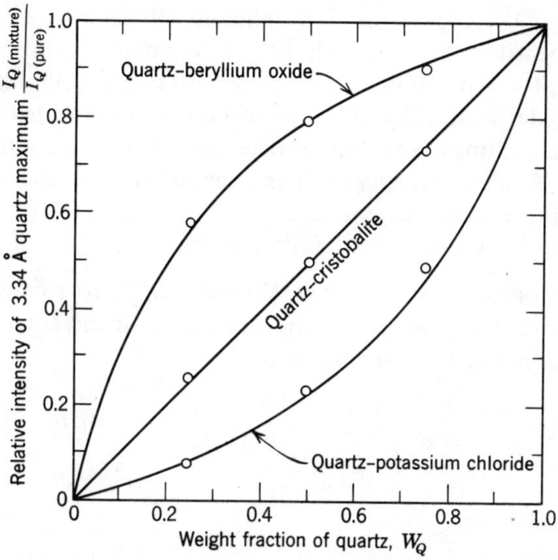

Fig. 5-12. Intensity concentration curves calculated from the relation $I_Q/I_{Q(\text{pure})} = W_Q\mu_Q^*/W_Q(\mu_Q^* - \mu_2^*) + \mu_2^*$ for three binary mixtures with quartz (μ_2^* refers to the second component). Open circles represent experimental measurements (from Klug and Alexander[8]).

the mass absorption coefficients $\mu_1{}^*$ and $\mu_2{}^*$ are equal. The relation is demonstrated in Fig. 5-12 by intensity-concentration curves calculated for three mixtures: quartz (SiO_2) and Be_2O, quartz and cristobalite (SiO_2), quartz and KCl. The mass absorption coefficients for these compounds are $Be_2O = 8.6$, quartz and cristobalite $= 34.9$, and $KCl = 124$. The curves further illustrate that the intensity of quartz is enhanced in the presence of a weaker absorber (Be_2O) and attenuated when mixed with a stronger absorber (KCl). The validity of the curves is proved by experimentally determined points, represented by circles on the figure, for different ratios of the three mixtures.

Preliminary to an analysis, x-ray powder films are prepared of the mixture to be analyzed, of a number of synthetic mixtures of its two components, and of the pure component which is the subject of the analysis. The intensity of a selected reflection is measured on each film with a microphotometer or is determined by comparison with a calibrated intensity strip. An intensity-concentration curve is then plotted from the intensity measurements of the known mixtures in accordance with Eq. 2. Finally, the weight fraction of the unknown is obtained by referring its intensity ratio to this curve.

This method suffers from an inherent disadvantage. Because the intensity measurement on each film is compared with the intensity measurement on the film of the pure component, all films must be prepared under identical conditions of sample size and packing density, instrumental settings, and film processing. In practice this is difficult to achieve. A more practical scheme for analysis is one that depends on comparative measurements made on the same film. This is the basis of the method described in the following section.

Analysis of the Ratio of the Weights of Two Components in a Mixture. According to Eq. 1, the intensities of any two components a and b in a mixture of n components are

$$I_a = K_a \frac{W_a/\rho_a}{\sum_1^n W_n \mu_n^*} \quad \text{and} \quad I_b = K_b \frac{W_b/\rho_b}{\sum_1^n W_n \mu_n^*}$$

Then

$$\frac{I_a}{I_b} = \frac{K_a}{K_b}\frac{W_a}{W_b}\frac{\rho_b}{\rho_a} = K\frac{W_a}{W_b}$$

Thus the intensity ratio I_a/I_b is independent of the absorption coefficient of the mixture and is proportional to the weight fraction ratio W_a/W_b. To make use of this relation, the I_a/I_b ratios are measured on films obtained from known mixtures and then plotted against their W_a/W_b ratios to give a line to which intensity ratios of unknown mixtures are referred. In contrast to the preceding method, the two measurements that give an intensity ratio are made on the same film. Because the experimental conditions are identical for the two components in a sample of a mixture, each film is, in a sense, self-calibrating and provides an independent measurement of the intensity ratio which can be compared with a measurement made on another film. If the unknown mixture contains only the two components, the intensity ratio immediately gives the absolute concentrations of *a* and *b*. In a multicomponent mixture the intensity ratio gives the relative concentrations of the two components. The relative values may be converted to absolute values, if the combined weight fraction of the two components or the weight fraction of one of the components can be determined by another method.

ANALYSIS OF A COMPONENT IN A MIXTURE OF UNKNOWN COMPONENTS. In the most general type of analysis it is necessary to determine the absolute concentration of a compound in a mixture of unknown components. In these circumstances the absorption coefficient of the mixture ($\mu_M{}^*$) is not known, and it becomes necessary to adopt a procedure similar to the one described in the preceding section to eliminate the need to consider this factor. The final statement in the preceding section suggests a modification whereby this method can be applied generally to the analysis of one component of a mixture. If the weight fraction of one component *s* in the unknown mixture is predetermined by the addition of an amount W_s, the weight fraction W_a of a component *a* can be found by referring the I_a/I_s ratio to a reference curve which gives W_a/W_s plotted against I_a/I_s for a range of W_a values and the fixed value of W_s. This is the basis of the internal standard method.

According to Eq. 1, if a known weight of a standard *s* is added to a mixture that contains a component *a* to be analyzed, then

$$I_a = K_a \frac{W_a'/\rho_a}{\sum_1^n W_n \mu_n{}^*}$$

where W_a' is the weight fraction of *a* after adding the standard. Also

$$I_s = K_s \frac{W_s/\rho_s}{\sum_1^n W_n \mu_n{}^*}$$

where W_s is the weight fraction of the standard. Then

$$\frac{I_a}{I_s} = \frac{K_a(W_a'/\rho_a)}{K_s(W_s/\rho_s)} \tag{3}$$

The weight fraction W_a of component *a* in the original sample is related to the weight fraction W_a' in the mixture which contains the standard as

$$\frac{W_a}{W_a'} = \frac{1}{1 - W_s}$$

Therefore $W_a' = W_a(1 - W_s)$. Substituting this value for W_a' in (3) and simplifying, we obtain

$$\frac{I_a}{I_s} = \frac{K_a W_a(1 - W_s)\rho_s}{K_s \rho_a W_s}$$

and from this

$$W_a = \frac{I_a \rho_a W_s K_s}{I_s \rho_s (1 - W_s) K_a}$$

Thus in a series of mixtures in which W_a varies but W_s remains constant

$$W_a = K\frac{I_a}{I_s}$$

that is, W_a is related in a linear manner to I_a/I_s.

Table 5-9. Data for Quartz Calibration Curve (Fig. 5-14)
(After Ballard, Oshry, and Schrenk[7])

Weight in Grams				
Quartz	Calcite	Fluorite	Quartz Per Cent	Ratio I_Q/I_F
2.00	0.00	1.00	100	2.61
1.80	0.20	1.00	90	2.36
1.60	0.40	1.00	80	2.08
1.40	0.60	1.00	70	1.81
1.20	0.80	1.00	60	1.55
1.00	1.00	1.00	50	1.27
0.80	1.20	1.00	40	1.03
0.60	1.40	1.00	30	0.78
0.40	1.60	1.00	20	0.53
0.20	1.80	1.00	10	0.28
0.16	1.84	1.00	8	0.23
0.12	1.88	1.00	6	0.20
0.08	1.92	1.00	4	0.14
0.04	1.96	1.00	2	0.11
0.02	1.98	1.00	1	0.08

This relation is used to determine the weight fraction of a compound in a mixture in the following way. A series of reference mixtures is prepared in which the weight fraction of a chosen standard remains constant, the weight fractions of the compound to be analyzed vary over an appropriate range, and a suitable diluent has been added to bring the weights of the mixtures to the same total. To a sample of the material to be analyzed is added the same weight fraction of standard. X-ray powder patterns are obtained of all the samples under identical

Table 5-10. X-Ray Diffraction Analysis of Samples Containing Known Amounts of Quartz

(After Ballard, Oshry, and Schrenk[7])

	Quartz, Per Cent						
	Individual Analyses						
Diluent	1	2	3	4	Average	Actual	Error Per Cent
$CaCO_3$ (calcite)	1.3	0.6	0.9	1.0	0.77[a]	0.59	31
$H_4Al_2Si_2O_9$ (kaolinite)	1.5	0.3	1.5	0.6	0.98	1.0	2
$CaSO_4.2H_2O$ (gypsum)	1.2	2.1	1.6	1.5	1.6	1.7	6
K_2SiF_6 (hieratite)	3.0	2.8	2.0	2.6	2.6	2.1	24
$KAl(SiO_3)_2$ (leucite)	2.6	3.8			3.2	3.4	6
$H_4Al_2Si_2O_9$ (kaolinite)	4.9	5.6	4.6		5.0	5.2	4
Vermiculite	5.1	6.3			5.7	5.3	8
$NaAlSi_3O_8$ (albite)	5.9	6.7			6.3	6.2	2
K_2SiF_6 (hieratite)	9.2	9.6			9.4	8.5	11
$ZrSiO_4$ (zircon)	8.9	9.6			9.3	9.1	2
W	10.3	10.7			10.5	10.1	4
SiC	9.9	8.7			9.3	10.3	10
$CaTiSiO_5$ (titanite)	13.4	12.9			13.2	12.1	9
$BaCl_2.2H_2O$	14.5	13.0			13.8	15.2	9
Cr_2O_3	18.5	18.0			18.3	17.0	8
CdO	18.6	18.4			18.5	20.7	11
Se	19.6	23.7			21.7	22.3	3
Coal	25.7	26.9			26.3	23.8	11
MgO	25.6	24.6			25.1	24.9	1
Fe_2O_3 (hematite)	25.3	26.9			26.1	27.2	4
$CaTiSiO_5$ (titanite)	29.3	24.4			26.9	28.9	7
Fe_2O_3 (hematite)	44.7	44.6	39.9	39.3	42.1	44.1	5
$H_4Al_2Si_2O_9$ (kaolinite)	50.7	46.1			48.4	53.4	9
Coal	72.1	73.0	74.2		73.1	78.6	7

[a] Including a fifth and sixth analysis which gave 0.5 and 0.3 per cent, respectively.

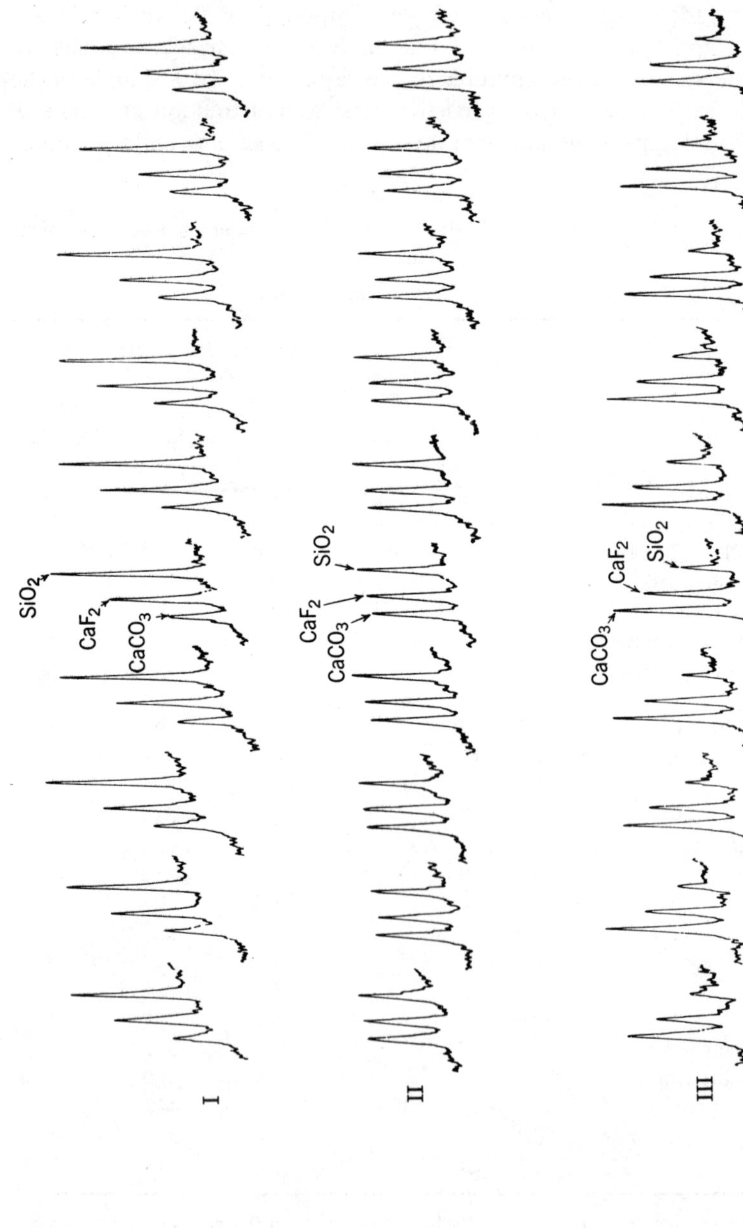

Fig. 5-13. Microphotometer records of ten traverses across selected calcite, fluorite, and quartz lines obtained in mixtures with 60 (I), 40 (II), and 20 (III) per cent quartz (from Ballard, Oshry, and Schrenk[7]).

experimental conditions, and the intensities of selected reflections of the compound in question and the standard are measured in the usual way. The intensity ratios I_a/I_s for the reference samples are plotted against composition to give a calibration curve. The weight fraction in the unknown mixture is then deduced from this curve by interpolation.

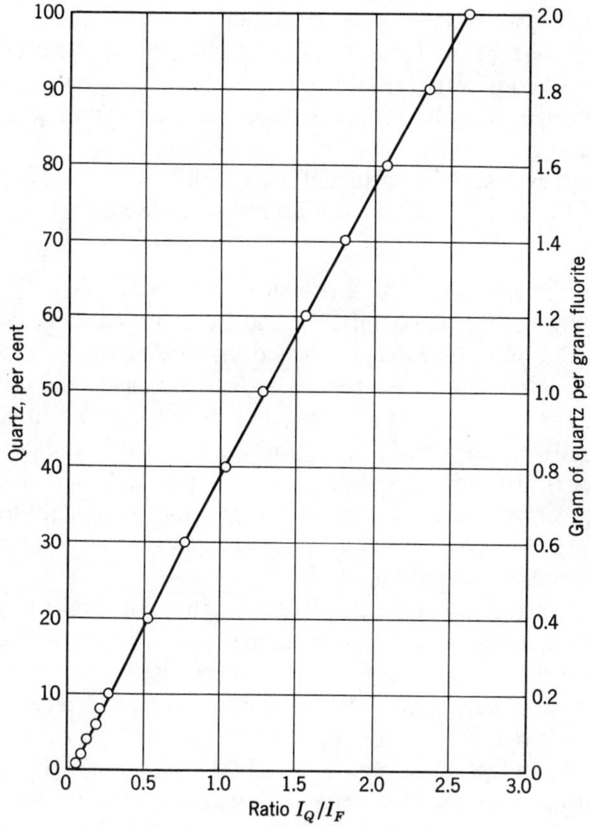

Fig. 5-14. Ballard, Oshry, and Schrenk's standard curve for the determination of quartz.

Table 5-9 illustrates the method of preparing the reference mixtures. It shows the composition of mixtures prepared by Ballard, Oshry, and Schrenk[7] for the determination of quartz. Fluorite (CaF_2) was chosen as the internal standard, and calcite ($CaCO_3$) was used as the diluent. The final column in the table gives the intensity ratios I_Q/I_F obtained by

[7] J. W. Ballard, H. I. Oshry, and H. H. Schrenk, *U.S. Bureau of Mines Report of Investigations* 3520, June 1940.

averaging ten micrometer traverses across the $10\bar{1}1$ quartz and the 111 fluorite reflection arcs (Fig. 5-13). The quartz calibration curve, plotted from these ratios, is reproduced in Fig. 5-14. To test the curve Ballard, Oshry, and Schrenk determined the quartz content of synthetic mixtures prepared in a manner similar to that followed for the standard series, except that various materials were substituted for the calcite diluent. Table 5-10 compares the values obtained by analysis with the actual quartz content in the mixtures. Although individual analyses of the same mixture vary considerably, the averaged values show a maximum percentage error of only 11 per cent for mixtures with more than 3 per cent quartz.

The accuracy which is attained in quantitative analysis by x-ray diffraction depends largely on the attention focused on two factors of the procedure:

1. *Sample Preparation.* A satisfactory reproducibility of intensity measurements on the same mixture can be achieved only if the components are homogeneously distributed and randomly oriented in the sample. The importance attached to proper mixing of the components is illustrated by the work of Ballard, Oshry, and Schrenk in connection with the analyses described. After grinding and sizing the components, the authors transferred their mixtures to a ball mill half filled with a dispersing agent (methanol) and rotated the mill for twenty-four hours. Efforts to achieve the perfectly random sample also serve to reduce the effect of secondary extinction.

Standard mixtures prepared from batches of uniformly ground components carefully mixed into homogeneous samples can be expected to give reproducible intensity ratios that precisely outline a reference curve. If the curve fails to give satisfactory results when intensity measurements from other mixtures are applied to it, the difficulty can usually be attributed to a difference in mosaicity, and therefore to a difference in the effect of primary extinction, between the component in the standards and in the mixtures to be analyzed. As noted in Chapter 3, differences in primary extinction can be eliminated by grinding standards and unknowns to a particle size that is at least as small as the smallest mosaic block size. Klug and Alexander[8] recommend grinding to less than 5 μ.

2. *Measurement of Intensities.* The measurement of intensities on powder films is neither so accurate nor so convenient as might be desired. Intensity measurements have been materially improved by the development of the diffractometer described later. The x-rays reflected by the

[8] H. P. Klug and L. E. Alexander, *X-Ray Diffraction Procedures*, John Wiley and Sons, New York (1954), p. 726.

sample are received by a Geiger, proportional, or scintillation counter and are converted to electrical pulses which can be electrically counted to give a more reliable estimate of the intensity of reflected x-rays than is possible with the film method. As a result of the convenience and precision afforded by the diffractometer, the film method has been largely superseded for quantitative studies.

The Determination of Crystallite Size

Although the diffraction behavior of ideally perfect and ideally imperfect (which is equivalent to the perfect powder specimen) crystals was discussed in some detail in Chapter 3, the reader was left with the

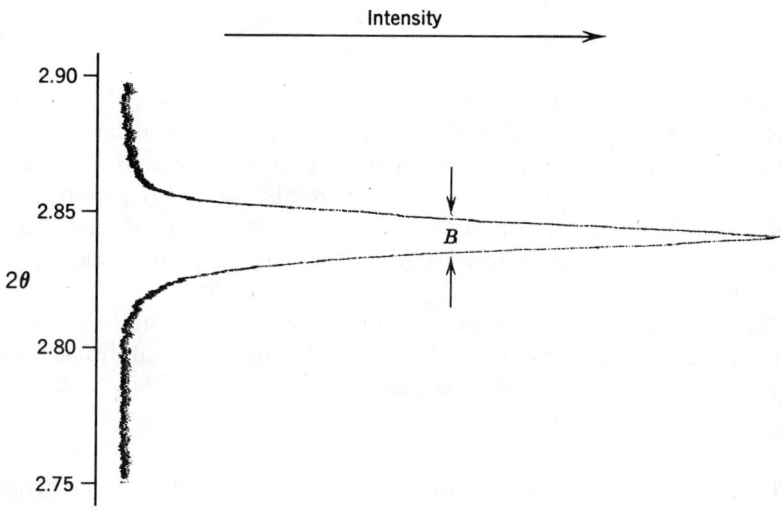

Fig. 5-15. A plot of intensity against diffraction angle for the 111 reflection from a specimen of powdered silicon.

impression that reflected rays, including the rays emanating from the various mosaic blocks of an imperfect crystal rotating through a Bragg condition, form an extremely narrow parallel beam. Actually, a typical plot of intensity against diffraction angle 2θ has the appearance of Fig. 5-15, indicating that diffraction lines have a certain breadth B (defined here as the angular width at half-maximum intensity). A number of instrumental factors contribute to the breadth of lines: the incident beam diverges somewhat and is not strictly monochromatic; the specimen has an appreciable diameter; the chemical composition may vary slightly from crystallite to crystallite with accompanying small variations in lattice dimensions; the structure may be distorted by strain. Line broadening caused by these factors is called instrumental broadening. The breadth at half-maximum intensity resulting from instrumental broadening is designated here as b.

When the thickness of crystallites is less than about 1000 Å (0.1 μ) another factor, called pure diffraction broadening, contributes to the total breadth B and causes line broadening beyond that expected from the instrumental arrangements mentioned in the preceding paragraph. Pure diffraction broadening increases with decreasing crystallite size, and the lines eventually become so broad and diffuse that they are virtually indistinguishable from the background. Scherrer[9] showed that this type of broadening β is related to the thickness t of the crystallites as

$$\beta = \frac{K\lambda}{t \cos \theta}$$

where K is a constant that depends on the way in which β and t are defined and on the shape of the crystallites. The constant has a value of 9.0 when β is defined as the angular width at half-maximum intensity of that portion of the reflection due to pure diffraction broadening, and the crystallites are platelets with a thickness t measured normal to the reflecting surface but whose lateral dimensions are not taken into account.

Scherrer assumed that the pure diffraction broadening can be obtained by subtracting the line breadth caused by instrumental broadening from the line breadth of the material under study

$$\beta = B - b$$

This relation has been found to have no general validity. A correction first used by Warren and Biscoe[10]

$$\beta = \sqrt{B^2 - b^2}$$

gives values which are in substantial agreement with more elaborate methods. According to Klug and Alexander,[8] the correction is quite satisfactory for comparing the crystallite sizes of a number of samples belonging to a related series—it yields satisfactory values of the *relative* sizes but leaves the *absolute* sizes uncertain. The reader in search of a complete treatment of the subject is advised to consult the Klug and Alexander reference.

In order to deduce the crystallite size (often wrongly called the particle size—particles may be aggregates of crystallites) from pure

[9] P. Scherrer, *Nachr. k. Ges. Wiss. Göttingen*, **98**, (1918).

[10] B. E. Warren and J. Biscoe, *J. Am. Ceram. Soc.*, **21**, 49 (1938).

diffraction broadening, it is necessary to know the breadth of lines given by crystallites which do not show the effect—that is, by crystallites with a thickness of 10,000 Å (1 μ) or more. The best method of determining this is to mix the material under study with another material having crystallites of the requisite size. Under these circumstances the instrumental effects are almost identical for the two compounds. The breadths of the sharp lines give b, and those of the material under study give B. The Scherrer relation indicates that β varies with θ. It is consequently necessary to compare lines in the same region of the film and to obtain by interpolation values for b that correspond to the positions of the broadened lines.

The accuracy with which pure diffraction broadening can be measured evidently decreases with increasing size of crystallites—that is, as the ratio b/B approaches unity. Careful experimental techniques are required to measure the effect on Debye–Scherrer films when the crystallites have a size of about 1000 Å. The limit can be extended to about 2000 Å with the use of a diffractometer.

It is worth noting that the Scherrer relation leads to a determination of the *average* crystallite size. The method does not provide the means for deducing the distribution of sizes in the sample.

The Attainment of Precise Lattice Dimensions with the Debye–Scherrer Method

Refinements in the determination of lattice constants with the powder method have advanced to the point at which, in the opinion of some scientists, the precision will soon be limited by the precision with which the x-ray wavelengths are known. Estimates of the highest accuracy obtainable vary from three to "probably as high as one" part in 100,000. The cell constants of many compounds change by detectable amounts with variations of little more than 1°C. Consequently, controlled-temperature cameras must be used to attain the highest accuracy. Furthermore, special measuring devices are required to read the films. Accuracy of at least five parts in 100,000 is possible with the apparatus available in most modern x-ray diffraction laboratories on powder patterns that show sharp, well-resolved reflections in appropriate regions of the film.

The positions of the reflections on a powder film are subject to a number of systematic errors. Both systematic and random errors are introduced in measuring the positions. To take full advantage of the powder method and to gain the precision of which it is capable, it is necessary to understand the source and nature of the errors, to know the techniques for keeping the errors at a minimum, and finally, to be able to compensate for errors when they are present.

The Source and Correction of Errors

FILM SHRINKAGE. The diameter of a Debye–Scherrer camera is easily and accurately measured, but the film shrinks during the process of developing, fixing, washing, and drying, and its length changes with the humidity and with the temperature of the measuring screen. It follows that the value of the film diameter, which is used in converting film measurements to θ angles, cannot be considered equal to the measured camera diameter. If accuracy is of prime importance, the effect of

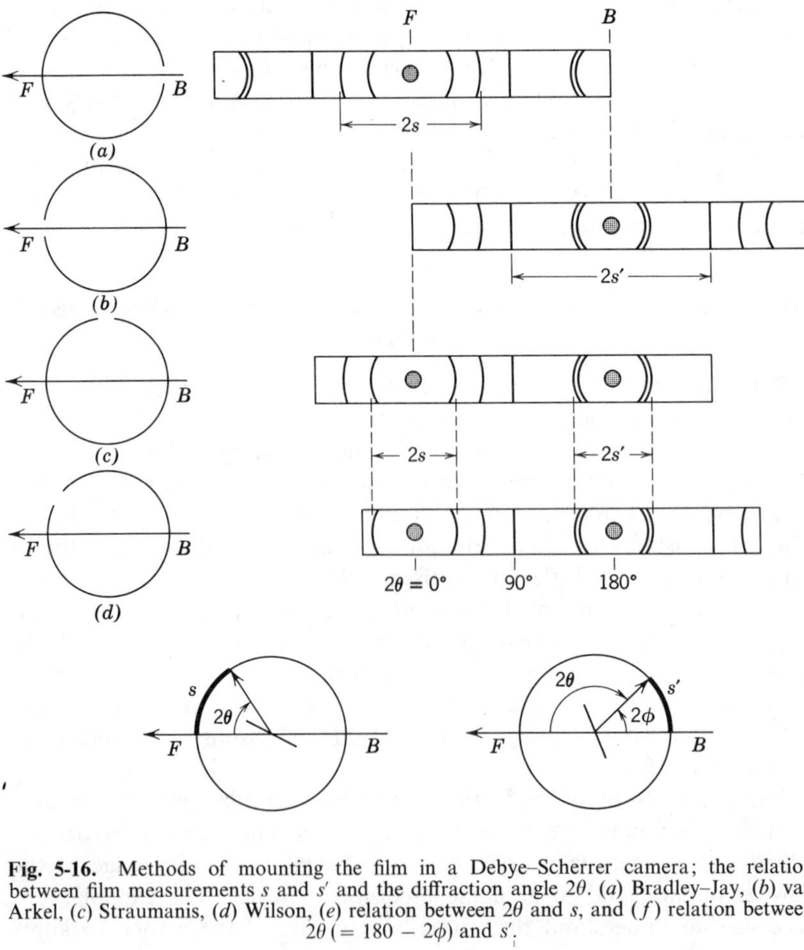

Fig. 5-16. Methods of mounting the film in a Debye–Scherrer camera; the relation between film measurements s and s' and the diffraction angle 2θ. (*a*) Bradley–Jay, (*b*) van Arkel, (*c*) Straumanis, (*d*) Wilson, (*e*) relation between 2θ and s, and (*f*) relation between $2\theta\ (= 180 - 2\phi)$ and s'.

shrinkage should be taken into account on each occasion when the film is measured.

The regular or Bradley–Jay instrumental arrangement described in connection with Fig. 5-2 is only one of four methods of positioning the film in a Debye–Scherrer camera. The four methods are compared in Fig. 5-16. In the precision or van Arkel arrangement the beam enters through a hole in the center of the film and passes out of the camera between the ends of the film, in direct contrast to the Bradley–Jay position. The back-reflection arcs are then curved about the film center, the curvature decreasing to zero at $2\theta = 90°$. The front-reflection arcs are curved about the ends of the film. In the asymmetric or Straumanis position the beam passes through the camera by way of two symmetrically situated and diametrically opposite holes in the film. As a result, the front- and back-reflection arcs are separately grouped on the film—the front reflections about one hole (F in Fig. 5-16*c*) and the back reflections about the other (B). The Wilson method is a modification of the Straumanis mount.

Bradley–Jay Arrangement. The shrinkage correction for a film mounted according to the Bradley–Jay arrangement can be determined with the aid of fiducial marks on the film. The marks can be produced by knife edges which are in contact with the film and cast sharp shadows on it during the exposure. In another procedure two small holes drilled in appropriate places on the circumference of the camera are opened for a few seconds in a weak light. The result is two tiny spots superimposed on the powder pattern after the film is processed. For greatest efficiency the fiducial marks should be placed near the ends of the film. The angle subtended by the knife edges or the fiducial holes (the so-called camera angle α) can be accurately determined by direct measurement on a spectrometer or similar instrument.

Figure 5-17 shows the relation between the true angles 2θ and α and the corresponding film measurements s and s_F. Assuming that the film shrinks uniformly, $2\theta/s = \alpha/s_F$, and therefore

$$2\theta = \left(\frac{\alpha}{s_F}\right) s$$

The factor α/s_F naturally varies for the same film from one day to the next. The ratio s/s_F increases with the angle 2θ, and therefore the error in 2θ due to an error in measuring α is smallest for the front (low 2θ) reflections.

Van Arkel Arrangement. To gain accuracy, 2θ measurements obtained from a film in this position must also be referred to fiducial marks,

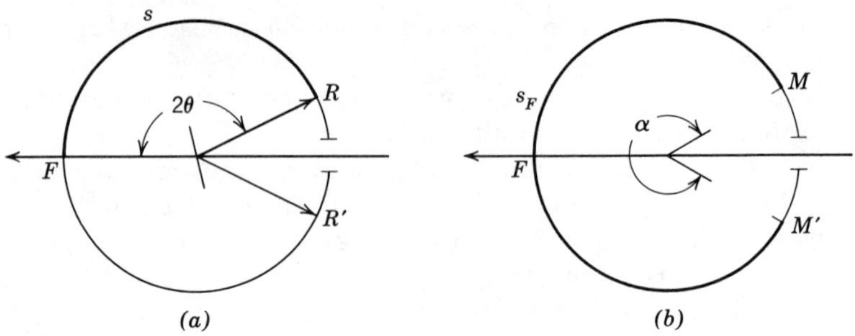

Fig. 5-17. The Bradley–Jay arrangement. (*a*) 2θ and s, half the measured film distance between reflections R and R'; (*b*) α, the camera angle, and s_F, the measured film distance MFM' between fiducial marks M and M'.

preferably placed near the ends of the film. From Fig. 5-18, $2\phi/s' = \alpha/s_F$, $2\theta = 180 - 2\phi$, and therefore

$$2\theta = 180 - \frac{\alpha}{s_F} s'$$

provided the film shrinks uniformly. The ratio s'/s_F decreases with increasing 2θ. Thus an error in measuring α has the least effect on the back (high 2θ) reflections. The van Arkel arrangement is useful for determining precise 2θ values from the back reflections; the Bradley–Jay arrangement is better for the front reflections.

Straumanis Arrangement. This method has the advantage that no fiducial marks are required—the film is self-calibrating. The midpoints of the front and back reflections (F and B, respectively, in Fig. 5-16) correspond to $2\theta = 0°$ and $2\theta = 180°$, respectively. The distance FB between these points is half the effective circumference of the film.

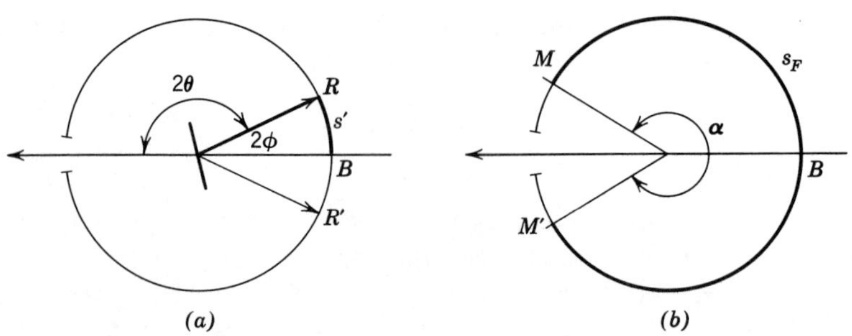

Fig. 5-18. The van Arkel arrangement. (*a*) 2θ and s', half the measured film distance between reflections R and R'; (*b*) α, the camera angle, and s_F, the measured film distance MBM' between fiducial marks.

Ideally, FB is 90 mm for the small standard camera (radius $= 360/4\pi$ mm) and 180 mm for the large camera. According to Fig. 5-16*e*, $2\theta/s = 180/FB$, and therefore in terms of s measurements

$$2\theta = \left(\frac{180}{FB}\right)s$$

From Fig. 5-16*f*, $2\phi/s' = 180/FB$, and because $2\phi = 180 - 2\theta$ measurements of s' lead to the relation

$$2\theta = 180 - \left(\frac{180}{FB}\right)s'$$

One method of measuring a Straumanis-type film, which establishes the FB length, and obtains accurate 2θ values, is illustrated by the data assembled in Table 5-11. The measurements were made on the powder pattern of silver which is reproduced in Fig. 5-19.

Table 5-11. X-Ray Powder Data for Silver Corrected for Film Shrinkage

Camera radius: $360/4\pi$ mm; Nickel-filtered copper radiation

	λ	Left	Middle	$\frac{L+M}{2} = F$	$\frac{M-L}{2} = s$	$\frac{180}{FB}s = 2\theta$	d
1	α	31.45	69.85	50.65	19.20	38.46	2.34
2	α	28.45	72.80	50.62_5	22.17_5	44.42	2.04
3	α	18.35	82.95	50.65	32.30	64.70	1.441
4	α	11.90	89.40	50.65	38.75	77.62	1.230
5	α	9.80	91.50	50.65	40.85	81.84	1.177
			Weighted average =	50.65			

	λ	Middle	Right	$\frac{M+R}{2} = B$	$\frac{R-M}{2} = s'$	$180 - \frac{180}{FB}s' = 2\theta$	d
6	α	99.60	181.40	140.50	40.90	98.06	1.021
7	α	105.90	175.05	140.47_5	34.57_5	110.74	0.9369
8	α	108.10	172.90	140.50	32.40	115.10	0.9135
9	α_1	118.05	162.95	140.50	22.45	135.02	0.8337
10	α_2	118.40	162.65	140.52_5	22.12_5	135.68	0.8338
11	α_1	128.90	152.10	140.50	11.55	156.86	0.7862
12	α_2	129.60	151.40	140.50	10.90	158.16	0.7864
			Weighted average =	140.50			

$FB = 140.50 - 50.65 = 89.85$

Wavelengths: Cu$K\alpha_1 = 1.54050$, $\alpha_2 = 1.54434$, $\alpha = (2\alpha_1 + \alpha_2)/3 = 1.5418$.

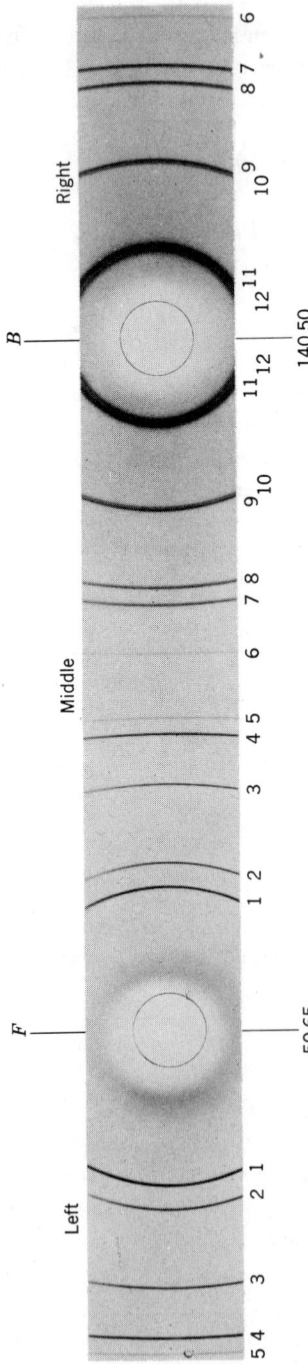

Fig. 5-19. A Straumanis-type x-ray powder pattern of silver; Ni-filtered Cu radiation; 1 mm $s = 1°\theta$. The lettering refers to the measurements in Table 5-11.

Wilson Arrangement. Powder films show a general diminution of intensity from low- to high-angle 2θ with the result that the back reflections of many compounds are too weak and diffuse to be measurable or are entirely absent. The self-calibrating feature of the Straumanis method fails under these circumstances. Wilson[11] has pointed out that the difficulty may be avoided by arranging that the free ends of the film lie in the low-angle region (Fig. 5-16*d*). Low-angle reflections then appear on each end of the film, and the position for F can be obtained from some of the front reflections, the position for B, from the remaining front reflections.

SAMPLE ECCENTRICITY. Sample eccentricity refers to a displacement of the sample from the axis of the film. This condition causes the reflection arcs to be displaced from their theoretical positions. If the sample displacement is normal to the x-ray beam direction (Fig. 5-20*a*), the $2s$ film distance is almost entirely unaffected, but the arcs are unsymmetrically disposed about the beam direction. A displacement parallel to the beam (Figs. 5-20*b* and *c*) gives $2s$ values which are either too small or too large. The diagrams indicate that the error in $2s$ becomes vanishingly small as 2θ approaches 180°.

Sample eccentricity is the result of faulty construction, whereby the manufacturer has failed to make the rotation axis of the sample coincident with the camera axis. The condition is nearly absent when normal care has been exercised in the workmanship.

DIVERGENCE OF THE BEAM. The effect of a diverging beam is to cause an uneven reduction of the intensity of the reflected beams. The net effect is to shift the positions of the reflections on the film. Fortunately, the effect is small compared with the shift caused by sample absorption and is almost eliminated in modern powder cameras by the use of pinhole collimators with small apertures.

ABSORPTION OF X-RAYS BY THE SAMPLE. The absorption of x-rays and its effect on the intensities of reflections has been described in connection with Fig. 3-18. Absorption also has an effect on the positions of reflections, no matter whether the specimen is a single crystal or a powder sample. Briefly, the reflections are displaced to give film measurements which do not accurately represent the diffraction angles, unless the most intense portion of each reflected beam arises from the center of the specimen. The proper conditions are realized only in the ideal specimen which is almost transparent to x-rays. In most real specimens the incident and reflected beams are absorbed to a degree that varies

[11] A. J. C. Wilson, *Rev. Sci. Instr.*, **20**, 831 (1949).

with the nature of the compound under investigation and with the shape, thickness, and packing density of the powder specimen. Rays which must travel the relatively long distances through the center of the sample are suppressed, and reflection occurs mainly from or near the surface. The effect on the positions of reflection arcs is shown in Fig. 5-21. At low diffraction angles either one or two arcs are formed, depending on the opacity of the sample. At higher angles only one arc is formed, which is displaced from the ideal position to give an s measurement which is too large, hence a d value which is too small. The diagrams

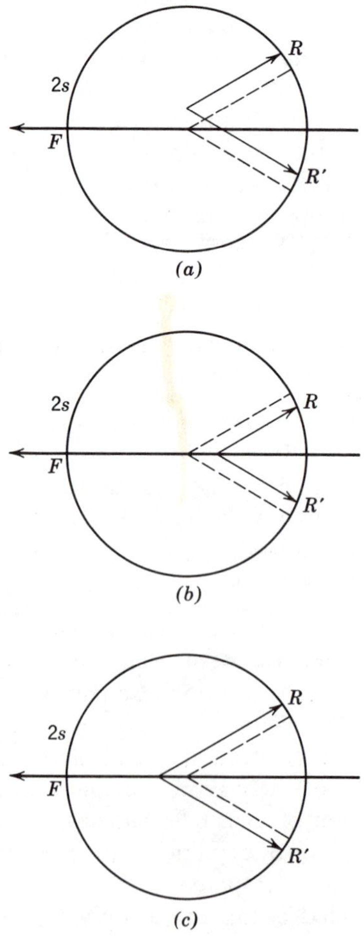

Fig. 5-20. Sample eccentricity and its effect on the film distance $RFR' = 2s$.

indicate that the displacement error in s decreases as θ increases and, like the eccentricity error, vanishes at $\theta = 90°$. The procedure for correcting errors due to absorption is to derive apparent cell dimensions from the reflections on the film, to plot the values against some function of θ, and to extrapolate to a value corresponding to $\theta = 90°$. Various functions have been proposed, but the practice is to plot against $\cos^2\theta$ (Bradley and Jay[12]) or against $\frac{1}{2}[\cos^2\theta/\sin\theta + \cos^2\theta/\theta]$ (Taylor and Sinclair[13]).

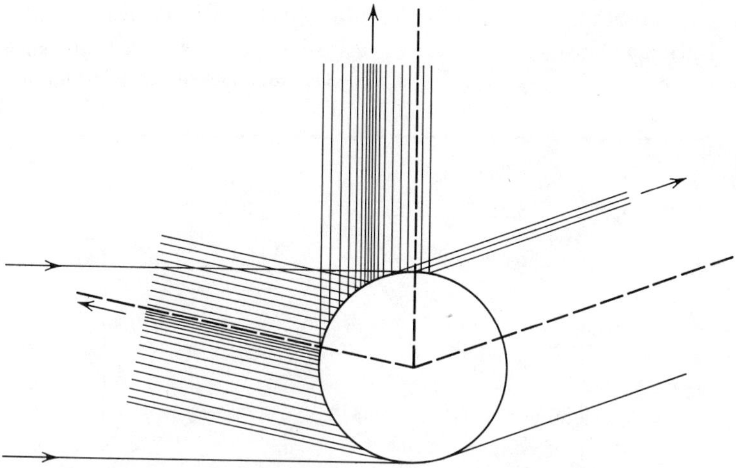

Fig. 5-21. The effect of absorption on the positions of x-ray reflections.

Nelson and Riley[14] investigated various extrapolation functions on powder data obtained with a 19-cm-diameter, constant-temperature camera designed to minimize sample eccentricity. Four powder specimens of the cubic alloy γCu_9Al_4, varying in thickness and dilution, were photographed (Table 5-12). The films were carefully measured to ensure highly accurate readings. The apparent a values, derived from the measured spacings, were then plotted against a number of function of θ. The results indicated the following:

1. When absorption is the greatest source of error, as in specimens 2 and 4, the function $\frac{1}{2}(\cos^2\theta/\sin\theta + \cos^2\theta/\theta)$ gives the best over-all linear plot (Fig. 5-22*a*). The linearity extends to θ angles of less than 15°. The plot against $\cos^2\theta$ is linear only above $\theta = 60°$ (Fig. 5-22*b*). This

[12] A. J. Bradley and A. H. Jay, *Proc. Phys. Soc. (London)*, **44**, 563 (1932).
[13] A. Taylor and H. B. Sinclair, *Proc. Phys. Soc. (London)*, **57**, 126 (1945).
[14] J. B. Nelson and D. P. Riley, *Proc. Phys. Soc. (London)*, **57**, 160 (1945).

Table 5-12

Specimen	Specifications
1 Medium thickness, diluted	A rod, 0.59-mm diameter, diluted with gum tragacanth
2 Thick, diluted	As above, 1.46-mm diameter
3 Medium thickness, undiluted	Undiluted and densely packed in a thin-walled silica tube measuring 0.45-mm inside diameter
4 Thick, undiluted	As above, in thin-walled Lindemann glass tube measuring 1.35-mm inside diameter

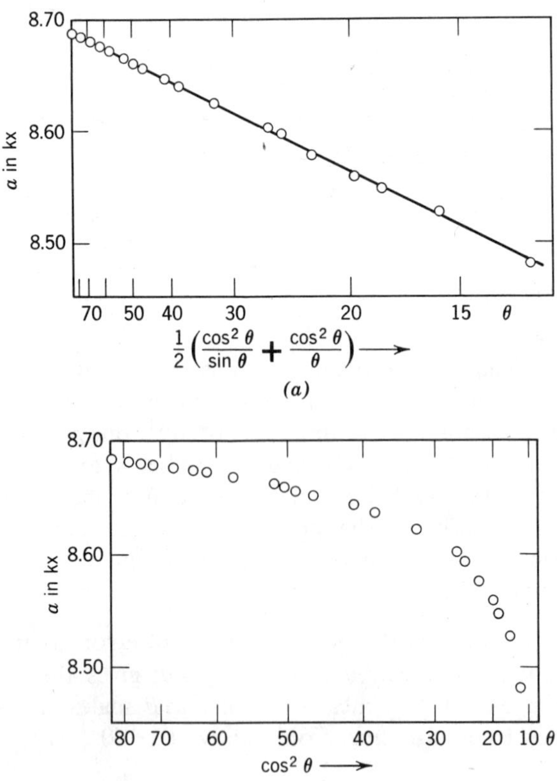

Fig. 5-22. Plots of apparent a when absorption is the dominant source of error, as in specimen 2 (after Nelson and Riley[14]).

Table 5-13

Specimen	a	Temperature
1 Medium thickness, diluted	8.6863 kx	15.8°C
2 Thick, diluted	8.6861	15.4°
3 Medium thickness, undiluted	8.6866	16.2°
4 Thick, diluted	8.6864	16.4°

Mean value of $a = 8.6864 \pm 0.0003$ kx

constitutes a disadvantage for the $\cos^2\theta$ function when relatively few appropriate reflections are present in the $\theta = 60$ to 90° region. The disadvantage is especially evident for noncubic crystals. In tetragonal crystals, for example, the extrapolation curves for the two cell dimensions are derived from plots of a and c values determined, respectively, from $hk0$ and $00l$ reflections. These reflections are comparatively few in number and are generally spaced over large angular ranges. Consequently, it is necessary to rely heavily on values obtained at low θ angles where the shape of the curve changes rapidly. A straight-line plot evidently requires fewer reflections to obtain the same accuracy given by $\cos^2\theta$ curves.

2. When absorption is slight, as in specimen 1, and observational errors are dominant, $\cot\theta$ gives the best overall linear plot (Fig. 5-23), although the plot against $\frac{1}{2}(\cos^2\theta/\sin\theta + \cos^2\theta/\theta)$ is linear above $\theta = 30°$.

3. Specimen 3 represents an intermediate case in which the effects of absorption and observational errors are comparable. Nelson and

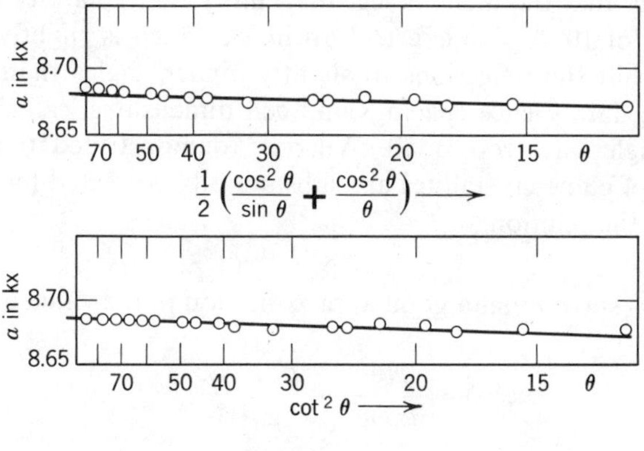

Fig. 5-23. Plots of apparent a when absorption is slight (as in specimen 1) and observational errors are dominant (after Nelson and Riley[14]).

Riley found that the plot was not strictly linear against either $\frac{1}{2}(\cos^2\theta/\sin\theta + \cos^2\theta/\theta)$ or $\cot\theta$ but was nearly linear against a mean of the two functions.

4. The function $\frac{1}{2}(\cos^2\theta/\sin\theta + \cos^2\theta/\theta)$ gave the best high angle ($\theta > 30°$) plot for all specimens. Despite a considerable variation in slope (Figs. 5-24*a* and *b*), this function gave reasonably consistent extrapolated values for the four specimens, the maximum deviation being about one part in 30,000 from the mean value (Table 5-13).

To obtain high accuracy, Nelson and Riley recommended the following:

1. The use of a thin specimen of optimum absorption with the object of reducing the slope of the extrapolation in order to give greater certainty to the final extrapolated value.

2. A careful choice of the wavelength of the x-radiation to give high-angle reflections; in particular, to ensure the appearance of at least one reflection with $\theta > 80°$. The closer the final reflection is to $\theta = 90°$, the greater the accuracy, no matter what extrapolation function is used.

3. To plot the apparent values against $\frac{1}{2}(\cos^2\theta/\sin\theta + \cos^2\theta/\theta)$ and to extrapolate linearly to zero value of this function. Reflections with θ values as low as 30° can be included in the plot. Values of this function are given in Appendix 4.

THE REFRACTION OF X-RAYS. The first experimental evidence that the index of refraction of x-rays differs from unity (i.e., that x-rays are refracted) appeared in 1919. Stenström found that the Bragg law was not accurately obeyed in his experiments and showed that the discrepancy could be explained by assuming that the index was slightly less than unity. Theoretical considerations and more recent experimental work indicate that the index is less than unity by a quantity δ which is of the order of 10^{-6}. The effect of an index which is slightly less than unity is to shift the reflections to slightly higher angles than predicted by the Bragg law. Hence spacings and cell dimensions, calculated from observed angles, are too small. Wilson[15] demonstrated that the cell dimensions of cubic crystallites are substantially corrected for refraction by applying the relation

$$a = a_{\text{meas}} \times (1 + \delta)$$

Sproull[16] has shown that a good approximation of the value of δ is given by

$$\delta = \frac{Ne^2}{2\pi mc^2} \times \frac{\rho\Sigma AN\lambda^2}{\Sigma AW}$$

[15] A. J. C. Wilson, *Proc. Cambridge Phil. Soc.*, **36**, 485 (1940).

[16] W. T. Sproull, *X-Rays in Practice*, McGraw-Hill, New York (1946).

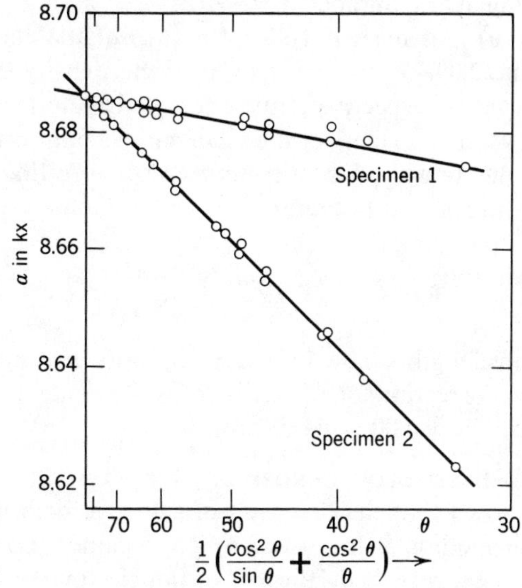

(a)

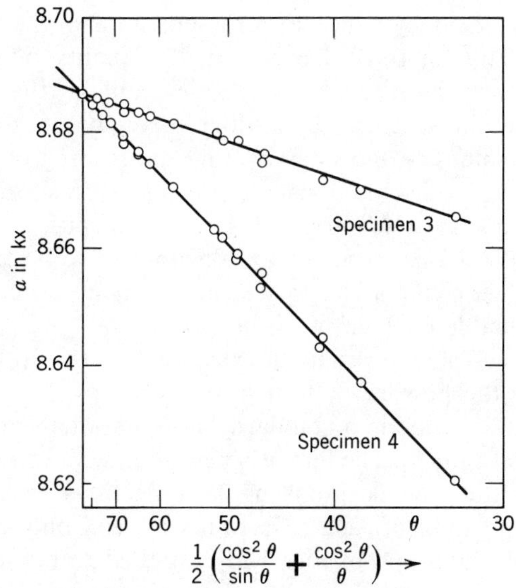

(b)

Fig. 5-24. Large-scale plots of apparent a against $\frac{1}{2}(\cos^2\theta/\sin\theta + \cos^2\theta/\theta)$ (after Nelson and Riley[14]).

where N is Avogadro's number (6.02×10^{23}), e and m are the charge (4.80×10^{-10} esu) and mass (9.107×10^{-28} g) of the electron, c is the velocity of light (2.998×10^{10} cm/sec), ρ is the density of the crystals, ΣAN and ΣAW are, respectively, the sum of the atomic numbers and the sum of the atomic weights of the elements per unit cell, and λ is the wavelength of the x-rays (in centimeters). After inserting the values of the constants, the relation becomes

$$\delta = 2.701 \times 10^{-6} \times \frac{\rho \Sigma AN \lambda^2}{\Sigma AW}$$

in which the wavelength is now in Angstrom units. It is evident from the value of δ that a correction for refraction must be made if accuracy of the order of one part in 50,000 is significant.

RANDOM AND SYSTEMATIC ERRORS IN MEASUREMENT. The preceding sections have shown that the precision of cell-edge determinations with the powder film method is improved by good camera construction and careful sample preparation, by the use of fiducial marks or special film mounts to provide a correction for film shrinkage, and by extrapolating apparent cell dimensions against a function of θ to minimize errors caused by sample absorption. The success of the last two procedures depends on the accuracy with which the positions of the reflections can be measured. This, in turn, depends on the quality of the measuring device, particularly its hairline and vernier, and on the care and skill of the observer in setting the hairline precisely on the position of maximum line intensity and in estimating the reading of the scale with the vernier. The scales on most commercial measuring devices are provided with a vernier reading to 0.1 mm by means of which the hair line setting can easily be estimated to 0.05 mm and with some experience, to 0.03 mm. A vernier reading to 0.02 mm, which permits estimates to 0.01 mm, is valuable for high-precision work.

Figure 5-25 depicts the results of extrapolating the apparent a values obtained from the powder pattern of a species of the mineral garnet. The exposure was made in a standard 114.6-mm-diameter, Straumanis-type commercial camera. The film was measured with the aid of a vernier reading to 0.1 mm; the positions of the reflections were estimated to 0.03 mm. The position of each reflection was read only once. To show sufficient detail, the graph of Fig. 5-25 is limited to points representing apparent values calculated from θ angles higher than 55°.

The points on the graph are scattered in random fashion, suggesting random errors in setting the hairline on the position of maximum line intensity. An obvious feature of the plot is the decrease in scatter as θ

approaches 90°. This is predicted by the relation between θ and the spacing d shown in Fig. 5-6. The curves of the figure indicate that for a given precision in measuring θ the error in d, hence in the cell edge a, decreases as θ increases. This can also be shown analytically by differentiating the Bragg equation with respect to θ:

$$n\lambda = 2d \sin \theta$$

$$0 = 2\frac{\Delta d}{\Delta \theta} \sin \theta + 2d \cos \theta$$

$$\frac{\Delta d}{d} = -\frac{\cos \theta}{\sin \theta} \Delta \theta = -\cot \theta \cdot \Delta \theta$$

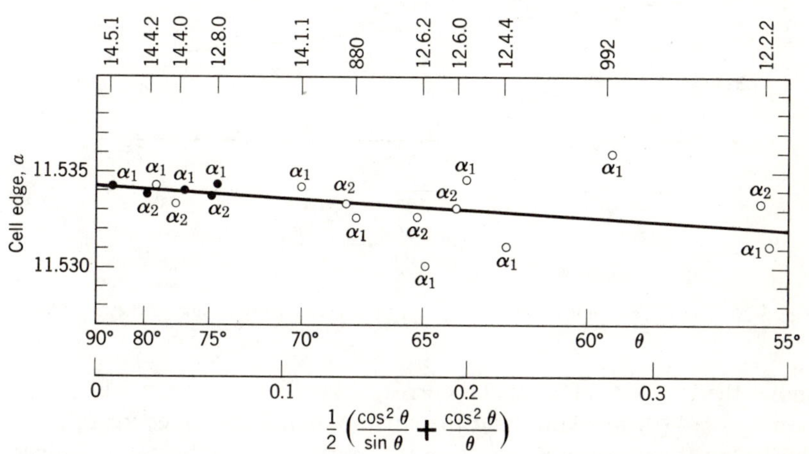

Fig. 5-25. Garnet. Plot of apparent a against $\frac{1}{2}(\cos^2 \theta/\sin \theta + \cos^2 \theta/\theta)$. The filled circles represent reflections that were remeasured to minimize random errors of measurement and then plotted in Fig. 5-26.

As θ increases to 90°, $\cot \theta$ approaches zero and the error $\Delta\theta$ produces ever smaller errors Δd in measured d. This explains why the function $\cot \theta$ gives the best over-all linear plot (Fig. 5-23) when absorption is slight and observational errors are dominant.

It is clear from Fig. 5-25 that considerable reliance has been placed in the accuracy of the final point, $\alpha_1(14 \cdot 5 \cdot 1)$, in deriving the extrapolated cell edge. Because of the random scatter in high-angle points, the exact position of the extrapolation line is in doubt. The uncertainty extends to the third decimal position of the cell-edge value, hence the accuracy is only of the order of one part in 10,000. Figure 5-26 shows the remarkable decrease in scatter that can be achieved by repeatedly determining the effective film circumference and reading the positions of selected reflections to obtain a weighted average of the best measurements. In preparing this extrapolation, only the seven reflections in Fig. 5-25 corresponding to θ angles higher than 74° were considered. One reflection, $\alpha_1(14 \cdot 4 \cdot 2)$, was judged to be too broad to permit a reliable setting of the hairline and was immediately omitted from further consideration. Four separate sets of measurements were then made of the film circumference and the positions of the remaining six reflections. Table 5-14

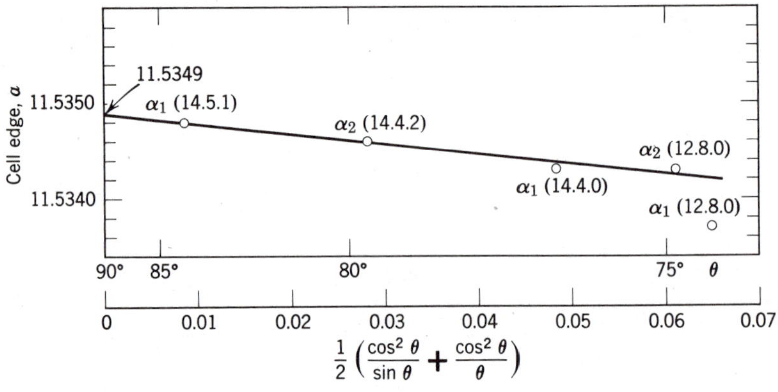

Fig. 5-26. Large-scale plot of apparent a for five selected high-angle garnet reflections.

shows the four sets of 2θ angles corrected for film shrinkage. The reflection $\alpha_2(14 \cdot 4 \cdot 0)$ was faint, and the corresponding 2θ values for this line illustrate the uncertainty in setting the hairline; the values were judged to be too inconsistent for inclusion in the extrapolation plot. The weighted averages of the remaining five reflections were converted to spacings, from which the cell edges were calculated and then plotted against $\frac{1}{2}(\cos^2\theta/\sin\theta + \cos^2\theta/\theta)$, as shown in Fig. 5-26. The new points offer more assurance for positioning the extrapolation line. The uncertainty in the extrapolated value of the cell edge now lies in the fourth decimal position, and the accuracy is probably of the order of one part in 50,000.

It is well known that measurements obtained on the same film by two independent observers often show systematic differences. Some observers have a tendency to set the hairline consistently to one side of the correct position. Thus, although repeated readings by the same observer can remove random errors in measurement and result in consistent values, the measurement may nevertheless be in error. It follows, then, that for

Table 5-14. Repeated Determinations of 2θ for Six Selected High-Angle Reflections from a Garnet Sample

(camera diameter 114.6 mm; Ni filtered Cu radiation)

	$\alpha_1(12\cdot 8\cdot 0)$	$\alpha_2(12\cdot 8\cdot 0)$	$\alpha_1(14\cdot 4\cdot 0)$	$\alpha_2(14\cdot 4\cdot 0)$	$\alpha_2(14\cdot 4\cdot 2)$	$\alpha_1(14\cdot 5\cdot 1)$
	148.78	149.82	152.97	154.22	[159.45]	168.48
2θ corrected for	148.80	[149.76]	152.97	154.20	159.39	168.46
film shrinkage	148.79	149.80	152.94	154.26	159.38	[168.54]
	148.78	149.82	152.99	154.27	159.40	168.48
Weighted average	148.79	149.81	152.97	...	159.39	168.47
d	0.79972	0.79976	0.79218	...	0.78483	0.77416_5
a	11.5337	11.5343	11.5343	...	11.5346	11.5348

The values in brackets were not used in deducing the weighted averages.

accuracy of the highest order the film should be measured independently by a number of experienced observers, unless the competence of the observer has been established by measurements on reference films.

THE ATTAINMENT OF HIGH PRECISION—A SUMMARY. The following points summarize the requirements for attaining high-precision cell dimensions with the powder film method:

1. The camera should (*a*) provide the means for determining shrinkage errors, (*b*) have a fine collimating system to reduce x-ray beam divergence errors, (*c*) provide for accurately centering the sample, and (*d*) have a diameter of at least 114.6 mm.
2. The sample should (*a*) be thin and symmetrically shaped to aid in accurate sample centering and (*b*) be sufficiently diluted to give a low slope to the absorption extrapolation line.
3. The wavelength of the x-rays should be selected to give appropriate reflections in the $\theta = 75$ to 90° region.
4. A weighted average should be made of the measurements of several observers to reduce systematic and random errors of measurement.
5. The apparent values should be extrapolated against $\frac{1}{2}(\cos^2\theta/\sin\theta + \cos^2\theta/\theta)$ to minimize absorption errors.
6. The extrapolated value should be corrected for refraction.

MINIMIZING SYSTEMATIC ERRORS BY THE METHOD OF COMPARISON. One method of compensating for systematic errors is to mix with the material a calibrating substance, the cell constants of which are accurately known. The powder patterns of the standard and the unknown are superimposed on the resulting film, and accurate 2θ values may be derived for the material under investigation by simple interpolation. To minimize the effects of physical differences between the two components, the sample should be ground to a very fine intimate mixture.

The reflection angles for a number of substances, suitable for calibration purposes, have been determined to a high level of precision with the aid of the methods already outlined and with the diffractometer (described below). These substances include high-purity silicon. Clear, well-crystallized quartz (SiO_2), is suitable, provided that the characteristics of the reference "batch" are first determined; the cell dimensions of this mineral vary with the locality. NaCl, which is often quoted as a standard substance, gives rather widely differing values for the cell edge. The reflections from the standard should naturally be sharp and well-distributed with respect to the reflections of the material under investigation. Superposition or even near superposition of reflections is, of course, undesirable. Obviously, then, no one calibrating substance can serve for all cases, and in practice it is often necessary to use a standard that does not meet all the requirements. It must also be remembered that the accuracy of a determination by the method of comparison is limited to the accuracy with which the standard is known. For these reasons most crystallographers prefer to minimize the effects of errors on powder films by the methods which have already been described. In recent years, however, the use of a calibrating pattern has gained some prominence in single-crystal x-ray studies, mainly because some of the powder techniques for the elimination of errors cannot be applied or because they require special costly apparatus.

Special Powder Techniques and Apparatus

Parafocusing Cameras

Parafocusing methods refer to geometrical arrangements that cause identical reflections from an extended specimen to converge to a point on the film. The methods are based on the theorem that all angles standing on an arc are identical and equal to half the angle subtended at the center of the same arc. Thus, in Fig. 5-27*a*, $\angle ABD = \angle ACD = \frac{1}{2}\angle AOD$. Hence, if XS and XS' in Fig. 5-27*b* represent rays of an x-ray beam diverging from X to a sample SS' and a set of planes at S reflects to R on the film, the same set of planes at S' will also reflect to R. The angles standing on the arc $XSS'R$ are equal to $180 - 2\theta$. Consequently, $\angle XOR$, subtended at the center, is equal to $2(180 - 2\theta) = 360 - 4\theta$. The value of θ is easily deduced from the length of the arc $XSS'R$:

$$\frac{\text{length of } XSS'R}{4\theta} = \frac{2r\pi}{360}$$

where r is the radius of the film.

A focusing camera has the advantage that it gives about twice the resolution of a Debye–Scherrer camera of the same radius. However, the useful angular range of 2θ that can be recorded on the film is correspondingly reduced by half. Furthermore, the sharpness of the reflections

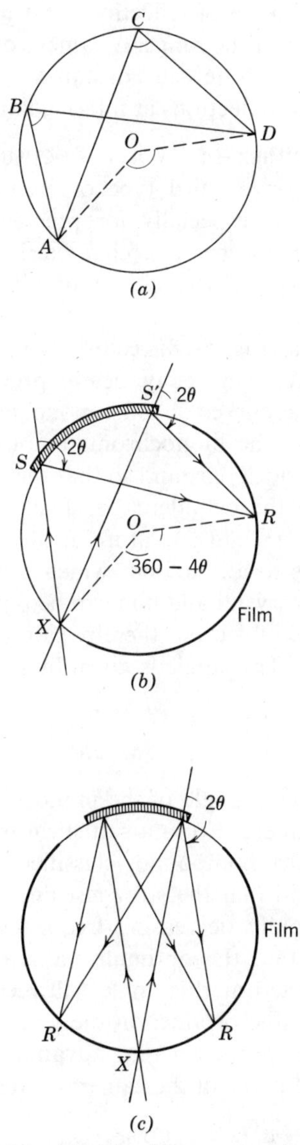

Fig. 5-27. The principle of the parafocusing camera.

decreases as the angle of incidence on the film becomes more oblique. These two factors dictate that no one design can meet the special needs of a variety of problems. Figure 5-27 shows two kinds of parafocusing geometry—namely, symmetric and asymmetric—both applicable to back-reflection cameras. The symmetric arrangement is designed to record the high-angle range of reflections and to give maximum sharpness to the extreme end of the range. A camera of this design is useful for the determination of accurate cell constants. The asymmetric arrangement gives the sharpest reflections at intermediate back-reflection angles.

THE MULTIPLE GUINIER–DE WOLFF FOCUSING CAMERA. In recent years cameras based on a third type of parafocusing geometry have gained wide acceptance, especially for precise studies of compounds whose back-reflection spectra are characteristically absent or weak. The cameras are based on the design of Guinier,[17] modified by de Wolff.[18]

The Guinier camera (Fig. 5-28) records the front-reflection region of the spectrum. A converging x-ray beam, produced by reflecting the incident beam from a curved crystal monochromator, is transmitted through the specimen. The monochromator on this camera consists of a quartz crystal cylindrically ground so that the strongly reflecting $(10\bar{1}1)$ planes are parallel to the cylinder axis. The crystal is bent about the axis to give the planes a slight curvature. It then converges the radiation reflected by the planes to a line focus which is received on the film. De Wolff showed that the crystal and film can be positioned to eliminate the separation of the $K\alpha$ doublet and thereby gain maximum line sharpness at a chosen value of θ. This angle is given by the relation

$$\tan\theta = \frac{PO\,(\tan\xi)}{2(SM)\cos\gamma}$$

in which ξ is the glancing angle of the monochromator crystal and SM is the radius of the camera. From a statistical investigation of the x-ray powder data of the 1000 compounds classified by Hanawalt, Rinn, and Frevel,[5] de Wolff found that the strongest lines have a maximum value at θ15–20° and a rapidly decreasing frequency on both sides of this value. He concluded that the optimum value of θ was about 15°. To ensure that rays reflected at this angle will meet the film with normal incidence for sharpest line production the γ value of the camera should be 30°. De Wolff also proposed that advantage be taken of the large vertical (normal to the plane of the camera) divergence of the beam from

[17] A. Guinier, *Ann. Phys., Paris*, **12**, 161 (1939).

[18] P. M. de Wolff, *Acta Cryst.*, **1**, 207 (1948).

the monochromator crystal to mount a number of cameras, one on the other, and to record several patterns on a single film. A commercial camera embodying these features is known as the Guinier–de Wolff quadruple focusing camera.

In summary—the Guinier–de Wolff camera records the reflection range θ0–45°. It has exceptionally high resolving power, centered at about 15°. The monochromator eliminates the white radiation in the

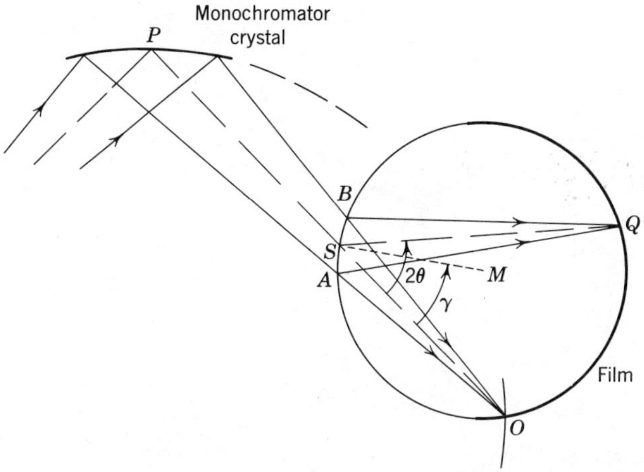

Fig. 5-28. Schematic drawing of Guiner-de Wolff focusing camera.

beam, with the result that patterns have a low background intensity. The possibility of recording several patterns on the same film is useful for comparison studies or for adding a reference pattern.

A focusing camera is not essential to the efficient operation of the average diffraction laboratory. Adequate resolution and accuracy can be gained under favorable circumstances with a large-diameter Debye–Scherrer camera at a cost, of course, of a longer exposure. A focusing camera requires a relatively large sample to ensure smooth lines, and therefore it cannot be used for studies in which the material under investigation is too precious or too rare to afford an adequate size sample. But the well-diversified laboratory and the laboratory that is concerned especially with the analysis of powder spectra will profit by the addition of a Guinier–de Wolff camera.

Pinhole Techniques

FORWARD-REFLECTION OR TRANSMISSION METHOD. This method uses a beam of characteristic x-rays and a flat film oriented normal to the beam. The sample is placed between the x-ray source and the film (Fig. 5-29*a*). The beam, collimated by a set of pinholes, generates reflection cones as it passes through the sample and is absorbed by a lead button in front of the film or is allowed to pass through a hole punched in the film. Reflection cones that are intercepted by the film are recorded

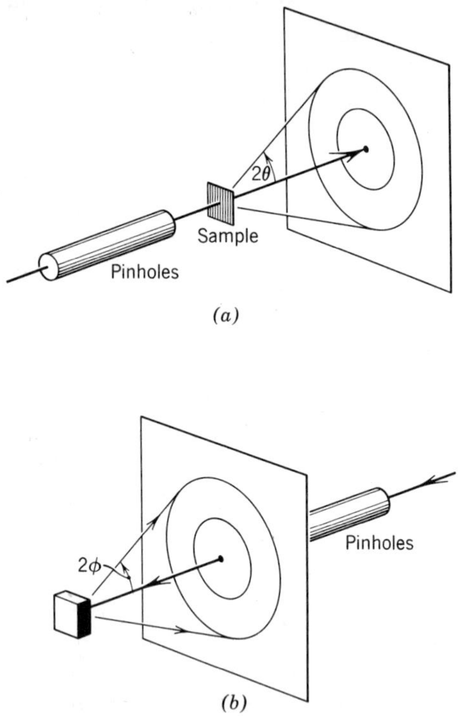

Fig. 5-29. Pinhole methods: (*a*) front reflection, (*b*) back reflection.

as circles. To prevent β reflections from reaching the film, either the beam is filtered before it reaches the collimator or the film is covered with a sheet of the filter material.

The usefulness of the method is limited by the restriction that only reflections of considerably less than $2\theta = 90°$ can reach the film. However, it has one advantage not afforded by the Debye–Scherrer method. Because entire reflection circles are recorded, it is possible to detect differences in intensity not apparent in ordinary powder photographs. The method is valuable, for this reason, in the study of preferred orientation and lattice strains in thin sheets of metals.

BACK-REFLECTION METHOD. In this method the film is placed between the x-ray source and the sample (Fig. 5-29*b*). As a result of this arrangement, a number of the back-reflection cones, corresponding to the smaller $2\phi = 180 - 2\theta$ reflections, are recorded as circles on the film. The thickness of the sample is immaterial, and consequently the polished surface of a metal or alloy which has been studied under a microscope can be used as an x-ray specimen to obtain correlative diffraction data. Because there are special advantages in the use of back reflections for spacing measurements, the method is suitable for detecting small changes in the cell constants of the metal or alloy. To aid these studies certain commercial back-reflection cameras are equipped with a shield, by means of which three patterns can be successively recorded on different regions of the same film for direct comparison.

High-Temperature Cameras

In order to investigate the crystal structures of phases stable only at elevated temperatures, special cameras have been devised to hold the specimen at a prescribed temperature during the exposure. In one of the simplest designs the powder is mounted on a wire which is centered in the camera and heated by an electric current. The temperature of the wire can be accurately measured with a thermocouple, but the temperature of the sample will be somewhat lower, and it is necessary to calibrate the thermocouple in terms of the actual sample temperature.

More elaborate equipment is necessary to reach temperatures well above 100°C. The heating unit is usually a small electric furnace which surrounds the specimen. In one commercially available camera that operates successfully to temperatures above 600°C the furnace is slotted to permit the passage of incident and reflected beams, the openings are covered with thin aluminum foil to prevent dissipation of the heat, and the body of the camera is cooled by a flow of water to protect the emulsion of the film. Materials that oxidize within the temperature range of this camera are sealed in thin-walled silica tubes. This procedure cannot be followed in cameras that are designed for considerably higher temperatures, because the silica reacts with many materials well below its melting point. Consequently, the very high-temperature camera must be evacuated or filled with an inert gas to prevent oxidation of susceptible compounds.

References

Tables of Spacing (d) Against θ Angles

Tables for Conversion of X-Ray Diffraction Angles to Interplanar Spacings, National Bureau of Standards, Applied Mathematics Series 10, U.S. Dept. of Commerce (1950).

Tables of d Spacings for Angle 2θ, by G. Switzer, J. M. Axelrod, M. L. Lindberg, and E. S. Larsen III, Geol. Surv. Circular 29, U.S. Dept. of the Interior (1948).

Analytical Methods of Indexing Powder Photographs

Azaroff, L. V., and M. J. Buerger, *The Powder Method in X-Ray Crystallography*, McGraw-Hill, New York (1958).

Bunn, C. W., *Chemical Crystallography*, Second Edition, Clarendon Press, Oxford (1961).

Henry, N. F. M., H. Lipson and W. A. Wooster, *The Interpretation of X-Ray Diffraction Photographs*, The Macmillan Co., London (1960).

Klug, H. L., and L. E. Alexander, *X-Ray Diffraction Procedures*, John Wiley and Sons, New York (1954).

Graphical Aids for Indexing Powder Photographs

Bunn charts available from Polycrystal Book Service, G.P.O. Box 620, Brooklyn 1, New York.

Identification of Materials by X-Ray Powder Data

American Society for Testing Materials: *The A.S.T.M. Index and the X-Ray Powder Data File*.

U.S. Dept. of Commerce, National Bureau of Standards, *Standard X-Ray Diffraction Powder Patterns*; NBS Circular 539, 1–10 (1953–1960); NBS Monograph 25, Sections 1–3 (1962–1964).

Geological Society of America, Memoir 85 (1962), *X-Ray Powder Data for the Ore Minerals*, by L. G. Berry and R. M. Thompson.

Part. 2. The Diffractometer

The diffractometer consists of a goniometer for measuring diffraction angles and a number of electronic circuits for determining the intensity of diffraction at any angle. The goniometer employs a large flat sample combined with a parafocusing arrangement to increase the intensity of diffraction and an x-ray counter tube in place of film to detect the diffracted radiation. The counter transforms the radiation spectrum emitted by the sample into a pulse spectrum which, after suitable amplification, is fed into the circuit panel and measured in one of a number of ways. The pulses, smoothed into a fluctuating current, may be caused to actuate a pen on a strip chart and to produce a permanent graphic record of intensity against diffraction angle. Alternatively, the pulse rate may be measured directly with a scaling circuit.

The Geometry and Optical System of the Goniometer

The parafocusing arrangement of a modern goniometer is illustrated in Fig. 5-30. An x-ray beam diverges from the focal spot F on the target

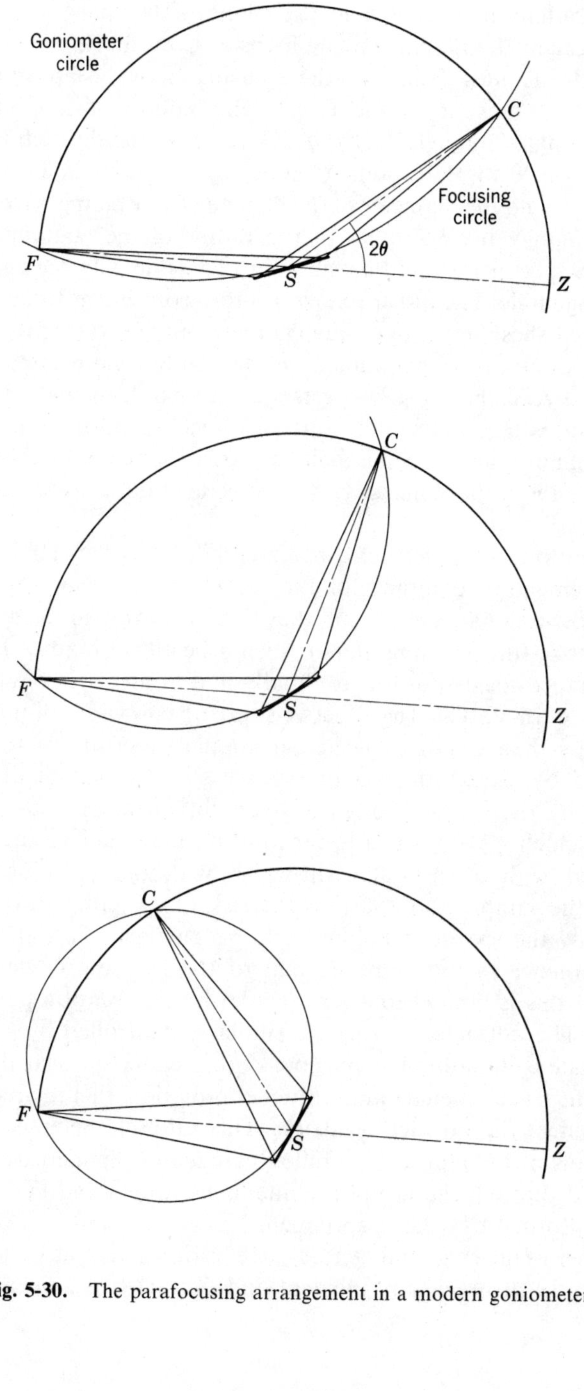

Fig. 5-30. The parafocusing arrangement in a modern goniometer.

to the flat surface of the specimen S which rotates about the goniometer axis. The radiation diffracted in the plane of the page at a given 2θ angle is brought to an approximate focus at C on the circumference of a circle which includes F and S—the focusing circle—by positioning the sample-surface tangent to the circle. The sample surface bisects the diffraction angle 2θ (angle CSZ) in this case. As the diffraction angle is increased from 0 to 180°, point C sweeps out a half-circle—the goniometer circle—and the diameter of the focusing circle changes from infinity to a minimum value FS equal to the radius of the goniometer circle. Ideally, the specimen should be curved to coincide with the curvature of the focusing circle. The flat specimen is more convenient to use, but only a fraction satisfies the parafocusing geometry and diffracts precisely to C. The x-ray counter tube is mounted on the arm SC and rotated about the specimen to scan the powder spectrum. The parafocusing arrangement, which requires that the sample surface bisect the diffraction angle and thereby remain tangent to the focusing circle, is maintained by rotating the counter about the same axis as the specimen but at twice the angular speed.

The x-ray optical system of a commercial diffractometer is shown in the complementary diagrams of Fig. 5-31. The focusing error is kept small by the use of an x-ray beam that is very narrow in the plane of the focusing circle (the focusing plane). Such a beam is obtained by viewing the line-shaped focal spot longitudinally and at an angle of about 3° to the surface of the target. The effective source-beam width of this arrangement is less than $\frac{1}{10}$ mm. The actual angular aperture of the beam is determined by the width of a divergence slit. The length of specimen irradiated by the beam through a given slit increases with decreasing angle of incidence. This sets a lower limit on the angular range that can be scanned with a particular slit-width without exceeding the total length of the sample and courting the risk of detecting stray radiation scattered by the specimen holder. A line-focus beam has an inherently great divergence in the plane normal to the focusing circle (the axial plane), but this is limited to a few degrees by the use of a parallel-plate assembly. The diffracted beams are similarly controlled by a slit and a parallel-plate collimator. The purpose of the scatter slit near the window of the counter is to exclude nondiffracted radiation, thereby reducing the background of the powder spectrum. This diffractometer has a goniometer radius of 170 mm to take full advantage of the increase in resolution gained through the use of the line-focus source and the increase in intensity afforded by a large specimen.

A number of instrumental factors, such as diffraction at various depths in the sample, divergence of incident and diffracted beams, and the use

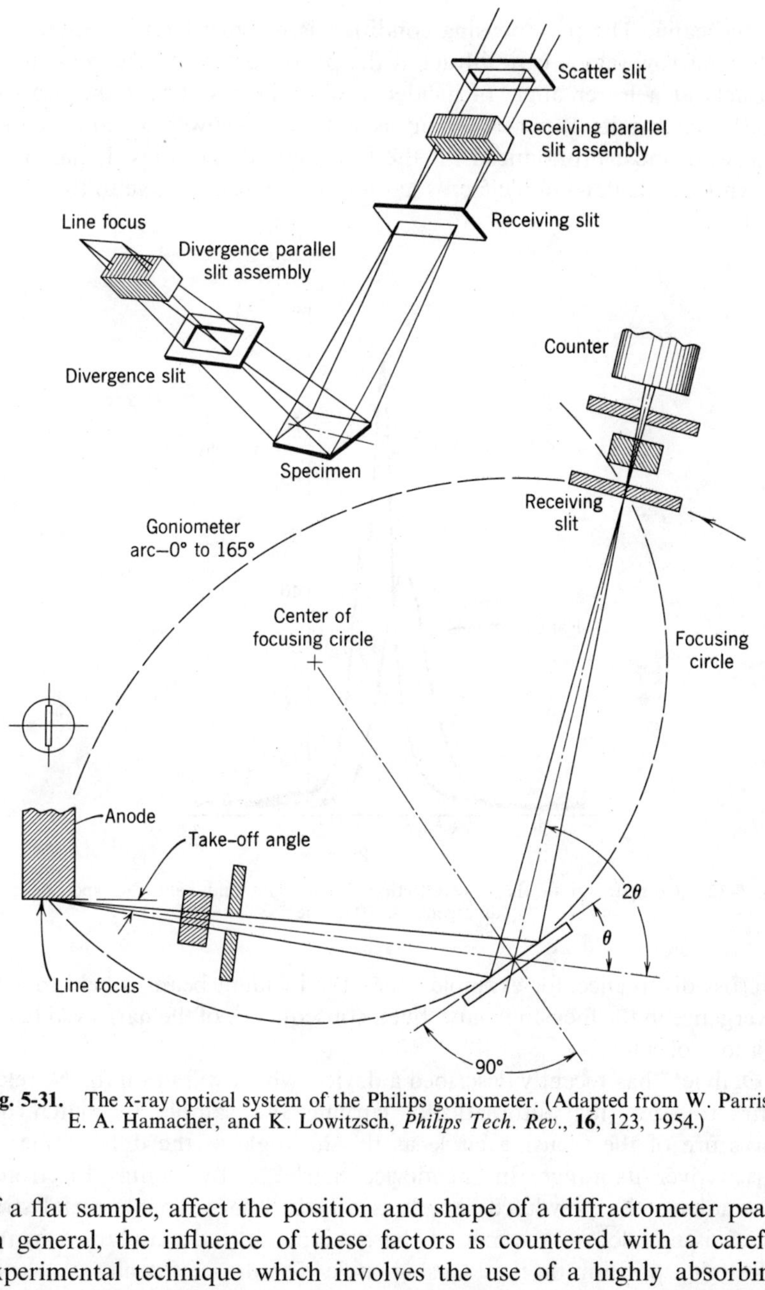

Fig. 5-31. The x-ray optical system of the Philips goniometer. (Adapted from W. Parrish, E. A. Hamacher, and K. Lowitzsch, *Philips Tech. Rev.*, **16**, 123, 1954.)

of a flat sample, affect the position and shape of a diffractometer peak. In general, the influence of these factors is countered with a careful experimental technique which involves the use of a highly absorbing sample to limit diffraction to the sample surface, and the use of the slits and parallel-plate assemblies already described to reduce the divergence of the beams. The parafocusing condition is not met by a flat sample—only a narrow region can diffract at the proper 2θ angle. The remainder diffracts at a lower angle of incidence with the result that the peak is broadened and its center of gravity is shifted to a lower 2θ value. These effects are most pronounced for the low-angle diffractions. It has been shown that the flat-sample factor has a minor effect, relative to the effect

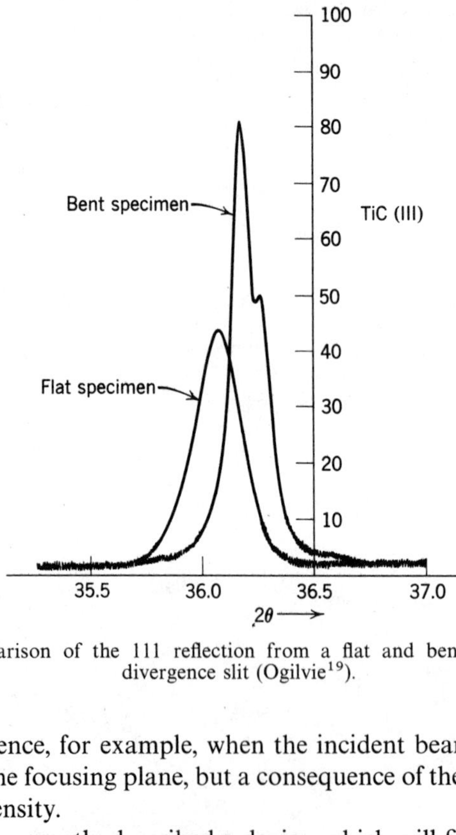

Fig. 5-32. Comparison of the 111 reflection from a flat and bent TiC specimen; 4° divergence slit (Ogilvie[19]).

of cross-divergence, for example, when the incident beam is held to a 1° divergence in the focusing plane, but a consequence of the narrowed beam is a loss of intensity.

Ogilvie[19] has recently described a device, which will fit on the Norelco diffractometer, for automatically curving the sample to match the curvature of the focusing circle as the 2θ angle of the diffractomer is varied over its range. In the device marketed by Philips Electronic Instruments the powdered sample is attached to a flexible metal strip which is coupled to the axis of the goniometer and maintained in the diffracting position on the focusing circle. The device permits the use of a more widely diverging incident beam at low-diffraction angles to gain

[19] R. E. Ogilvie, *Rev. Sci. Instr.*, **34**, 1344 (1963).

intensity while maintaining sharp resolution and accuracy of peak position. Ogilvie compared the 111 diffraction peaks of TiC, obtained with a 4° divergence slit, from a flat specimen and a specimen mounted in the bending device. Figure 5-32 shows that some resolution of the $K\alpha_1$–$K\alpha_2$ doublet is possible with the 4° beam at $2\theta = 36°$. The angular displacement and loss of intensity for the flat specimen is evident. Ogilvie found that with a 1° divergence slit the resolution from the flat specimen was about as good as that obtained with a 4° slit and the bending device, but the intensity was reduced by a factor of four.

X-Ray Counters

The detectors used to measure the intensities of reflections in powder diffractometry are of two kinds: the gas-filled detector tubes (Geiger–Mueller and proportional counters) which make use of the initial ionization that follows absorption of x-rays by a gas and the photoelectric detectors (scintillation counters) which utilize the light produced when x-rays are absorbed by a phosphor. Both types of detectors have the property of emitting electrical pulses when stimulated by x-ray quanta.

Gas-Filled Detectors

Gas-filled detectors consist essentially of a grounded cylindrical cathode and a coaxial wire anode. The assembly is enclosed in a gas-filled tube which has a window to admit x-rays. When an x-ray quantum enters the counter, it ionizes atoms of the gas, liberates electrons, and forms positive gas ions. The free electrons and positive ions are together called *ion pairs.* They separate in the presence of an electrical field and collect on the electrodes of the detector. The strength of the field depends on the voltage applied to the anode and determines the extent to which the initial ionization spreads through the tube, which establishes the character of the detector.

If the applied voltage is just high enough to prevent the electrons and ions from recombining (about 200 volts), the positive ions move to the cathode under the influence of the electrical field, and the electrons are drawn to the anode where they discharge to send an electrical pulse to the external circuit. The amplitude or height of this pulse is fixed by the initial ionization and is proportional to the energy of the x-ray quantum that initiated the ionization. Thus an x-ray beam of constant intensity generates a small constant flow of current from the detector. This is the principle of the ionization chamber used in the Bragg spectrometer.

At higher voltages the height of the pulse is affected by the phenomenon known as multiple ionization or gas amplification. Under the influence of a more intense electrical field than is maintained in an ionization chamber, the electrons liberated in the initial ionization are accelerated and acquire energy to ionize more gas atoms in their movement to the anode. This liberates electrons which then ionize more gas atoms. As a result, the ionization initiated by the original x-ray quantum is amplified, and when a detector that is designed to take advantage of the amplification effect is operated at proper voltages (800 to 1700 volts, depending on the design) the amplification factor has a value of about 10^4. The anode is bombarded by a hail of electrons in the vicinity of the original ionization, and a pulse of easily detectable size (of the order of 1 mV) is produced. The height of the pulse for a given applied voltage is directly proportional to the number of ion pairs formed in the initial ionization, which is proportional to the energy of the initiating x-ray quantum (i.e., it is inversely proportional to the wavelength of the x-radiation that enters the counter). A detector designed to be operated under these conditions is aptly called a proportional counter.

The gas amplification in a proportional counter is moderate compared with the amplification obtained in a Geiger counter operating under optimum voltages. The electrical field maintained in a Geiger counter is so intense that the electrons liberated from the gas acquire sufficient energy to cause the initial ionization to spread throughout the tube. As a result, the gas amplification factor reaches a value of 10^7 or more, and the anode is bombarded along its entire length by an avalanche of electrons, which yields a comparatively large pulse (of the order of 1 to 10 volts) when they discharge. The initial energy of the x-ray quanta is unimportant—the original quanta act merely as triggers. A discharge of about the same size is built up each time, and the pulses delivered by the counter vary only slightly in size (height or voltage). Hence there is no relation between pulse height and the energy of the x-ray quanta (or the wavelength of the radiation) that enter a Geiger counter.

PLATEAUS OF THE GAS-FILLED DETECTORS. Ordinarily, the radiation diffracted by a nonfluorescing sample is not strictly monochromatic, even though it is passed through a β filter, because of the ever-present white radiation spectrum in the primary beam. As a consequence, the x-ray quanta that enter a detector vary somewhat in energy, and the pulses emitted by a proportional counter exhibit a range of amplitudes. At a certain applied voltage the amplitudes of some of the pulses exceed the minimum sensitivity level of the measuring equipment and are detected by the counting circuits. An increase in the applied voltage raises the amplitudes of more pulses above the detection level, and the counting rate (intensity) is increased. Eventually the heights of all pulses exceed the level. Then all the pulses are counted, and the counting rate has reached a constant value or plateau. A further increase in the counter voltage merely increases the height (voltage) of the pulses without affecting the counting rate. The voltage at the threshold of the plateau is therefore the minimum voltage required to raise the height of all pulses,

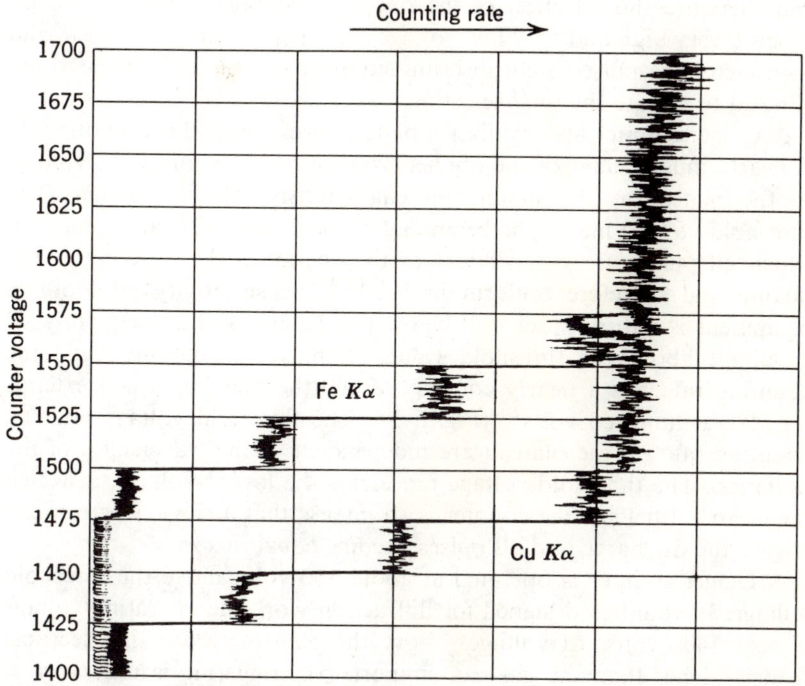

Fig. 5-33. Experimentally determined relation between the voltage applied to a proportional counter (base line of pulse-height analyzer set at three volts) and the counting rate for Fe$K\alpha$ and Cu$K\alpha$ radiation. Primary beam intensities adjusted to give identical counting rates on the plateau.

corresponding to a particular wavelength, above an instrumental sensitivity level. The level can be varied, as explained later, by changing the baseline of the pulse-height analyzer. Because pulse height is inversely proportional to wavelength, more voltage must be applied to raise the long wavelength (small-size pulses) above the sensitivity level. Hence the threshold voltage increases with the wavelength as demonstrated by Fig. 5-33, which shows the experimentally determined relation between the voltage applied to a proportional counter and the counting rate for Fe$K\alpha$ (1.9373 Å) and Cu$K\alpha$ (1.5418) radiation. Thus the threshold voltage for a proportional counter plateau varies with the wavelength and with an instrumental setting; the counting rate on the plateau naturally depends on the intensity of the beam admitted into the counter.

A proportional counter is operated at some voltage on the plateau. Counter power supplies normally have a voltage stability of the order of 0.1 per cent in the 1500-volt range. It is evident from Fig. 5-33 that a voltage drift of 0.1 per cent will have little effect on the counting rate, and therefore the selection of the operating voltage is not critical. In general, very high and very low voltages on the plateau are avoided, and such factors as pulse-height discrimination (described subsequently) are allowed to dictate the final selection.

A Geiger counter also exhibits a plateau over which the counting rate is nearly independent of the applied voltage, but the threshold voltage to the plateau is a fundamental characteristic of the counter. The threshold voltage marks the beginning of the Geiger discharge region in which all pulses are independent of the energy of the absorbed x-ray quanta and therefore uniform in height. The sensitivity level of the equipment is naturally set well below this height, and consequently, at a voltage above the threshold value, all pulses are counted and the counting rate has a nearly constant value (the plateau) which extends for several hundred volts[20] (Fig. 5-34). The threshold voltage and the counting rate on the plateau are independent of the wavelength of the radiation. The threshold voltage represents the lowest voltage, at which the electrical field in the counter is so intense that a single electron can trigger the discharge and all pulses become equal in size.

A Geiger counter is operated at about 100 volts above the threshold voltage. In counters designed for diffraction work the operating voltage is near 1400 volts. At voltages above the plateau region the electrical field is higher than the gas can support. No triggering is necessary to start the electron avalanche, and the counter goes into a state of continuous discharge. A counter may be permanently damaged if exposed to a continuous discharge for even a short time.

THE GEIGER COUNTER. It is useful now to present the design and characteristics of one commercial Geiger counter in some detail. The Philips counter has been described in several papers.[21] This counter (Fig. 5-35*a*) consists of a chrome–iron cathode cylinder measuring 2 cm in diameter by 10 cm in length with a 0.75-mm tungsten anode wire mounted along its axis. The gas is argon mixed with about 0.4 per cent

[20] Spurious counts, which appear in the form of "after-discharges" or trains of counts following a valid count and are attributed to gaseous impurities in the tube, give a slope to the plateau.

[21] P. H. Dowling, C. F. Hendee, T. R. Kohler, and W. Parrish, *Philips Tech. Rev.*, **18**, 262 (1956/1957), for example.

chlorine at a pressure of 55 cm Hg. The counter is operated at about 1400 volts; its plateau is more than 300 volts wide. X-rays enter the tube axially through an end window of mica, 0.013 mm thick. The "active"

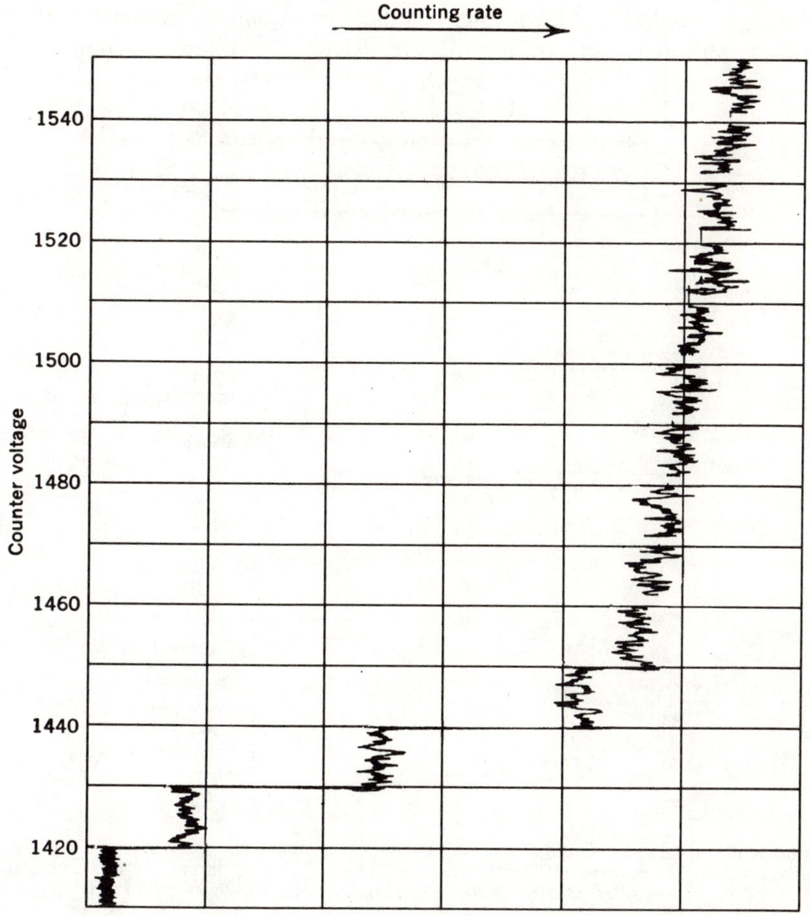

Fig. 5-34. Experimentally determined plateau for a Geiger counter tube.

volume of gas in which an x-ray quantum may be absorbed and give rise to a discharge pulse is 1.2 cm in diameter in this tube.

Most of the ionization into free electrons and positive ions takes place near the wire anode where the electrical field strength is most intense. The electrons are quickly collected at the anode, leaving the positive ions as a sheath about the wire. This effectively lowers the field strength in the gas below the value required to produce a pulse, and for an interval (the "dead" time) the counter cannot detect an entering x-ray quantum. The field strength increases as the positive ion sheath drifts toward the cathode. It attains the threshold value in 170 μsec in the Philips counter. This is the end of the dead time, and incoming quanta can now produce pulses, although not yet of full size. The normal field strength and

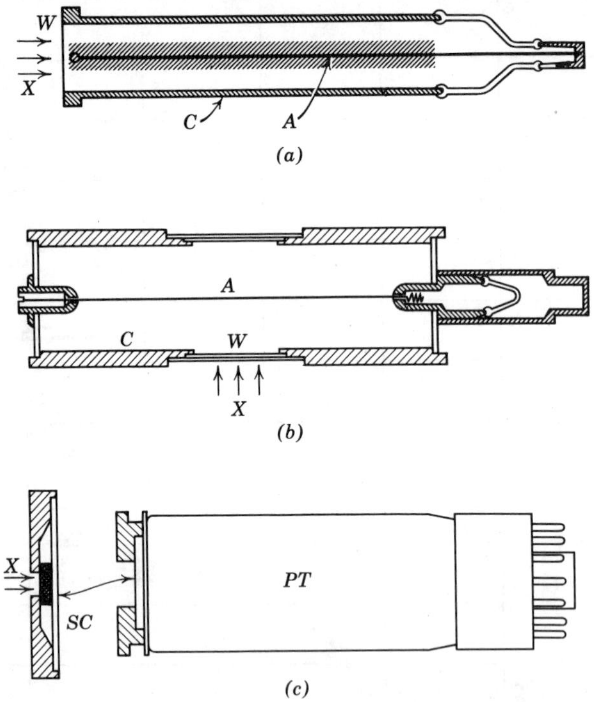

Fig. 5-35. (*a*) Geiger counter and (*b*) proportional counter: *A*—anode, *C*—cathode, *W*—window, *X*—x-rays. (*c*) Scintillation counter: *SC*—scintillating NaI · Tl crystal, *PT*—photomultiplier tube. (From Dowling, Hendee, Kohler, and Parrish.[21])

pulse height are not recovered until the sheath reaches the cathode. The period of total recovery is near to 300 μsec. The comparatively long dead time of the Geiger counter means that the response to incoming quanta becomes nonlinear at rather low counting rates. At rates as low as 500 cps[22] the deviation from linearity approaches 10 per cent.

The purpose of the small amount of chlorine mixed with the main gas of the counter can now be explained. If argon is the only gas in the tube,

[22] cps = counts per second.

electrons are ejected from the cathode on the arrival of the positive argon ions that form the space-charge sheath. These electrons, accelerated toward the anode, initiate a new chain of ionization which causes another electron avalanche onto the anode and starts the movement of a new positive ion sheath to the cathode. The result is a continuous state of discharge, and no newly arriving quanta can be counted. The discharge can be "quenched" as soon as the sheath reaches the cathode by an external circuit that momentarily drops the counter voltage to a value that is too low to maintain the discharge but is high enough to collect all the ions on the cathode. A Geiger counter tube can be made self-quenching by the addition of a small quantity of a gas such as a halogen, which prevents the initial avalanche from becoming a continuous discharge. The positive argon ions of the sheath collide with chlorine atoms, acquire electrons, and are neutralized. The ionized chlorine atoms are neutralized at the cathode without releasing electrons.

THE PROPORTIONAL COUNTER. Proportional counters normally differ from Geiger counters in two important respects—they have a side window and are filled with a heavier gas, such as xenon or krypton (Fig. 5-35*b*). To ensure that the pulse size is proportional to the energy of absorbed quanta, the spatial distribution of the electrical field should be the same for all quanta, no matter where they are absorbed. This is more fully realized in a tube that admits the x-ray quanta normal to the axis of the anode (i.e., in a side-window tube). The absorption path is much shorter in such a tube, and to keep the unabsorbed quanta at a minimum it is necessary to raise the pressure of the gas or use a heavier gas. Higher gas pressures are unsatisfactory because undesirably high operating voltages are required. Heavier gases, on the other hand, satisfactorily compensate for the shorter absorption path. Figure 5-36 shows that the detecting efficiency of the Philips side-window proportional counter, filled with xenon to a pressure of 32 cm Hg, is almost identical with the argon-filled Geiger counter already described.

The long recovery period of a Geiger counter is due to the discharge spreading throughout the volume of the tube. Because the voltage of a proportional counter is maintained below the threshold value of the Geiger plateau, the discharge is confined essentially to a radial area or plane normal to the axis of the tube, located where the original x-ray quantum was absorbed. As a result, only that small portion of the anode is affected by the ion sheath. The remainder is active and can detect a second quantum absorbed shortly after the first. The recovery time is approximately 0.2 μsec; hence very high detecting rates are achieved. The counter is capable of counting rates as high as 10^7 cps without loss

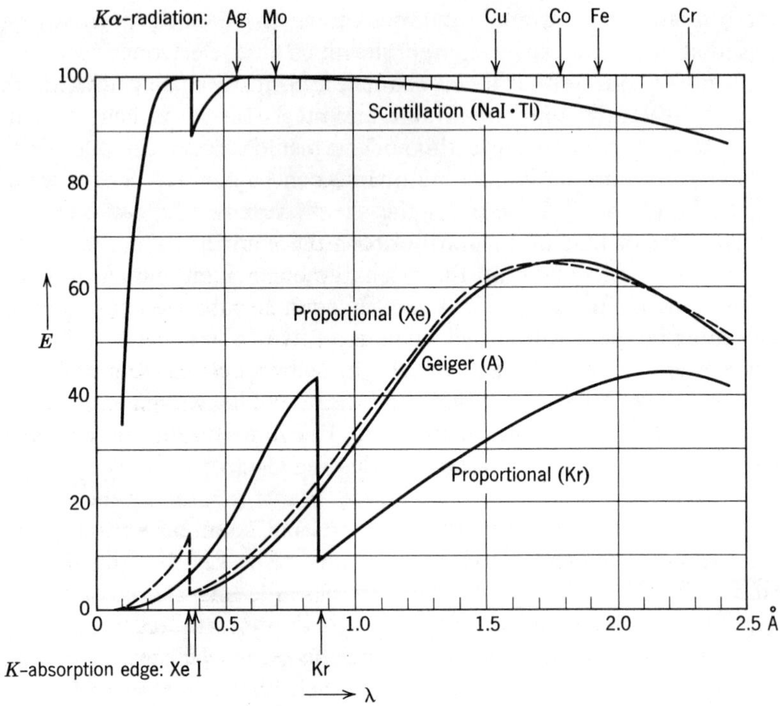

Fig. 5-36. The calculated quantum counting efficiency E as a function of wavelength for a number of counters. E changes abruptly at the absorption edge of the gas or scintillation crystal (the Xe K- and Kr K-edges of the proportional counters and the I K-edge of the scintillation crystal). (Modified from Dowling, Hendee, Kohler, and Parrish.[21])

of linearity, and consequently the practical limits of linearity are set by other circuits in the diffractometer.

Photoelectric Detectors

THE SCINTILLATION COUNTER. A scintillation counter (Fig. 5-35*c*) has two basic elements—a crystal that fluoresces visible light or scintillates when struck by x-ray quanta and a photomultiplier tube that converts the light to electrical pulses. The crystal is usually NaI activated by thallium. When an x-ray quantum is absorbed by an atom of the crystal, it produces a photoelectron and usually one or more Auger electrons which energize a large number of the fluorescent centers located at the sites of thallium ions in the crystal. Each energized fluorescent center may emit a visible photon. With the aid of reflecting surfaces near or on the crystal, the light is directed onto the cathode of the photomultiplier tube. For approximately ten of these photons, one electron is liberated from the cathode. These electrons are drawn by a potential difference of about 100 volts onto the first of a series of electrodes known as dynodes which, as the name implies, perform two functions—they collect and amplify the stream of electrons from the cathode. For each electron that strikes the first dynode, a definite number of secondary electrons is produced and drawn to the second dynode where the process is repeated. The multiplication factor or gain is about four at each dynode. Thus the total gain in a photomultiplier tube with ten stages of dynodes is of the order of 4^{10} or 10^6, and the anode receives an avalanche of electrons, which results in a pulse that can be amplified to operate the counting circuits in the diffractometer.

The scintillation counter is similar in two respects to the proportional counter. First, the number of fluorescent centers energized by an incoming quantum depends on the energy of the quantum; hence the pulse delivered by the photomultiplier is also proportional to the quantum energy. Second, the fluorescent decay constant of NaI crystal is about 10^{-7} sec; this results in a dead time of about 0.2 μsec and ensures linear counting rates up to a very high intensity.

The scintillation counter, too, exhibits a plateau over which the counting rate does not vary appreciably with the applied voltage. This plateau, like the plateau shown by the proportional counter, owes its existence to an adjustable sensitivity level below which the counting equipment cannot detect the pulses; the threshold voltage to the plateau varies with the radiation, increasing with the wavelength.

One disadvantage of the scintillation counter is its high background count. Ohmic leakage, thermionic and field emission of electrons, and gaseous ionization cause noise pulses of relatively large amplitudes. High counter voltages favor the generation of these pulses and, of course, cause an increase in the size of all pulses. High amplifier gains also increase the size of all pulses, including the noise pulses generated between the photomultiplier tube and the counting circuits. The strategy, therefore, to attain a high ratio of relevant-to-noise pulses is to suppress the formation of noise pulses with low counter voltages and to increase the size of relevant pulses to detectable levels with moderate amplifier gains. The Philips scintillation counter, for example, is designed to be operated at less than 1000 volts. The exact choice of voltage is discussed in the section on pulse-height discrimination.

A Comparison of the Three Types of Counter

Each of the three counters has special advantages for powder diffractometry which derive from the characteristics that have been discussed in the preceding pages. It is useful now to describe these advantages and to evaluate the counters for various diffraction purposes.

THE QUANTUM COUNTING EFFICIENCY. This factor has been defined by Taylor and Parrish[23] as the ratio of the number of x-ray photons counted by the circuits to the total number of photons that strike the window of the detector (assuming that there are no counting losses due to dead time). The value of the function E may be calculated from the expression

$$E = f_T f_A$$

where f_T is the fraction of radiation transmitted through the window (and through the inactive volume of gas in a Geiger counter) and f_A is the fraction absorbed in the active material. The values calculated by Taylor and Parrish for commercial models of the three types of counter are plotted as a function of wavelength in Fig. 5-36. The figure shows that the efficiency of the scintillation counter is outstanding compared with both Geiger and proportional counters and is nearly 100 per cent across the diffraction wavelength range (Ag$K\alpha$ to Cr$K\alpha$). The counter shows a marked superiority as a detector of Mo$K\alpha$ and Ag$K\alpha$ radiation.

THE ENERGY RESOLUTION. The pulses obtained from counters show a range in pulse height, varying about an average value, even when the incoming radiation is monochromatic. This variation in pulse height limits the usefulness of a counter for distinguishing and separating radiation of closely spaced wavelengths. Figure 5-37 shows pulse-height distribution curves determined experimentally for Cu$K\alpha$ radiation with two kinds of counter. The separating power or energy resolution is expressed by the ratio W/A (usually as a per cent) where W is the width of the curve at half-height and A is the average pulse height (roughly equal to the value of the abscissa at maximum curve height). The smaller the ratio, the better the energy resolution. Commercial proportional counters have an energy resolution of about 20 per cent compared with 50 per cent for scintillation counters.

The range in pulse height has been shown to be inversely related to the number of electrons per quantum of absorbed radiation which take part in the pulse-forming discharge. In the xenon-filled proportional counter one Cu$K\alpha$ quantum will produce, on the average, $m = 350$ electrons.[21] Owing to the fluctuation in m and a variation in the gas amplification factor for each electron, the resulting pulse height varies with a relative standard deviation σ/A, which is about equal to $1/\sqrt{m}$.

[23] J. Taylor and W. Parrish, *Rev. Sci. Instr.*, **26**, 367 (1955).

Taking $W = 2.36\,\sigma$[24]

$$\frac{W}{A} = \frac{2.36\,\sigma}{\sqrt{m\,\sigma}} = \frac{2.36}{\sqrt{350}} \sim 13 \text{ per cent}$$

The difference between the calculated and actual values in mass-produced counters is probably the result of variations in the gas amplification factor caused by irregularities in the anode wire and in its position in the sheath.

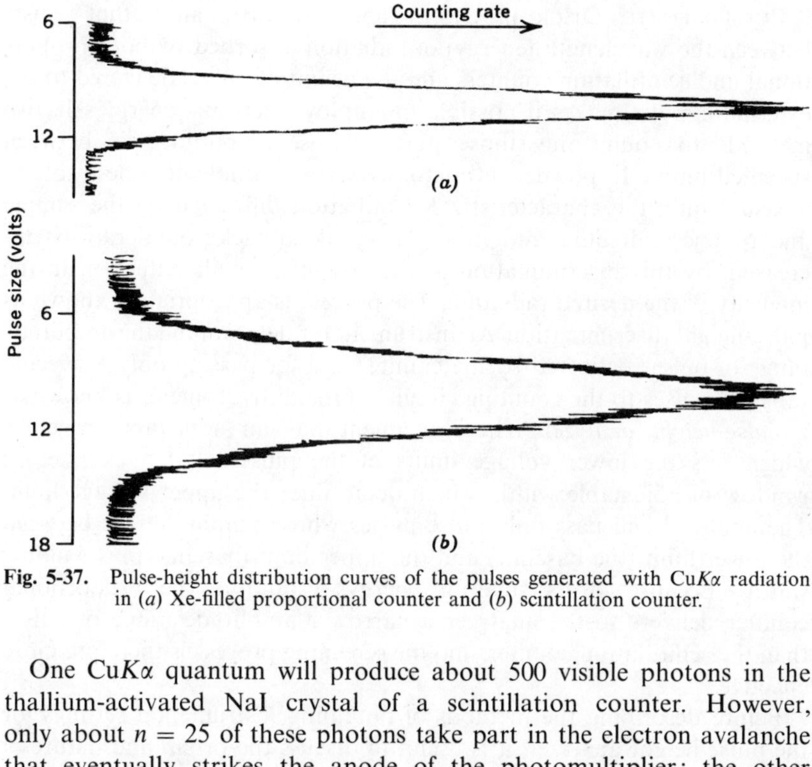

Fig. 5-37. Pulse-height distribution curves of the pulses generated with Cu$K\alpha$ radiation in (*a*) Xe-filled proportional counter and (*b*) scintillation counter.

One Cu$K\alpha$ quantum will produce about 500 visible photons in the thallium-activated NaI crystal of a scintillation counter. However, only about $n = 25$ of these photons take part in the electron avalanche that eventually strikes the anode of the photomultiplier: the other photons either do not reach the photocathode, or fail to liberate an electron, or the liberated electron fails to reach the first dynode.[25] Then

[24] In the Gaussian error curve the standard deviation σ is the horizontal distance from the mean to the point of inflection of the distribution curve; it is related to the width at half-height W as $2.36\,\sigma = W$.

[25] This does not detract from the quantum counting efficiency described above; nearly every absorbed quantum is counted, despite the loss of a large part of the energy in the crystal.

σ/A is about equal to $1/\sqrt{n}$, and

$$\frac{W}{A} = \frac{2.36\,\sigma}{\sqrt{n}\,\sigma} = \frac{2.36}{\sqrt{25}} \sim 47 \text{ per cent}$$

The cause of the variation in height of pulses produced by proportional and scintillation counters is fundamental in nature, and therefore it is unlikely that significant improvements in energy resolution will be forthcoming.

PULSE-HEIGHT DISCRIMINATION. The proportionality that exists between the wavelength (energy) of radiation absorbed by both proportional and scintillation counters and the height of pulses delivered to the external circuits makes it possible to employ electronic energy-selective methods to count only those pulses whose amplitudes lie between specified limits. In powder diffractometry, for example, it is desirable to measure only the characteristic $K\alpha$ radiation diffracted by the sample and to reject all other radiation. The peak-to-background ratio is increased, by this discrimination at the cost of a small reduction in the intensity of the desired radiation. The process is appropriately known as pulse-height discrimination. An instrument for determining the amplitude range of pulses delivered by the counter and for passing only a selected range of pulses to the counting circuits of the diffractometer is known as a *pulse-height analyzer*. The instrument has an adjustable baseline, which sets the lower voltage limits of the pulses, and a channel or window of adjustable width, which determines the upper voltage limit. The analyzer will pass only those pulses whose amplitudes lie between the lower limit (the baseline) and the upper limit (baseline plus window width). Because of its superior energy resolution, the proportional counter delivers to the analyzer a narrower amplitude range of pulses than the scintillation counter, and the screening process is therefore more effective.

Before describing the methods of obtaining instrumental settings for the pulse-height analyzer, it is useful to discuss the origin and nature of the unwanted radiation that contributes to the background in powder diffractometry.

1. A portion of the $K\alpha$ radiation is incoherently scattered by the sample. This radiation has the same wavelength as the diffracted radiation and consequently cannot be separated by pulse-height discrimination. It represents the minimum background.
2. The $K\beta$ component of the beam cannot be satisfactorily separated from the $K\alpha$ radiation by pulse-height discrimination because of

the small energy difference between the lines. It can be effectively removed by passing the primary beam or the reflected beam through an appropriate filter (Table 2-4).

3. The white radiation of the continuous spectrum is always present in the primary beam and is diffracted and incoherently scattered by the sample. The intensity of this undesirable radiation is reduced by the β filter, but a large proportion passes on to the counter and adds to the background. This radiation has a peak intensity near 0.5 Å at the voltage applied to x-ray diffraction tubes. Its wavelength is therefore close to the wavelength of Mo$K\alpha$ (λ 0.71) and Ag$K\alpha$ (λ 0.56) radiation but markedly different from the wavelength of Cu$K\alpha$ radiation (λ 1.54) and other radiation used in diffraction studies. The amplitude ranges overlap to such an extent that pulse-height discrimination is not particularly effective for Mo$K\alpha$ and Ag$K\alpha$ radiation, and peak-to-background ratios after discrimination are therefore much lower than those of the wavelengths characteristic of copper, iron, and chromium targets.
4. Fluorescent x-rays are generated in the sample if it contains an element that is excited by the incident beam. The presence of iron in a sample exposed to a Cu$K\alpha$ beam is a typical example in mineralogy. Fluorescent radiation always has a longer wavelength (lower-voltage pulses) than the generating radiation, but the energy difference is generally so small that the amplitude ranges overlap and electronic discrimination leaves much to be desired. It is best to avoid the problem by the use of primary radiation that is not absorbed by any element in the sample.
5. Strongly radioactive samples give high-voltage background pulses which may materially reduce the precision of measurements. Kohler and Parrish[26] have shown that pulse-height discrimination is effective in reducing the background and increasing the sensitivity of radioactive samples to the primary radiation.

Appropriate counter voltages and optimum analyzer settings can be determined from a study of either differential or integral pulse-height distribution curves. Differential curves are prepared by recording the counting rate of pulses passed by a narrow window as it is drawn across the pulse distribution of a monochromatic source. This is performed in practice by opening the window of the pulse-height analyzer to a width of $\frac{1}{2}$ or 1 volt and driving the baseline, in synchronism with the recording chart, down through the pulse voltage range. Figure 5-38 shows examples

[26] T. R. Kohler and W. Parrish, *Rev. Sci. Instr.*, **26**, 374 (1955).

of such curves obtained by viewing the Ni-filtered Cu$K\alpha$ 111 reflection of silicon powder with the xenon-filled proportional counter whose *counting rate/applied voltage* characteristics are reproduced in Fig. 5-33. Four curves, at operating voltages beginning immediately above the threshold voltage and spaced at intervals of 25 volts on the plateau, are reproduced. Each shows two peaks—the main pulse-height distribution peak due to Cu$K\alpha$ radiation and a minor peak at a lower voltage known as the escape peak.[27] The sharp upturn in the curves at very low voltages is due to counter noise. Figure 5-38 shows that the average amplitude of the relevant $K\alpha$ pulses is increased by raising the applied voltage. Pulse-height discrimination can be effective only if the counter voltage provides a distinct separation between the relevant $K\alpha$ peak and the escape peak or, if none is present, the counter noise. A separation of 3 volts is adequate for the proportional counter, and thus a satisfactory choice for the operating voltage of the counter described here is 1600 volts for Cu$K\alpha$ radiation. At this voltage the escape peak has maximum intensity at 4.8 volts, the main peak is at 11.4 volts, and an intensity low of about 3 volts separates the peaks. Because of its inferior energy resolution and greater counter noise, it is advisable to operate the scintillation counter at a voltage that isolates the main peak by an interval of at least 6 volts from all irrelevant low-voltage pulses.

Exactly the same information can be obtained from a series of integral pulse-height distribution curves. Such curves are prepared by plotting intensity against pulse voltage as the baseline is lowered at a constant rate with the window of the analyzer opened to its full width. Thus, as the baseline moves to lower voltages, ever more pulses are passed by the analyzer, and the counting rate increases until the full distribution,

[27] The primary process in the absorption of an x-ray quantum by one of the counters described here is the ejection of an inner-shell electron from a gas or crystal atom. If the wavelength of the absorbed radiation is shorter than the K absorption edge of the atom, the electron will originate from the K shell. "Falling in" from the L shell to replace the electron is associated with the ejection of one or more electrons (the Auger process) or the emission of a new x-ray quantum (fluorescence). Auger electrons, as well as ejected inner-shell electrons, are absorbed and contribute to the formation of ion pairs in the counter gas and to the excitation of fluorescent centers in a scintillation counter. However, some of the fluorescent x-ray quanta may not be reabsorbed by gas or crystal atoms and will "escape" as fluorescent radiation. Hence some of the original quanta contribute a lesser amount of energy and result in pulses of lesser size. This is of no consequence in the Geiger counter, but with other counters an additional peak—the escape peak—appears in the pulse-height distribution curve. The escape peak produced by a krypton-filled proportional counter with Mo$K\alpha$ radiation is especially large, constituting about 55 per cent of the total pulse energy and severely limiting the usefulness of this counter for measurements which require discrimination methods. W. Parrish and T. R. Kohler, *Rev. Sci. Instr.*, **27**, 795 (1956).

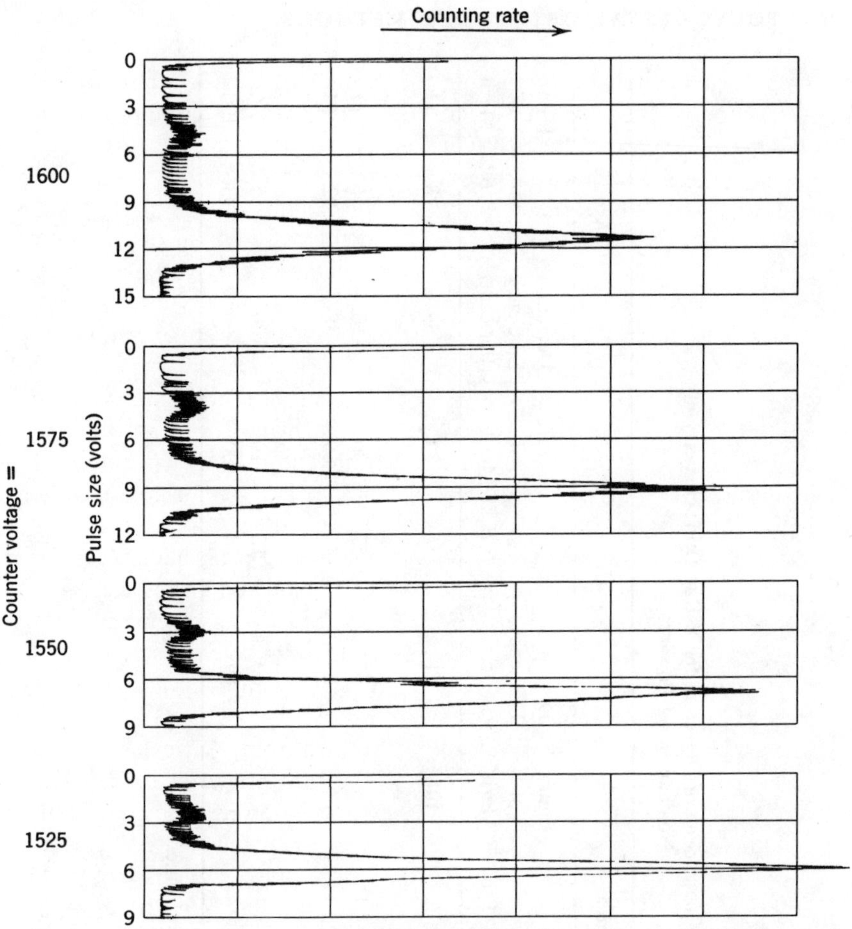

Fig. 5-38. Differential pulse-height distribution curves obtained by viewing the Ni-filtered Cu$K\alpha$ 111 reflection from powdered silicon through a $\frac{1}{2}$ volt window with a Xe-filled proportional counter. The average pulse voltage is shifted to higher values with an increase in the operating voltage.

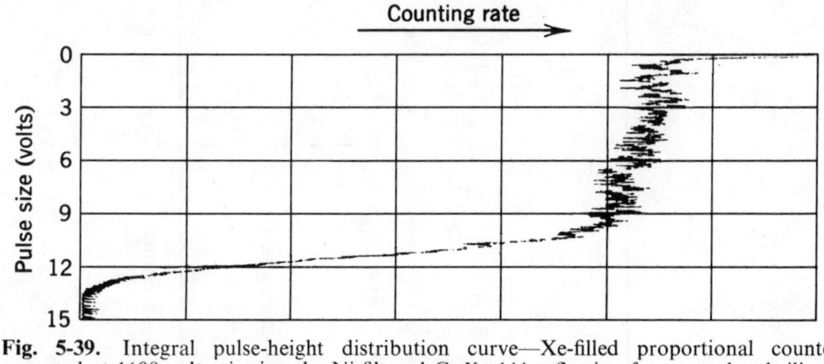

Fig. 5-39. Integral pulse-height distribution curve—Xe-filled proportional counter operated at 1600 volts viewing the Ni-filtered Cu$K\alpha$ 111 reflection from powdered silicon.

Table 5-15. A Comparison of Detectors for X-Ray Diffractometry of Nonfluorescing Powders

Counter	Mo$K\alpha$			Cu$K\alpha$			Cr$K\alpha$		
	Detection Efficiency (Per cent)	P/B	Figure of Merit	Detection Efficiency (Per cent)	P/B	Figure of Merit	Detection Efficiency (Per cent)	P/B	Figure of Merit
Geiger (argon)	14	27	10	51	46	20	59	18	13
Proportional (Xe)									
no PHD	13	16	5	63	57	30	60	16	11
PHD	12	44	13	56	146	69	58	92	64
Scintillation									
no PHD	100	10	25	100	12	10	100	5	6
PHD	90	44	100	89	134	100	90	93	100

Geiger-counter measurements corrected for dead time.

marked by a plateau, is received by the counting circuit. Figure 5-39 shows the integral distribution curve for Cu$K\alpha$ radiation obtained with the same proportional counter operating at 1600 volts. This curve should be compared with the upper curve in Fig. 5-38.

According to Parrish and Kohler,[28] the settings of the pulse-height analyzer that give the highest intensities and peak-to-background ratios with nonfluorescing samples are obtained with a window symmetrically placed across the pulse-height distribution and just wide enough to transmit 90 per cent of the characteristic radiation. These settings are found by raising the baseline, with the window wide open, until the total intensity is reduced by 5 per cent and then narrowing the window width until the intensity is reduced by another 5 per cent. The analyzer then discriminates against counter and electronic noise on the low-voltage side and against the short-wavelength radiation of the continuous spectrum on the high-voltage side. It is unwise to attempt a more severe discrimination by narrowing the window beyond these arbitrary limits: the peak intensity falls more rapidly than the peak-to-background ratio increases because the background contains, after discrimination, at least as much $K\alpha$ as noncharacteristic radiation. Furthermore, as the window is narrowed, the settings correspond to positions on the steep flanks of the pulse-height distribution curves, and small variations in the stability of the circuits become critical. The procedure of Parrish and Kohler leads to a baseline value of 9.5 volts and a window width of 3.5 volts for Cu$K\alpha$ radiation when a potential of 1600 volts is applied to the proportional counter described here.

FIGURE OF MERIT. Powder spectra are generally composed of relatively sharp reflections, many of very low intensity, superimposed on a background which may have various components as already described. To detect and measure the weak lines, an x-ray counter should have a high quantum counting efficiency and should yield a high peak-to-background ratio. No disadvantage, other than a decrease in quantum counting efficiency, derives from a nonuniform spectral response (Fig. 5-36). Linearity of counting rates is of significance in the measurement of intensities but of little importance in many routine applications of powder diffractometry. Giving equal importance to quantum counting efficiency and peak-to-background ratio, Dowling, Hendee, Kohler, and Parrish[21] derived a figure of merit for a number of Philips counters. All measurements were made on the filtered 111 reflection from a specimen of powdered silicon. This element does not fluoresce measurably with the x-radiations used in powder diffractometry and, of course, is not

[28] W. Parrish and T. R. Kohler, *Rev. Sci. Instr.*, **27**, 795 (1956).

radioactive. Table 5-15 gives, for three useful wavelengths, the authors' values for

(1) *detection efficiency (per cent) = quantum counting efficiency* × *transmission factor* (the last factor is the percentage of relevant pulses passed by the pulse-height analyzer—arbitrarily set to pass 90 per cent of proportional and scintillation counter pulses);
(2) *measured peak-to-background* (P/B) *ratio* based on x-ray tube voltages chosen to give the highest P/B ratio for each wavelength;
(3) *figure of merit = detection efficiency* × P/B (all figures adjusted to reduce the highest value for each wavelength to 100).

Table 5-15 shows that the scintillation counter, with pulse-height discrimination, clearly outranks the other counters for Mo$K\alpha$ radiation. This is evidently because of its markedly superior quantum counting efficiency for radiation of this wavelength (Fig. 5-36). It also rates as the best detector for Cu$K\alpha$ and Cr$K\alpha$ radiation, but its superiority over the xenon-filled proportional counter is less pronounced for these wavelengths.

Dowling, Hendee, Kohler, and Parrish[21] also derived a figure of merit for the proportional and scintillation counters in the presence of a strongly fluorescing specimen. The fluorescent background given by a sample of iron with CuKα radiation was chosen as a discrimination test for the counters. The evaluation, in this case, is the question which counter is more effective at pulse-height discrimination between the relevant Cu$K\alpha$ and the fluorescent FeK radiation. Hence the product, *detection efficiency of CuKα × ratio of detection efficiencies CuKα/FeK*, serves as a useful figure of merit for comparing the two detectors. Figures 5-40*a*, and *b* show the superimposed integral curves for Cu$K\alpha$ (obtained by viewing the 111 reflection of powdered silicon) and FeK (the fluorescent background of the iron pattern) for the xenon-filled

Table 5-16. A Comparison of Detectors for X-Ray Diffraction of a Fluorescing Powder (Iron Sample Irradiated by Cu*Ka* Beam)

	Symmetrical Window		Asymmetrical Window	
Counter	Ratio of Detection Efficiency Cu$K\alpha$/FeK	Figure of Merit	Ratio of Detection Efficiency Cu$K\alpha$/FeK	Figure of Merit
Proportional (Xe)	1.6	13.5	25	100
Scintillation	1.2	16.3	4.1	27.5

proportional counter and the scintillation counter, respectively. The vertical dashed lines indicate the limits of the conventional symmetrical setting of the pulse-height analyzer window narrowed to transmit 90

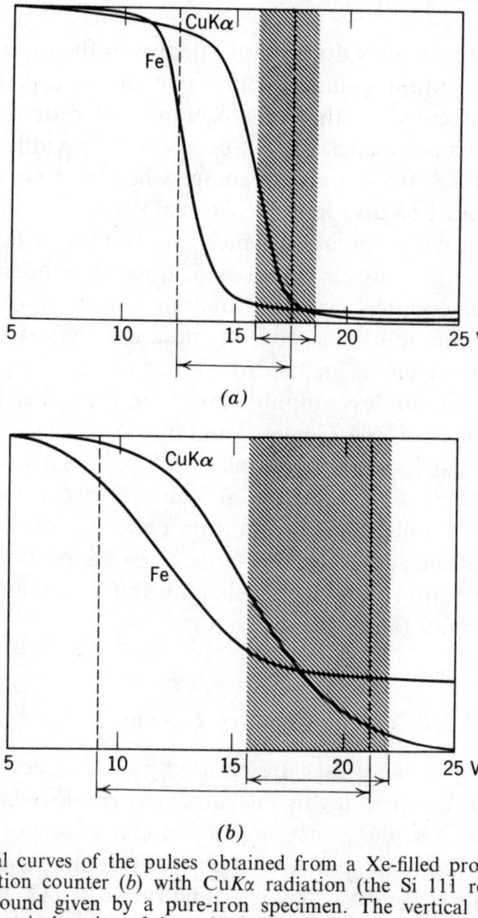

Fig. 5-40. Integral curves of the pulses obtained from a Xe-filled proportional counter (*a*) and a scintillation counter (*b*) with Cu*K*α radiation (the Si 111 reflection) and the fluorescent background given by a pure-iron specimen. The vertical lines indicate the conventional symmetrical setting of the pulse-height analyzer window to give a 90 per cent transmission of the Cu*K*α radiation. The cross-hatched area represents an asymmetrical setting to discriminate strongly against the Fe*K* radiation. Curves normalized to have same counting rate at 5 volts pulse amplitude. (From Dowling, Hendee, Kohler, and Parrish.[21])

per cent of the Cu*K*α pulses. The ratios of the detection efficiencies, Cu*K*α/Fe*K*, are a disappointing 1.6 and 1.2 for the two counters. It is not necessary, however, to use a symmetrical setting for the window. An asymmetrical setting selected to discriminate strongly against the fluorescent radiation at the expense of Cu*K*α detection (cross-hatched areas,

Figs. 5-40*a*, and *b*) raises the detection ratio to 25 and 4.1, respectively. The better performance of the proportional counter is a consequence of its superior energy resolution. Table 5-16 summarizes the evaluation of the two counters, with merit figures adjusted to give the highest value a figure of 100.

The preceding summary does not do justice to the inexpensive Geiger counter. The quantum counting efficiency of an argon-filled Geiger counter is about equal to that of a xenon-filled proportional counter (Fig. 5-36). It has a peak efficiency at about $\lambda = 1.75$ Å; the poor response in the vicinity of $\lambda = 0.5$ Å is an advantage when the counter is used with Cu$K\alpha$, Fe$K\alpha$, and Cr$K\alpha$ radiation, because it discriminates against the short-wavelength white radiation which contributes to the background. A study of the merit figures in Table 5-15 shows that unless proportional and scintillation counters are operated in conjunction with a pulse-height analyzer the additional cost of these counters is not warranted. The Geiger counter has a further advantage in that its pulses are relatively large and require less amplification than the pulses from the other two counters. In brief, the Geiger counter is a satisfactory counter for all routine and many specialized applications, especially when simplicity and cost of equipment is an important factor. Nevertheless, the scintillation counter has a much higher quantum counting efficiency, especially for Mo$K\alpha$ radiation, and when this counter or the proportional counter is used in conjunction with a pulse-height analyzer important gains in peak-to-background ratios are realized.

The Counting Circuits

The rate at which the pulses are emitted by the detector or passed by the pulse-height analyzer, if this instrument is employed, is analyzed on a scaling circuit. A scaler consists of a number of stages each of which transmits a fraction of the pulses to the next stage. If each stage transmits every second pulse, the stages have scaling ratios of 1:2, 1:4, 1:8, and so on. The output or *scale factor* of the scaler can be selected at any stage of the circuit to reduce the pulse rate to measurable proportions.

The intensity of the x-radiation received by the detector is measured in one of two ways:

1. *Chart recording.* The output of the scaling circuit is fed into a ratemeter which converts the digital information into a continuous electric current. The current can be indicated on an ammeter which then acts the part of a counting ratemeter. More importantly, it can be caused to actuate a pen on a strip-chart recorder. The pen responds to the radiation

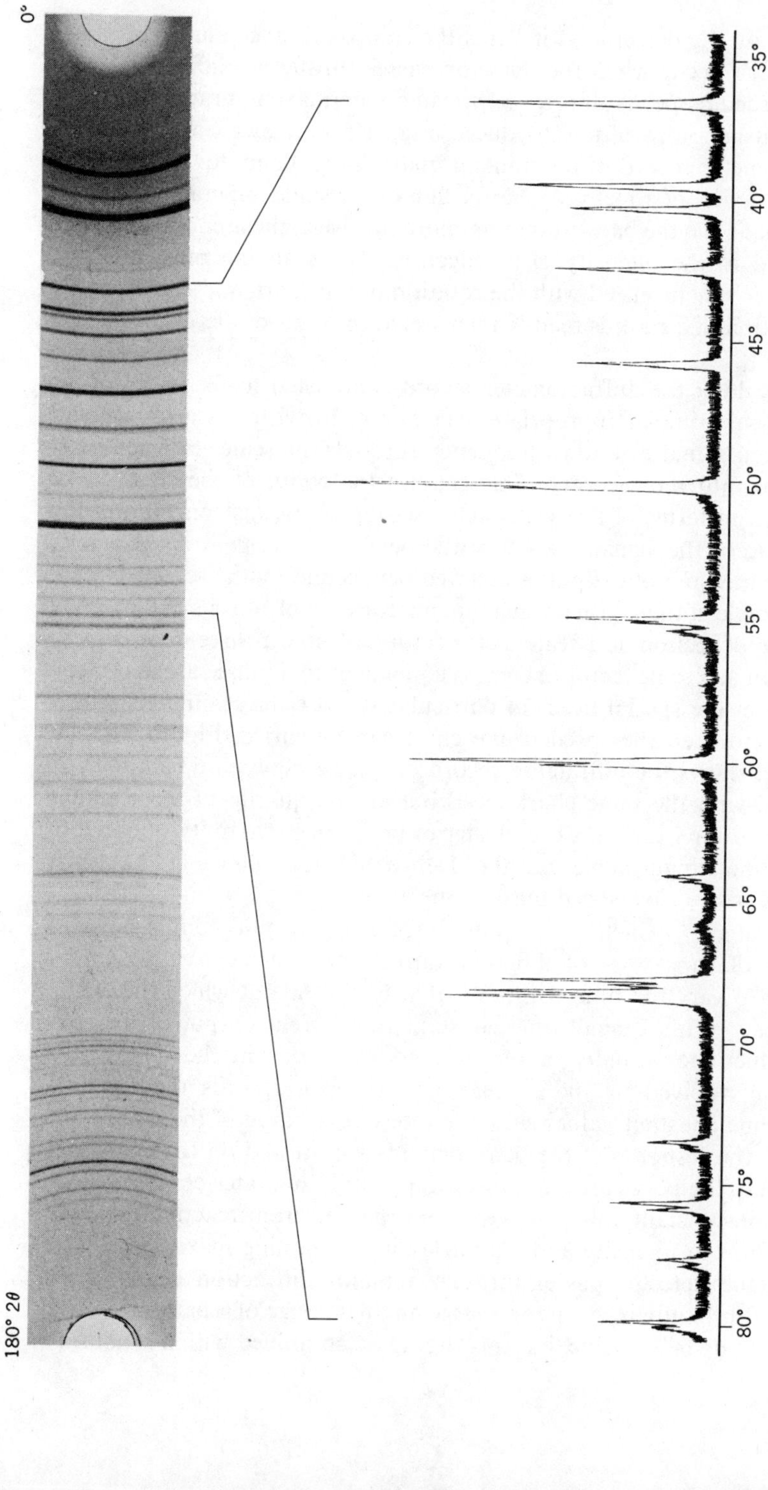

Fig. 5-41. Diffraction record of quartz; Ni-filtered Cu radiation. Diffractometer chart from 34 to 80° 2θ; 40 kV/15 mA; proportional counter 1600 V; PHA/BL 7.2 V, W 6 V for 90 per cent transmission; 1° slits; scan speed 1°/min; 7/24 actual size. Debye–Scherrer pattern from 0 to 180° 2θ; 33 kV/15 mA; 6-hour exposure; actual size negative print.

received by the detector as it scans the sample. Sudden surges of radiation, which occur when the detector passes through a diffracted beam, are recorded as peaks on the chart. Thus a permanent, graphical record of the diffraction pattern is produced. Figure 5-41 shows a portion of the diffractometer record of the mineral quartz (SiO_2) and, for comparison, a reproduction of a Debye–Scherrer film of the same mineral. The height of a peak above the background or, more precisely, the area under a peak is related to the intensity of a reflection. The scanning motion of the detector is synchronized with the rotation of the chart, and consequently the position of a peak is readily interpreted in terms of the 2θ diffraction angle.

The scale of the diffractometer record is adjusted to the width of the chart by selecting an appropriate scale factor. Provision is made to send an internal signal of known frequency through the scaler-ratemeter circuits and into the recorder. This enables the width of the chart to be calibrated in terms of the scale factor setting of the instrument and the pulse rate of the signal. Therefore the height of a peak is a reasonably accurate measure of the pulses counted per second by the scaling circuit. For example, if the internal signal has a frequency of 100 cps and causes a full scale deflection at a scale factor setting of one, a three-quarter scale deflection at a scale factor of eight is equivalent to a pulse rate of 600 cps.

To meet the special needs of particular studies, the scanning speed of the detector and the speed of the chart can be varied. Figure 5-42, for example, shows the diffraction record of quartz between 67 and 69°, 2θ obtained with the same chart speed but at one quarter of the scanning speed used for Fig. 5-41. The cluster of peaks in the vicinity of 68° 2θ is almost indistinguishable on the Debye–Scherrer film but is clearly resolved in the slow-speed diffractometer record.

The ratemeter includes a pulse-integrating circuit composed of a number of capacitors of different ratings, any one of which can be selected to vary the time constant—that is, the rate at which the circuit discharges. With a small time constant the current output of the ratemeter fluctuates rapidly in response to variations in the intensity of radiation received by the counter, and the pen records these details. Large time-constant values slow the rate of discharge of the circuit and dampen the response of the pen—detail is suppressed on the chart and, as a result, the peaks are smoothed. Rapid scanning rates combined with large time-constant values may prevent the pen from reacting to minor fluctuations in intensity and may delay it in attaining maximum deflection as the detector passes through a major diffraction beam. If the object of the study is to survey a large angular range of a pattern, a rapid scanning rate is desirable, but this should be combined with a small time constant to give a pen response that will keep pace with rapidly changing intensities and provide a reasonably satisfactory record of intensities and 2θ angles. If the object is to measure accurately the relative integrated intensities of a number of selected peaks, a slow scanning rate is indicated. This does not require a rapid pen response, and a large time constant may be selected to smooth the peaks and aid in the measurement of the areas they cover. Finally, a moderate scanning rate combined

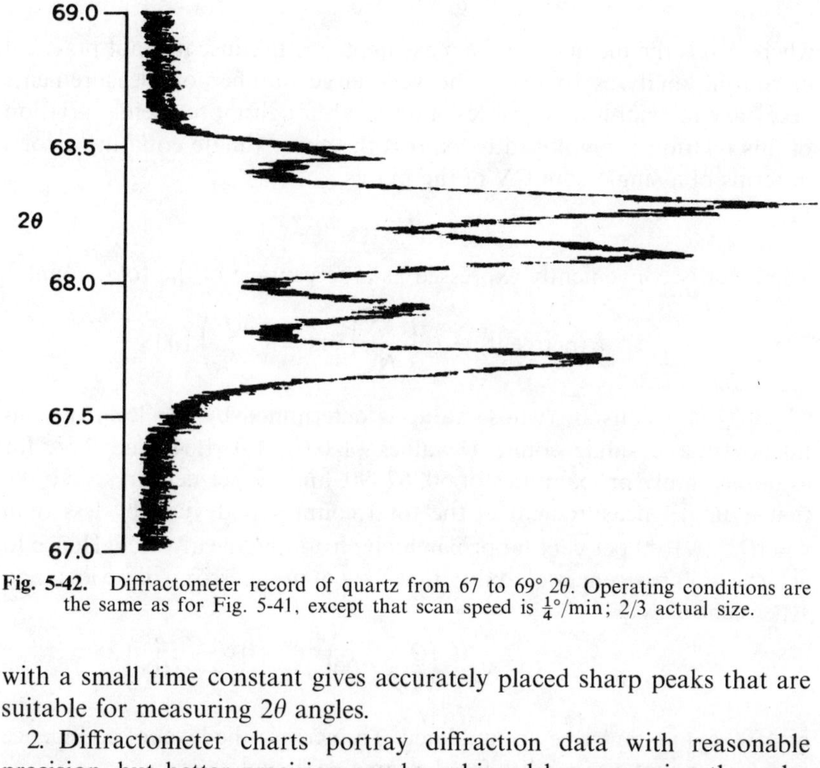

Fig. 5-42. Diffractometer record of quartz from 67 to 69° 2θ. Operating conditions are the same as for Fig. 5-41, except that scan speed is $\frac{1}{4}$°/min; 2/3 actual size.

with a small time constant gives accurately placed sharp peaks that are suitable for measuring 2θ angles.

2. Diffractometer charts portray diffraction data with reasonable precision, but better precision can be achieved by measuring the pulse output rate of the scaler on an electronic count register. Although the average intensity of a diffracted beam may be constant during a period of measurement, the x-ray quanta arrive at the counter in a random time sequence. Consequently, pulses are received and counted by the ratemeter in a random fashion, and for this reason statistical rules must be applied to the measurements. Theoretically, there is no limit to the precision of measurement if enough pulses are counted. Practically, however, the counting time is limited and, furthermore, the intensity of the beam emitted by an x-ray tube may vary over a period of time. It

is desirable, then, to choose a counting method which gives an acceptable precision in a minimum counting time.

Suppose that a very large number of counts is made of the total number of pulses received from a beam of constant intensity in a certain interval of time. Statistical analyses have shown that the range of measurements will be a Gaussian distribution with a standard deviation (standard counting error)

$$\sigma = \sqrt{N_M}$$

where N_M is the mean of all the measurements. Because it is not practical in routine analyses to make the very large number of measurements necessary to establish N_M, a less precise although more useful variation of this relation is employed to express the approximate counting error ε in terms of a single count N of the pulses

$$\varepsilon = \sqrt{N}$$

The error is conveniently expressed as a percentage of the total count

$$\varepsilon \text{ (per cent)} = Q\left(\frac{\sqrt{N}}{N}\right)100 = \left(\frac{Q}{\sqrt{N}}\right)100$$

where Q is a constant whose value is determined by the level of confidence in the single count. Q values of 0.67, 1.0, 1.64, and 2.58, for example, imply probabilities of 50, 67, 90, and 99 per cent, respectively, that a single measurement of the total count will deviate by less than $\varepsilon = (Q/\sqrt{N})100$ per cent (approximately) from the mean value. Hence to achieve a 50 per cent probability that the counting error will be less than 1 per cent of N_M

$$N = \left(\frac{Q}{\varepsilon} \times 100\right)^2 = \left(\frac{0.67}{1} \times 100\right)^2$$

or 4500 counts must be accumulated. To increase the level of confidence to a 67 per cent probability for the same percentage error

$$N = \left(\frac{1.0}{1} \times 100\right)^2$$

or 10,000 counts must be taken.

This expression of the counting error is valid only if the counting rate is very high compared with the counting rate of the background. If the background count is appreciable, the error in the net count will be higher. This follows from the rule that individual errors must be combined to find the error in a difference. If the standard error in the total (peak plus background) count is described as

$$\varepsilon_T = \sqrt{N_T}$$

and that in the background as

$$\varepsilon_B = \sqrt{N_B}$$

the standard error in one measurement of the net $(N_T - N_B)$ count is

$$\varepsilon_{T-B} = \sqrt{N_T + N_B}$$

The percentage error, relative to the net count, is therefore

$$\varepsilon_{T-B}\ (\text{per cent}) = Q\left(\frac{\sqrt{N_T + N_B}}{N_T - N_B}\right)100$$

where Q has the same meaning as in the preceding paragraph. Klug and Alexander[8] have demonstrated the effect of the background with a series of curves (Fig. 5-43) which give the total count required to attain a 50 per cent probability ($Q = 0.67$) for various percentage errors and total to background counting rates. The curves show, for example, that less than 200 counts need be accumulated to achieve a probable counting error of 5 per cent when the background is negligible (when the ratio $R = N_T/N_B = \infty$); almost 2000 counts are needed when the ratio R is 2 and almost 60,000 counts if the ratio is 1:1.

The counting rate is determined by one of two general methods—either the total number of pulses which arrive in a selected interval of time is counted (fixed-time method) or the time to receive a selected number of pulses is measured (fixed-count method). In both cases the data are reduced to give the counting rate as counts per second. The precision of the fixed-count method is independent of the counting rate, but the time required to accumulate a prescribed number of pulses, when the counting rate is very low, may be excessively long.

Integrated intensities are measured by counting at regular angular increments across a peak. The size of the increment evidently determines the precision of the data. Automatic devices are available to facilitate the measurement of integrated intensities. It is obviously impractical to establish the intensities of a complete pattern by point-to-point counting, and this procedure should be reserved for selected reflections and special studies.

The Reliability of Diffractometer Powder Data

The reliability of powder data, including measurements of relative intensities and diffraction angles, depends on the precision with which the goniometer has been aligned and on the care that has been exercised in the preparation of the specimen. In a properly aligned instrument the goniometer axis, which is a line in the plane surface of the specimen, lies midway between the x-ray source and the receiving slit (Fig. 5-31).

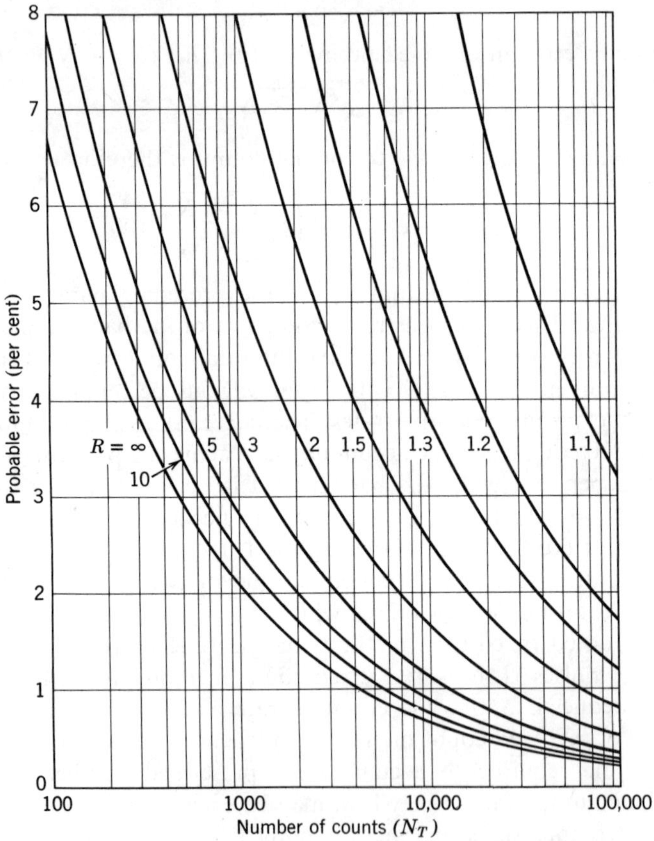

Fig. 5-43. The total number of counts (N_T) required to give a desired degree of accuracy for various ratios (R) of total-to-background counting rates (from Klug and Alexander[8]).

The specimen surface should include the central ray of the incident beam (*FZ* of Fig. 5-30) when the instrument is in the zero position. Then the coupling that maintains the 2:1 ratio between scanning arm angulation and specimen rotation ensures that the specimen surface will remain tangent to the focusing circle at all angular settings and will bisect the angle through which the receiving slit has rotated from the zero position. An error in the 2:1 relation causes a decrease in intensity, an increase in line breadth, and a shift in line position. If the specimen surface does not coincide with the circumference of the focusing circle, either because of an error in aligning the instrument or a fault in the preparation of the sample surface, line positions are shifted. The shift is to higher angles if the surface is within the focusing circle and vice versa.

An x-ray specimen should consist of tiny crystallites, randomly oriented to give a true representation of the relative intensities and fashioned into a form that will provide accurate measurements of diffraction angles. These conditions are easily satisfied in the Debye–Scherrer method. The shape of the specimen, which is a rod or a sphere, and the usual methods of fashioning it, reduce the possibility of a preferred orientation. Furthermore, the specimen is rotated continuously during the exposure with the result that in general, the film accurately portrays the relative intensities of the reflections. To avoid serious shifts in line position the specimen is diluted (thereby reducing the absorption) and accurately centered on the axis of the cylindrical film.

The shape of a diffractometer specimen, with its large flat surface, favors the alignment of crystallites into a preferred orientation. In fact, it is virtually impossible to attain a completely random orientation if the crystallites have a characteristic shape. Diffractometer and Debye–Scherrer specimens differ in another important way. The sharp resolution and accurate line position of which the diffractometer is capable depend in part on having the reflections emanate from the surface of the specimen. If the specimen is "transparent," the beam penetrates and is diffracted at various depths, with the result that reflections are broadened and shifted to lower angles. Hence, in contrast to the Debye–Scherrer method, it is desirable to have a highly absorbing specimen in powder diffractometry.

The size of crystallites is critical in diffractometry. Parrish[29] demonstrated, for example, that when silicon crystallites (which are irregular in shape and pack without a preferred orientation) are several times larger than 5 μ, they have a pronounced effect on the intensities of the reflections. Figure 5-44 shows this effect on the intensity of the 111 reflection for specimens of different crystallite sizes rotated at 77 rpm and then at 1/7 rpm about the axis normal to the specimen surface. The rapid rotation gives an average intensity whose fluctuations have an instrumental origin. When the crystallite sizes are less than 5 μ, there is no significant difference between fast and slow rotation. However, the slow-rotation pattern shows an increasing amount of fluctuation from the average intensity as the crystallite size increases, which must be attributed to an increasing variation in the number of diffracting

[29] W. Parrish, *Advances in X-Ray Diffractometry of Clay Minerals*, Proc. Seventh National Clay Conf., Washington (1958).

crystallites with azimuthal angle (ϕ). De Wolff, Taylor, and Parrish[30] made measurements of the peak intensities of the 111α and the $220\alpha_1$ lines given by a 30–50 μ fraction of silicon with the specimen set at ten different stationary azimuthal angles. The 111α measurements ranged from 1345 to 3330 cps and the $220\alpha_1$, from 956 to 1504 cps. The ratios varied from 1.11 to 2.48. When the same specimen was rotated rapidly, the values were $111\alpha = 2145$ and $220\alpha_1 = 1088$ cps; the ratio = 1.97. It is apparent from the work of Parrish and the others that intensity measurements on stationary specimens composed of large ($>5\ \mu$) crystallites are sensitive to the specimen's position in the beam and can show large variations in the relative intensities of the reflections. If stationary specimens must be used, the crystallites should be less than 5 μ in size to obtain an acceptable precision in measurement.

Figure 5-44 shows that rotating the specimen has little effect on the average total intensity. The 5–10 μ fraction gave the highest average intensity. Parrish attributed the lower intensity of the larger sizes to extinction and a poorer packing density. The lower intensity of the $< 5\ \mu$ fraction proved to be caused by a layer of amorphous silica around the grains; the intensity increased when it was removed by selective etching.

Specimen Preparation

The flat surface of the specimen must be of sufficient size to intercept the entire incident beam at all diffraction angles, if the measurement of intensities is an object of the study. The manufacturers of diffractometers supply proper specimen trays, including those that can be inserted in a "flat specimen spinner" and rotated about an axis normal to the specimen's surface. Satisfactory mounts are obtained by packing the trays with finely ground powder, which is wetted to prevent the material from pouring out at steep angles and smoothed to a flat surface level with the edges of the tray. If the material has a cleavage or characteristic grain shape, this procedure will lead to a preferred orientation and therefore to unreliable intensity measurements, whether or not the specimen is rotated. A rapid rotation improves the reproducibility of intensity measurements, as has been shown, but can have only a limited influence in improving the randomness of crystallites. It has been said that very fine grinding (to less than 5 μ) is the single most effective method of combatting preferred orientation; this has the added advantages of improving reproducibility and minimizing the effects of extinction. Alignment due to the smoothing of the surface can be avoided by mixing the powder with a viscous binder, overfilling the tray, and removing the

[30] P. M. de Wolff, J. M. Taylor, and W. Parrish, *J. Appl. Phys.*, **30**, 63 (1959).

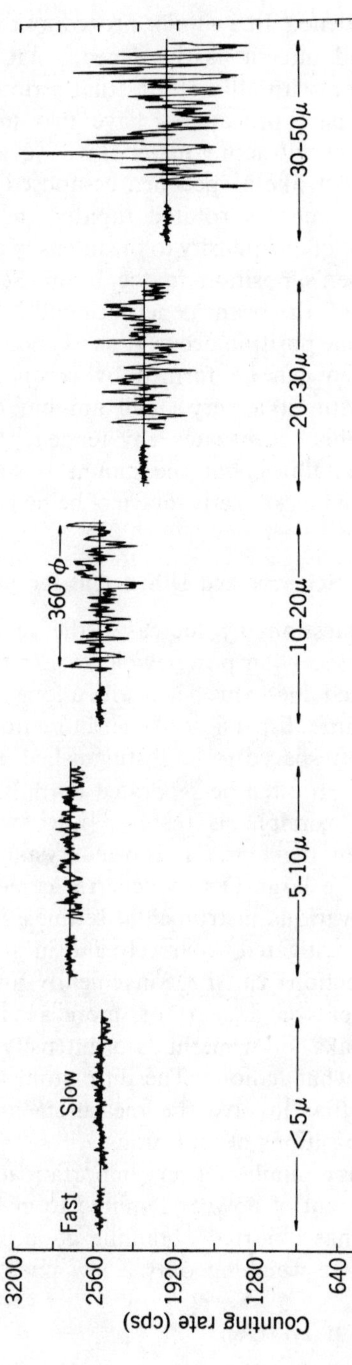

Fig. 5-44. The effect on the 111 reflection of slowly and rapidly rotating silicon crystallites of various sizes (from W. Parrish[29]).

excess after the mixture has hardened. In a similar procedure the mixture is poured onto a glass slide and, after it has hardened, a flat surface is ground on the mass. In common with all methods that seriously dilute the diffractometer specimen, these procedures have two undesirable results. First, the concentration of diffracting material will vary throughout the mixture; hence it will act like a specimen composed of coarse crystallites. Unless such a specimen is rotated rapidly, the absolute intensity of any line and the ratio of its intensity to the intensity of another line can vary with the specimen's position in the beam. Second, the specimen becomes "transparent," the beam penetrates and is diffracted at various depths, and shifts in line position occur. A nearly homogeneous and "nontransparent" specimen can be formed by compressing the powdered material, with or without a very small amount of binder, into a dense cake or pellet. This accentuates any tendency toward a preferred alignment of the crystallites, but the mount is suitable for determining diffraction angles and has the advantage of being permanent.

A Comparison of the Debye–Scherrer and Diffractometer Methods

The diffractometer has unquestionably increased the scope of the powder method. The excellent resolving power which arises from combining the use of a well-collimated line-focus beam with a long sample-to-detector distance results in a large dispersion of the diffraction pattern. Reflections which are too closely spaced to be distinguished on Debye–Scherrer films of standard size can often be separated on diffractometer charts. The $\alpha_1\alpha_2$ doublet, for example, is resolved into two distinct peaks at less than $35° \, 2\theta$ when the material is well-crystallized; this occurs at about $110° \, 2\theta$ with the large Debye–Scherrer camera. When full advantage is taken of the various instrumental settings, interesting sections of a pattern can be investigated separately and in great detail. The relative intensities of reflections can be conveniently and rapidly estimated by direct measurement on a chart or, more accurately, by statistically counting across peaks. Film methods of intensity measurements are elaborate and somewhat tedious. The diffractometer is particularly useful in all studies that involve the measurement of small changes in the intensities and positions of reflections.

The film method, however, has a number of very important advantages. Foremost is the very small amount of powder required to give a usable diffraction pattern. Donnay[31] has reported obtaining good patterns of polycrystalline samples of quartz weighing only a few micrograms. It

[31] J. D. H. Donnay, *Am. Mineralogist*, **31**, 191 (1946).

seems then that the practical minimum size is determined by the technician's skill in mounting and orienting the specimen. This is of inestimable value to the mineralogist who frequently must work with microscopic quantities of minerals. In general, film methods are better for the detection of weak lines. Another desirable feature is the size of the diffraction record. The entire pattern obtained with the small standard-size camera is recorded on a strip of film measuring roughly 1 by 7 inches. Thousands of these patterns can be stored and arranged for quick reference in a small cabinet. The small size of the film also makes it convenient to compare a number of diffraction patterns. Finally, the Debye–Scherrer method is capable of yielding accurate lattice spacings if the material is well-crystallized. The apparatus is uncomplicated, and the determination is straightforward. In contrast, the intensity peaks detected by the diffractometer are displaced from their theoretical positions by a number of instrumental factors, which include the use of a flat rather than a curved specimen, the divergence of the primary beam, and its penetration and diffraction at various depths in the sample. Practical difficulties are encountered in aligning the surface of the specimen exactly tangent to and on the focusing circle and in precisely locating the 0 and 180° 2θ positions. The geometrical aberrations and alignment errors combine in a complex manner to give cumulative errors, which exceed the sum of the individual errors and are difficult to evaluate. The measured 2θ values, for example, cannot be refined with the same assurance of reliability associated with the Debye–Scherrer method, in which a single parameter $[\frac{1}{2}(\cos^2\theta/\sin\theta + \cos^2\theta/\theta)]$ extrapolation function applies in the back-reflection region.

CHAPTER SIX

ORIENTATION AND PROJECTION OF MORPHOLOGICAL CRYSTALS

Orientation of Crystals by Optical Methods

Single-crystal x-ray methods require that a chosen direction in the crystal be precisely oriented in the x-ray beam. The direction may be the axis of a prominent zone,[1] the normal to a prominent crystal face or easy cleavage, an edge between two faces, a few ill-defined growth striations, or an optic axis apparent in transmitted polarized light. Because x-ray crystals measure between 0.1 and 0.5 mm in average dimension and need to be adjusted to within about three minutes of arc of the desired orientation, refined equipment is necessary to orient the crystal and to hold it in the beam. The adjustment can be made on an x-ray goniometer, but, if the crystal is reasonably well-developed and has light-reflecting faces, this can be done more rapidly on an optical goniometer. A poor crystal may allow only a preliminary optical adjustment, in which the final precision adjustment is made on the x-ray goniometer, but even this provides a considerable saving.

Optical Goniometers

Two types of optical goniometer are in common use for the adjustment and measurement of crystals. Figure 6-1 is a simplified plan drawing of a *one-circle* type. It consists of a horizontal circle (1), which carries the crystal holder (2), and rotates in a fixed stand (3). A collimator (4), which is provided with a signal slit, directs a beam of light onto the crystal. The telescope (5) contains cross-hairs and views the crystal or the light image reflected from the crystal. The circle is read on the vernier (6). Figure 6-2 is a similar drawing of a *two-circle* goniometer. It resembles a

[1] A zone comprises a set of planes whose intersecting edges are parallel. The direction defined by the edges is called the *zone axis*.

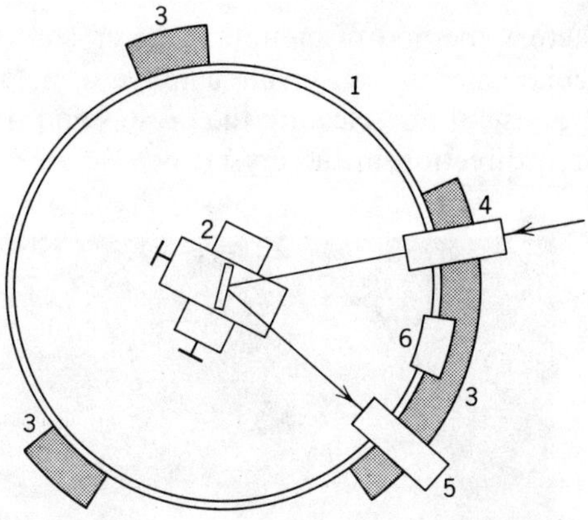

Fig. 6-1. A one-circle goniometer.

one-circle instrument (explanatory notes 1 to 6 as in Fig. 6-1) except that a vertical circle (7), rotates on the axis of the horizontal circle and carries the crystal holder. This circle is read on the vernier (8). Figure 6-3 is a photograph of a crystal holder or *goniometer head*, especially compactly built for x-ray work. The instrument is provided with two mutually perpendicular arc movements (and for this reason is sometimes called "a pair of arcs") and two translation movements for adjusting and centering the crystal on the optical goniometer.

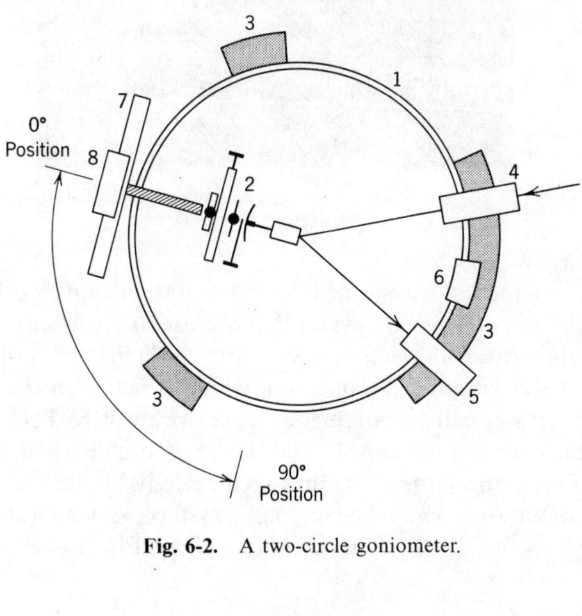

Fig. 6-2. A two-circle goniometer.

One satisfactory method of mounting the crystal is illustrated in Fig. 6-4*a*. A binocular microscope, giving a range of magnification of 10 to 50 diameters, is useful for selecting the crystal and fixing it to the glass rod. The chosen direction in the crystal needs to be parallel to the rod

Fig. 6-3. An x-ray goniometer head. (Courtesy Unicam Instruments Ltd., Cambridge, England.)

in some x-ray methods described here and normal to it in others. Figure 6-4 also shows two special devices that are useful for meeting this condition in certain cases. In the first device (Fig. 6-4*b*) the rod is attached with a length of soft wire to a second rod, which is seated in the goniometer head. The crystal can be rotated, and the wire can be bent to vary the orientation over a wide range. The device is useful when the angle of reorientation is too large to be managed entirely by the arc adjustments and for mounting poorly developed crystals in which the required orientation is in doubt. The second device (Fig. 6-4*c*), designed by

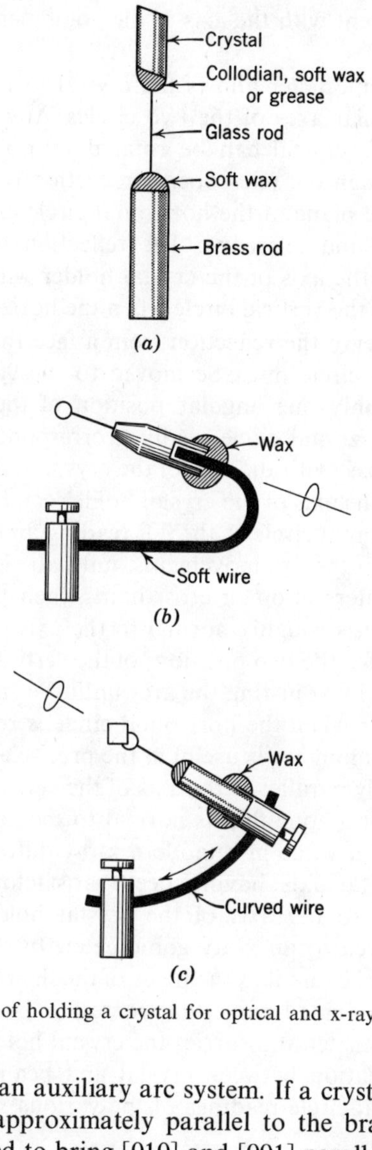

Fig. 6-4. Methods of holding a crystal for optical and x-ray goniometric study.

Grenville-Wells,[2] is an auxiliary arc system. If a crystal is mounted with an axis, say [100], approximately parallel to the brass rod, it must in general be remounted to bring [010] and [001] parallel to the axis of the goniometer head. Remounting can be avoided with this device by cementing the crystal with its [111] direction parallel to this axis. If the rod is then slid along the curved wire to about the 45° position, three orthogonal directions in the crystal can be successively brought into

[2] H. J. Grenville-Wells, *Acta Cryst.* **8**, 519 (1955).

approximate alignment with the axis of the goniometer head by rotating the rod through 360°.

The two-circle goniometer allows the crystal to be turned about the mutually perpendicular axes of the two circles. Any face except one on the underside of the crystal can be rotated into a position to reflect the light beam through the telescope. A reflection is observed when the face-normal is in the plane of the horizontal circle and bisects the angle between collimator and telescope. The reflection from a face that is precisely normal to the axis of the crystal holder will be observed at all angular positions of the vertical circle when the horizontal circle is in the 0° position. To observe the reflection from a face that is parallel to the axis, the horizontal circle must be moved to the 90° position; the reflection occurs at only one angular position of the vertical circle. (It follows that the horizontal circle reading corresponds to the angle that the face-normal makes with the axis of the crystal holder.) A zone that is roughly parallel to the axis of the crystal holder can be precisely oriented by setting the horizontal circle at the 90° reading and adjusting the slope of the zone axis with the arc movements, until all signals from the zone pass through the intersection of cross-hairs when the vertical circle is rotated. A zone that is roughly normal to the axis of the crystal holder can be precisely set for the two positions of the vertical circle in which the zone axis is vertical by adjusting the arcs until the signals remain on the horizontal cross-hair when the horizontal circle is rotated.

The one-circle goniometer is useful in the precise adjustment of a zone that is approximately parallel to the axis of the crystal holder but cannot be used on a face or a zone that is normal to the axis. This goniometer therefore is of limited value in a modern x-ray diffraction laboratory.

The chosen crystal axis having been satisfactorily oriented either parallel or normal to the axis of the crystal holder, the holder and crystal are transferred to an x-ray goniometer. In the operation of this instrument an x-ray beam takes the place of the light beam. It is generally equipped with a graduated circle that corresponds to the vertical circle of the optical goniometer and carries the crystal holder. To maintain the exact azimuthal relation between crystal and beam, it is necessary to calibrate the vertical-circle readings of the various x-ray goniometers in the laboratory.

Methods of Projection

The measurement of a well-formed crystal on an optical goniometer can lead to a determination of the symmetry of the crystal and its unit cell. The study can, of course, provide only the relative dimensions of the cell, but it may be of great value in planning the strategy of the x-ray attack on the absolute dimensions when there is doubt about how to orient the crystal in the beam.

The study of a morphological crystal is simplified by representing its faces with a projection that preserves the essential angular relations between the faces while eliminating such trivial features as the size and shape of the faces. The projections are equally useful for displaying the angular relations between reflecting lattice planes in some of the x-ray methods described in this book. The projections represent the *normals*

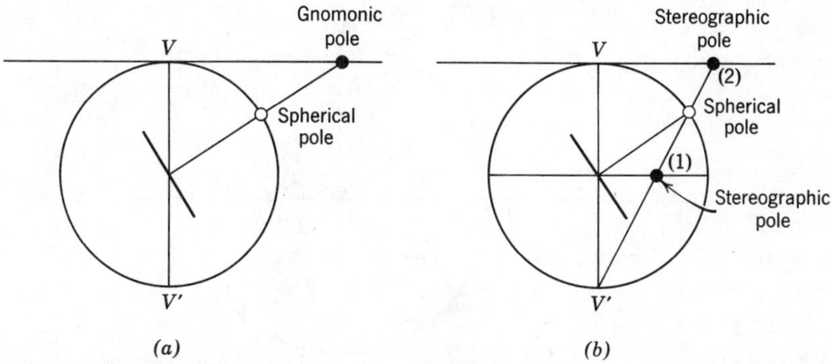

Fig. 6-5. The projection of a plane: (*a*) gnomonic projection; (*b*) stereographic projection showing alternate positions for the plane of projection—(1) through center of sphere and (2) tangent to sphere.

to reflecting faces, or lattice planes, on a plane. The center of the idealized crystal, or the origin of the lattice, coincides with the center of a reference sphere. Normals are drawn from the center of the sphere to intersect its surface in a set of *poles*. The poles are then projected onto a plane situated at right angles to a chosen reference axis (VV' of Fig. 6-5). The point of projection is on this axis and at the center of the sphere in the gnomonic projection (Fig. 6-5*a*) or at the south pole in the stereographic projection (V' of Fig. 6-5*b*). The stereographic plane of projection may be taken either through the center of the sphere or tangent to it. The first position is somewhat more convenient for projecting the poles corresponding to opposite ends of a normal—that is, the poles representing a pair of faces on opposite sides of a morphological crystal, or the opposite surfaces of a stack of lattice planes.

The stereographic projection is more useful for displaying the orientation of crystals and consequently only brief reference is made here to the gnomonic projection.

The Stereographic Projection

The stereographic projection is illustrated in greater detail in Fig. 6-6. The figure is a section through the center of the sphere and normal to the stereographic plane which here coincides with the equator of the sphere. PP' is one member of a stack of parallel and identical *hkl* planes which are also at right angles to the section. The normal to these planes intersects the upper hemisphere at N and the lower hemisphere at N'. The stereographic pole corresponding to the upper surface of the stack is found by joining N to V'. The join crosses the plane of the projection at

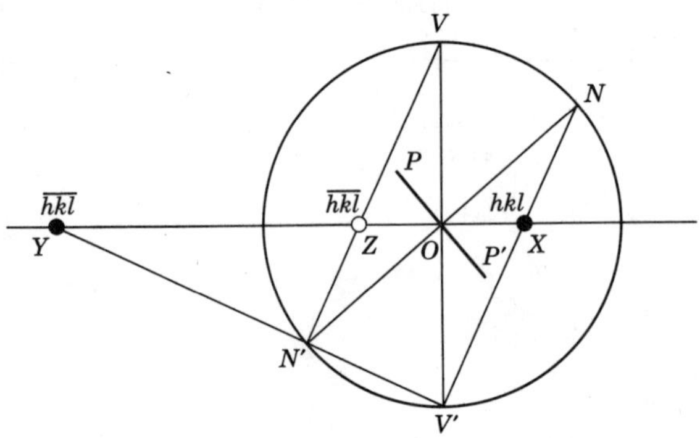

Fig. 6-6. The stereographic projection of both surfaces of a plane.

X, and this point, designated by a filled circle, is the stereographic pole *hkl*. The stereographic pole corresponding to the lower surface of the stack is found by joining N' to V' and extending the line until it intersects the projection plane. This point, Y, is appropriately designated by the indices $\overline{hkl}$. The area of the projection is unduly extended by this method of projecting planes whose normals intersect the lower hemisphere, especially when the planes are nearly parallel to the equator. A preferable method, which has almost universal acceptance, consists of joining N' to V and designating the point Z at which the join crosses the plane of projection by a distinguishing symbol, such as an open circle. In this manner all planes are projected within the confines of the equatorial or primitive circle.

This description has dealt with the projection of general planes that are inclined to both the vertical reference axis and to the plane of projection. Two specially-oriented types of plane are worthy of mention. The normals to planes that are parallel to the vertical axis lie in the plane of the projection. Hence the stereographic poles of these planes occur on

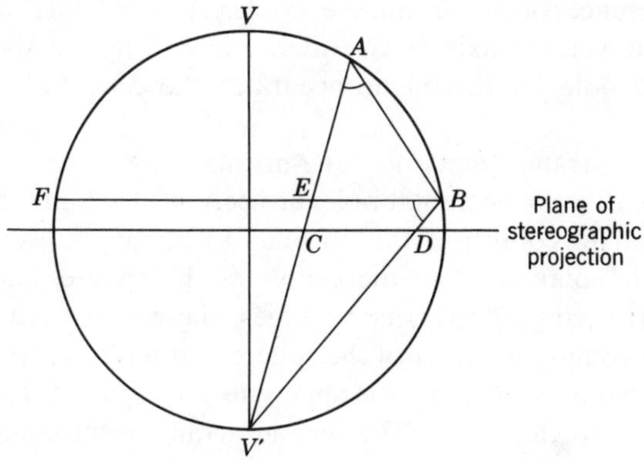

Fig. 6-7. The stereographic projection of a circle on the surface of the sphere.

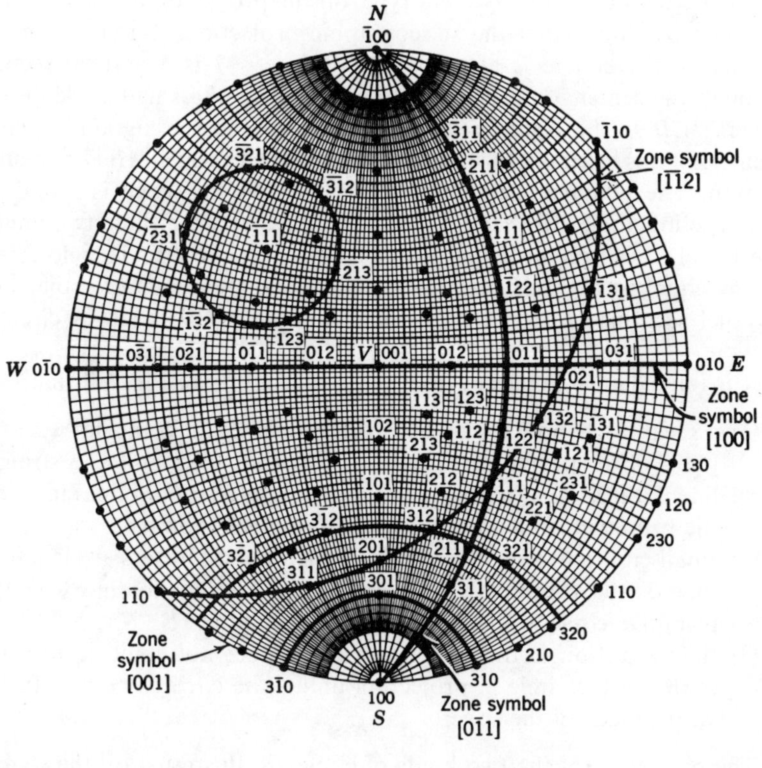

Fig. 6-8. The stereographic projection of cubic planes showing small and great circles.

the circumference of the primitive circle. A second special plane is normal to the vertical axis. A consideration of Fig. 6-6 shows that the stereographic pole for this plane occurs at the center of the primitive circle.

Two special arrangements of plane-normals are of particular interest in the application of stereographic projections to Laue photographs. One arrangement comprises the normals to planes in a zone. Because plane-normals emanate from the center of the sphere, and the planes in a zone have parallel intersection edges, these normals lie in a plane that passes through the center of the sphere and intersects the sphere in a circle of the same diameter as the sphere (a great circle). The axis of the zone is normal to this plane. The second arrangement comprises plane-normals which are equally inclined at an angle of less than 90° to an axis. The normals outline a cone that intersects the sphere in a circle whose diameter is less than that of the sphere (a small circle). Consequently, the stereographic poles of the first type fall on the projection of great circles and of the second type, on the projection of small circles.

It can be shown that the stereographic projection of a circle on the surface of the sphere is also a circle. Figure 6–7 is a vertical section through the center of the sphere. AB is the trace of a small circle on the sphere; CD is the trace of its projection on the stereographic plane. Hence $AV'B$ is a section through a general cone, of which AB is a circular section. The conjugate circular section, EB, in this cone is found by constructing angle EBV' equal to angle BAV'. Because these two angles are equal, they rest on equal arcs of the circle $ABV'F$. Therefore, arc FV' is equal to arc $V'B$, and EB is parallel to the plane of projection. Parallel sections of a cone are similar, and thus CD is also a circular section.

It is immediately apparent from the construction of Fig. 6-7 that

(1) a great circle projects as the primitive circle if it coincides with the plane of the projection, as an arc if it is inclined, and as a straight line if it is normal to this plane; the arcs and straight lines intersect the primitive circle at diametrically opposite points;
(2) small circles project as circles of which the diameters are less than that of the primitive circle or as arcs that do not intersect the primitive circle in diametrically opposite points;
(3) the projection of the center of a circle does not coincide with the center of the circle in projection unless the circle is normal to the vertical axis of the sphere.

The projection of the two kinds of circles is illustrated on the stereographic projection of cubic planes in Fig. 6-8. Four great circles in projection, including the primitive circle, marking the four zones of planes [001], [100], $[0\bar{1}\bar{1}]$, and $[\bar{1}\bar{1}2]$,[3] have been indicated on the figure. Three small circles in projection are also shown. Two of the small circles are normal to the [100] axis of the lattice; the third circle is normal

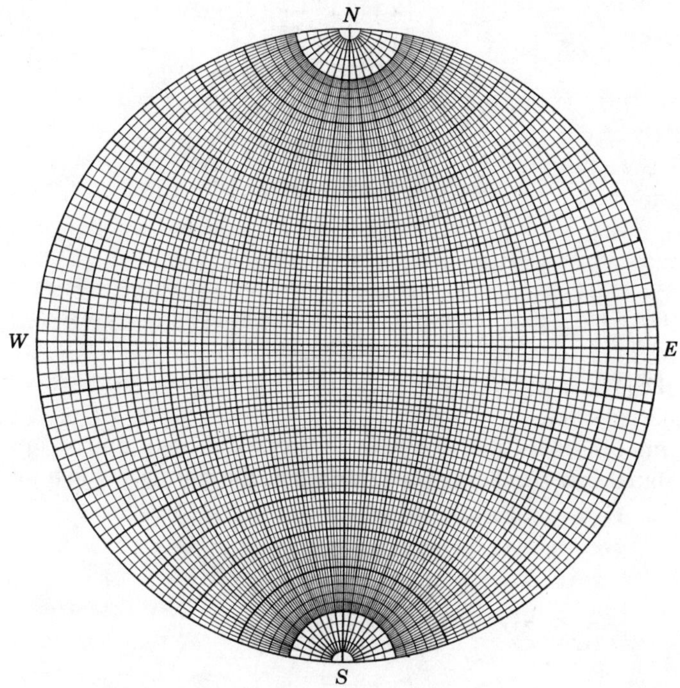

Fig. 6-9. The Wulff net consisting of the projection of a system of great circles having the *N-S* axis as common diameter, and a system of small circles whose centers lie on the same axis. The circles are spaced at 2° angular intervals.

to the $[\bar{1}\bar{1}\bar{1}]$ axis and most clearly shows that the projection of the center of the circle does not coincide with the center of the circle in projection, when the circle is inclined to the vertical axis. The center of this circle

[3] The zone symbol for two planes (e.g., $3\bar{1}1$ and 201) is found by a type of cross multiplication. Each index is written twice; the pairs joined by the strokes of a cross are then multiplied, and the product of an upper right and a lower left is subtracted from the other product:

$$\begin{array}{c|ccccc|c} 3 & \bar{1} & 1 & 3 & \bar{1} & & 1 \\ & & \times & \times & \times & & \\ 2 & 0 & 1 & 2 & 0 & & 1 \\ \hline & & [\bar{1} & \bar{1} & 2] & & \end{array}$$

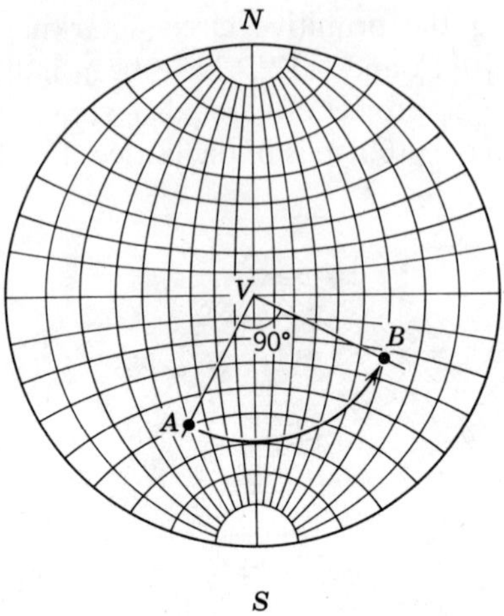

Fig. 6-10. To rotate a pole A 90° anticlockwise about the vertical axis.

lies on the $[\bar{1}\bar{1}\bar{1}]$ axis; the pole of the axis is obviously not at the center of the projected circle.

One important feature of circles in projection may not be apparent. It was stated in a preceding paragraph that the normals which outline a

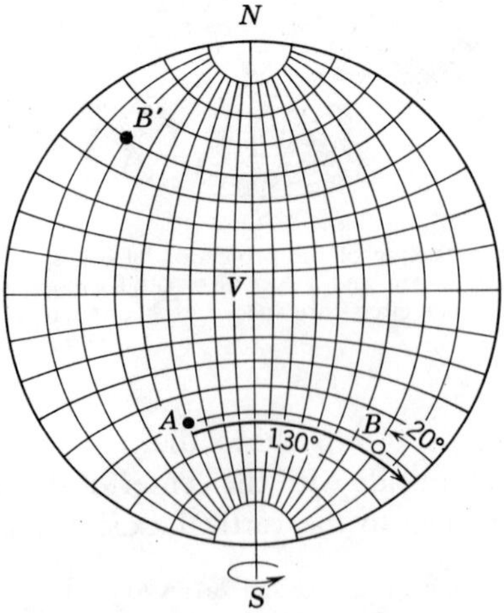

Fig. 6-11. To rotate a pole A 150° from west to east about the N-S axis. A rotation of 130° moves the pole along its small circle to the edge of the projection. At this point the plane-normal lies in the plane of the projection. A further rotation causes the normal to intersect the lower half of the stereographic sphere; the pole moves back along the small circle and reaches its position B within 20°. The position is appropriately shown by an open circle. Simultaneously, the pole representing the opposite end of the normal reaches B', diametrically opposite to B.

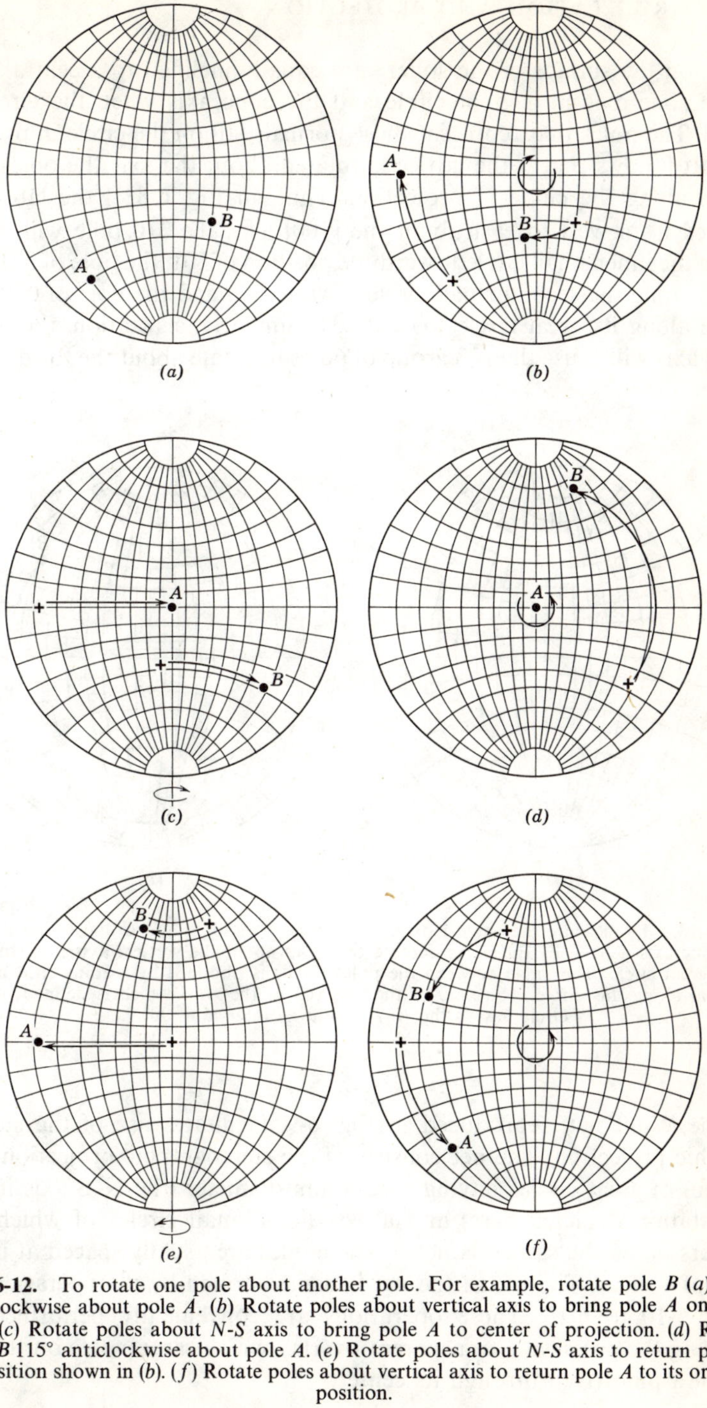

Fig. 6-12. To rotate one pole about another pole. For example, rotate pole *B* (*a*) 115° anticlockwise about pole *A*. (*b*) Rotate poles about vertical axis to bring pole *A* on *W-E* axis. (*c*) Rotate poles about *N-S* axis to bring pole *A* to center of projection. (*d*) Rotate pole *B* 115° anticlockwise about pole *A*. (*e*) Rotate poles about *N-S* axis to return pole *A* to position shown in (*b*). (*f*) Rotate poles about vertical axis to return pole *A* to its original position.

circle on the surface of the sphere are equally inclined to a central axis —the zone axis in a great circle and the cone axis when the circle is small. The poles that represent these normals fall on the circle in projection. It follows that if the lattice is rotated about the axis the poles will travel along the circle in projection. Thus, in Fig. 6-8, if the lattice is rotated from west to east about the [100] axis, the $3\bar{1}0$ pole will move along the inner small circle toward the position of 301, the pole of $3\bar{2}1$ will move along the second small circle toward $3\bar{1}2$, and the pole of $0\bar{1}0$ will move along its great circle toward $0\bar{3}1$. Similarly, a rotation about the $[\bar{1}\bar{1}1]$ axis will cause the $\bar{1}23$ group of poles to rotate about the third small circle.

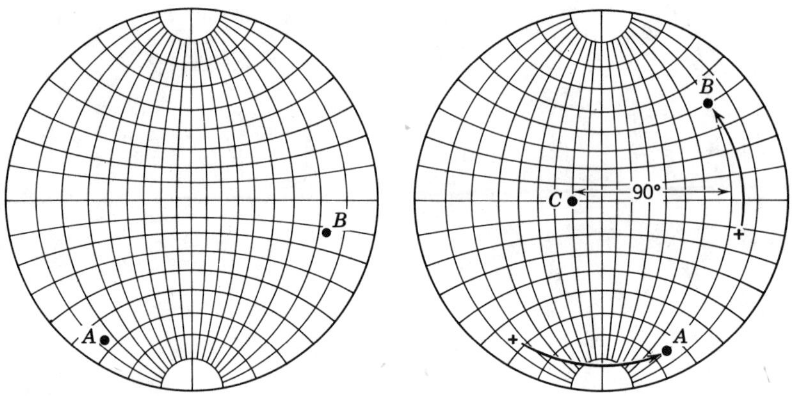

(a) (b)

Fig. 6-13. To measure the angle between two poles (*A* and *B*) and to locate the pole of the zone axis defined by the poles. Rotate the poles about the vertical axis until they lie on a great circle. The angle between the poles (105°) is obtained by reading the angular interval along this circle. The pole of the axis (*C*) of the zone which is defined by this circle lies on the *W-E* axis at an interval of 90°.

The Wulff net, reproduced in Fig. 6-9, facilitates use of the stereographic projection. The net consists of the stereographic projection of a system of great circles having a common diameter (the N–S axis in the orientation depicted here) and a system of small circles of which the centers lie on the same diameter. The circles are usually spaced at intervals of 1 or 2°. The use of the net is demonstrated by the operations in Figs. 6-10 to 6-13. These operations are conveniently performed by plotting the poles on a transparent overlay which can revolve on the net about a pin driven through its center.

The Gnomonic Projection

One distinctive feature of the gnomonic projection is immediately apparent. The normals to the faces of a zone lie in a plane, and when extended to intersect the plane of the gnomonic projection they give a set of collinear poles. Thus a zone of faces projects as a row of poles. This is in contrast to the stereographic projection, in which the poles of a zone outline a great circle.

The gnomonic projection is useful in morphological studies that seek to deduce the shape and symmetry of the unit cell and to index the crystal faces. A relation between this projection and the reciprocal lattice is described in chapter 8.

CHAPTER SEVEN

THE LAUE METHOD

The Laue method is the original x-ray diffraction method—it was used by Friedrich and Knipping under von Laue's direction, to test the diffraction of x-rays with a crystal of copper sulfate. The method makes use of a stationary single crystal, which is irradiated by an x-ray beam consisting of a range of wavelengths. The radiation from a tungsten target is well-suited to this purpose because it includes a high percentage of noncharacteristic radiation. However, the unfiltered beams from diffraction tubes used for powder studies contain enough white radiation to give satisfactory results and are often used for Laue work at the cost of a longer exposure.

Diffraction by a crystal structure is geometrically equivalent to reflection from lattice planes in accordance with the Bragg condition

$$n\lambda = 2d_{hkl} \sin \theta$$

Because a Laue crystal is stationary, the orientation of each lattice plane is fixed in the x-ray beam. Hence the reflection angle θ is fixed for each set of planes, and thus for a set of planes of spacing d_{hkl}, the value of $2d_{hkl} \sin \theta$ is also fixed and the only variables are n and λ. Now, n must be an integer, and therefore the set of planes can reflect only specific components of the white beam, of wavelengths $\lambda_1, \lambda_2, \ldots, \lambda_n$, such that

$$1\lambda_1 = 2\lambda_2 = n\lambda_n = 2d_{hkl} \sin \theta$$

This relation is designated as

$$\lambda_1 = 2d_{hkl} \sin \theta$$

$$\lambda_2 = 2d_{2h \cdot 2k \cdot 2l} \sin \theta$$

$$\lambda_n = 2d_{nh \cdot nk \cdot nl} \sin \theta$$

in the terminology of x-ray crystallography and is equivalent to stating that the various orders of reflections which emanate from a set of planes are given off at the same angle and form only one Laue spot. This limits the usefulness of the Laue method for the determination of systematically missing reflections, hence space group. Furthermore, each set of planes in the crystal is characterized, in general, by a different spacing and, as a consequence, reflects a different set of wavelength components. Because it is not practical to determine the values of the wavelengths for a particular reflection, it is not possible to calculate the spacings of the reflecting planes. The method gives only a series of θ values for the different crystal planes, and this leads to the shape, not the size, of the unit cell.

The usefulness of the method lies in the fundamental diffraction property that the directions of possible reflections depend on the crystal's orientation in the beam. A crystal that is symmetrically oriented to the primary beam gives rise to reflected beams which are symmetrically arranged about the beam. When the reflections are recorded on an appropriately situated film, the photograph portrays the orientation of the crystal and its diffraction symmetry in this orientation. A slight misalignment from a symmetrical position is readily detected. This application is especially valuable for determining the orientation of crystals. Usually a knowledge of the indices of the reflecting planes is a prerequisite to such a determination.

The pattern of reflections which emanates from a Laue crystal may be conveniently and usefully studied in terms of the reflections from zones of planes in the crystal. The reflection directions for the planes of a zone lie on the surface of a cone whose axis is the cone axis. Figure 7-1 illustrates reflection cones for zone axes inclined to the incident beam at less than 45°, at 45°, between 45 and 90°, and at 90°. The undeviated portion of the primary beam is always a line in the cone.

The apparatus for preparing Laue photographs is exceedingly simple. It consists essentially of a pinhole system for collimating the incident beam, a crystal holder, and a film casette. The components are normally mounted on a set of tracks to facilitate alignment. The reflections are usually recorded on a flat film oriented normal to the primary beam. This offers two kinds of instrumental arrangements.

The Transmission Method

In this arrangement (Fig. 7-2) the specimen is placed between the x-ray source and the film. A basic requirement of the method is a specimen of sufficiently low absorption to allow the reflected beams to pass through the specimen and reach the film. In the study of minerals a thin slice,

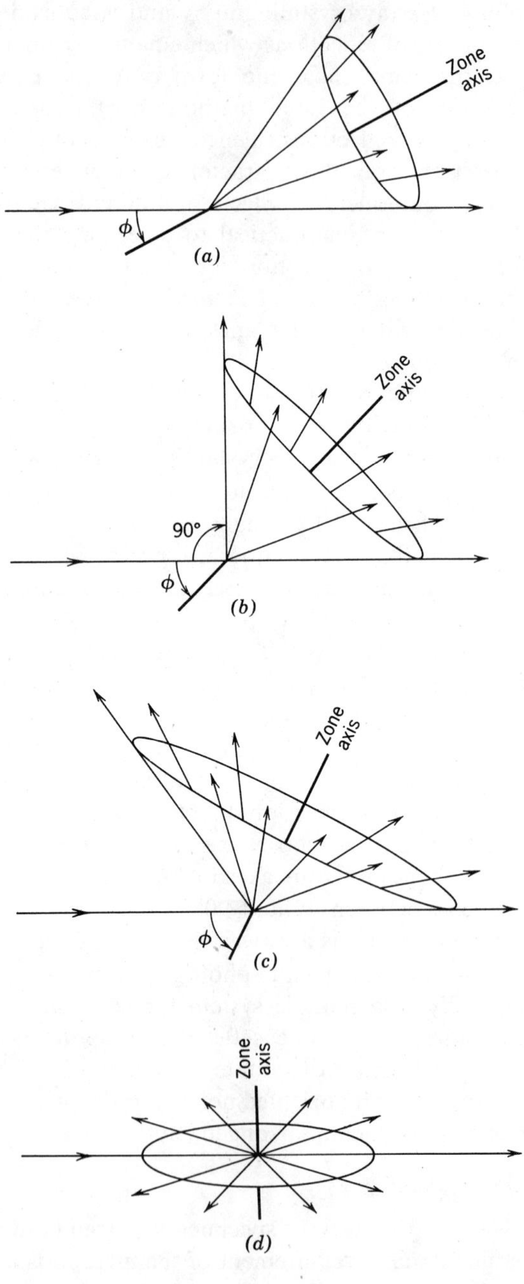

Fig. 7-1. Laue reflection cones from zones of planes of which the zone axes are inclined at various angles to the x-ray beam. (*a*) $\phi < 45°$, (*b*) $\phi = 45°$, (*c*) $\phi > 45°$, and (*d*) $\phi = 90°$.

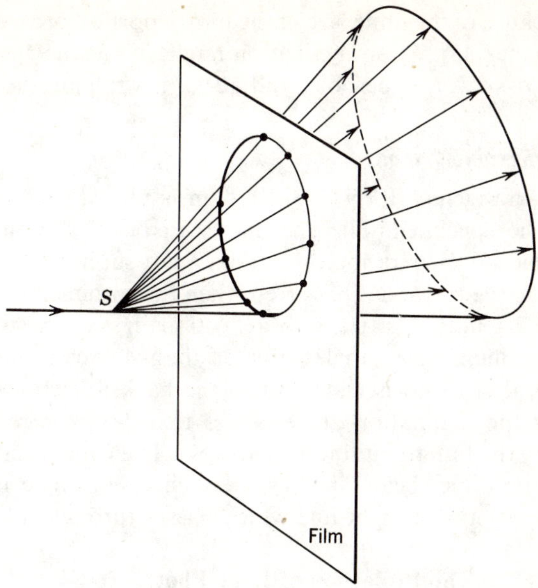

Fig. 7-2. The transmission Laue method—sample at *S*.

cleavage flake, or small crystal is usually chosen. The size depends on its absorptive properties. The various reflection cones illustrated in Fig. 7-1 are intercepted by the film in a series of conic sections, all of which pass through the direct beam spot on the film. When the zone axis is inclined at less than 45° to the incident beam, the diffraction spots outline an

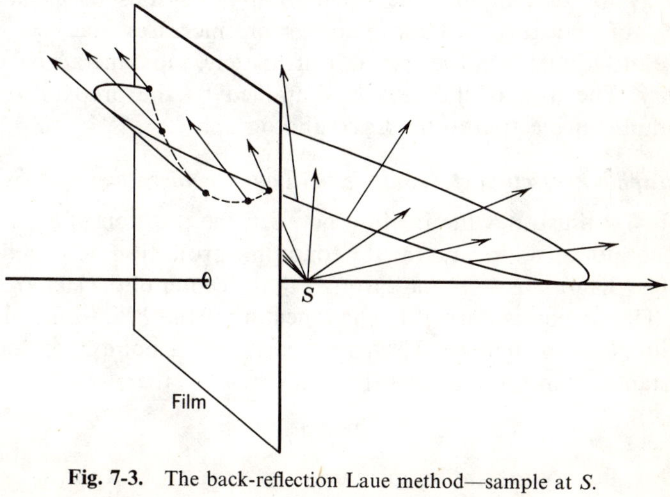

Fig. 7-3. The back-reflection Laue method—sample at *S*.

ellipse or, because of the finite size of the film, a portion of an ellipse as the angle approaches 45°. At an inclination angle of 45° the spots outline a parabola, above 45°, a hyperbola, and at 90° a straight line.

The Back-Reflection Method

With this arrangement (Fig. 7-3) the film is placed between the x-ray source and the specimen. The specimen can be of any thickness and requires no special preparation other than a flat surface facing the beam. For this reason the method is preferred when the specimen is, of necessity, large. Figure 7-1 indicates that only reflections from zones of which the zone axes are inclined at angles greater than 45° can reach the film. Furthermore, the record is restricted to the back reflections from these zones. When the inclination angle is less than 90°, the reflections are intercepted by the film to outline a hyperbola. The hyperbola flattens and approaches the direct beam spot as the inclination angle increases. At 90° the spots lie on a straight line which passes through the film center.

The Projection of Laue Photographs

In general, the task of indexing a Laue pattern is more difficult, and the results are less certain, than with the Weissenberg and precession single-crystal methods described in subsequent chapters. However, these methods require very small crystals, measuring about 0.2 mm in cross section. Because the thickness and opacity of the specimen is immaterial in the Laue back-reflection technique, this method is peculiarly suited to the study of large crystals; hence it is extensively used in the field of metallurgy for determining the orientation of crystals in metals. The difficulty of indexing a Laue photograph increases markedly with decrease of symmetry in the crystal. Fortunately, most metals have high symmetry. The task of indexing is simplified by the preparation of a stereographic projection of the Laue photograph.

Stereographic Projection of Transmission Laue Photographs

Figure 7-4 illustrates the relation between the position of a transmission Laue spot on a film and the stereographic projection of the reflecting plane. The diagram is a section through O, the center of the stereographic sphere. The section is normal to the reflecting plane PP', to the plane of projection, and to the film. The plane reflects to a point S on the film. The distance from the center of the film C to S is, therefore

$$F \tan 2\theta$$

where F is the crystal-to-film distance.

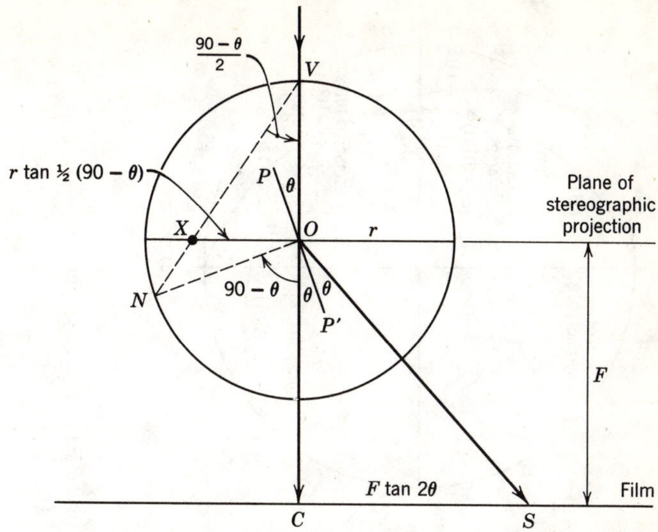

Fig. 7-4. Relation between a transmission Laue spot and the stereographic projection of the reflecting plane.

The normal to the reflecting plane intersects the sphere at N. Its stereographic pole X is found by joining N to the point V, at which the incident beam enters the sphere. Angle NOC is $(90 - \theta)$, and therefore angle NVC is $\frac{1}{2}(90 - \theta)$. Hence OX, the distance from the center of the projection to the pole, is

$$r \tan \tfrac{1}{2}(90 - \theta)$$

where r is the radius of the stereographic sphere.

The projection of a Laue photograph is conveniently performed with the aid of a stereographic scale (Fig. 7-5), which directly converts film measurements (values of $F \tan 2\theta$) to distances on a stereographic projection $[r \tan \frac{1}{2}(90 - \theta)]$. The scale is constructed by marking on the left of a central point values of $F \tan 2\theta$ for 2θ angles ranging from 0 to 60° (which is about the practical limit for a 5 × 7 inch film); values of the projection distance, $r \tan \frac{1}{2}(90 - \theta)$, calculated for an arbitrarily chosen value of r and the same range of 2θ angles, are marked to the right of the central point. Data for the construction of these scales are given

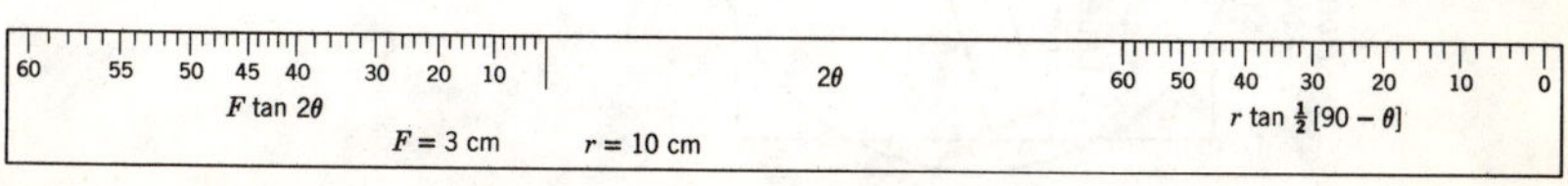

Fig. 7-5. Transmission Laue stereographic scale. $F = 3$ cm, $r = 10$ cm, 2/5 actual size.

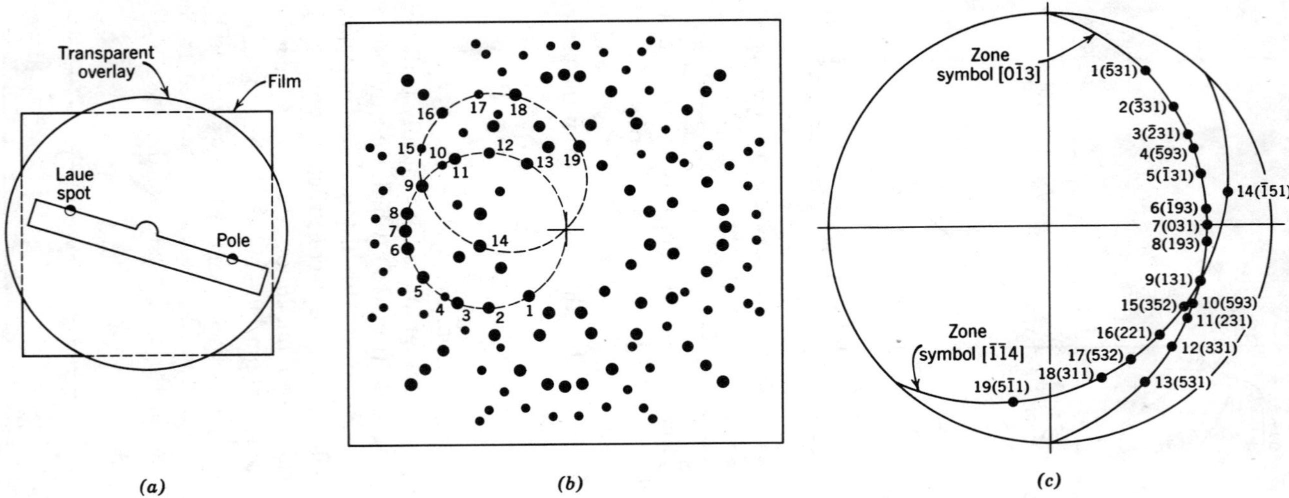

Fig. 7-6. Projection of a transmission Laue photograph with the aid of a stereographic scale.

in Table A5-1. The film and a transparent overlay are fastened to a viewing screen, and a pin is driven through the central point of the scale and the center of the film. The scale is then rotated until the left edge passes through the center of a Laue spot; the 2θ value is read from the scale, and the stereographic pole is marked on the overlay at the corresponding 2θ position on the right side of the scale, as indicated in Fig. 7-6*a*.

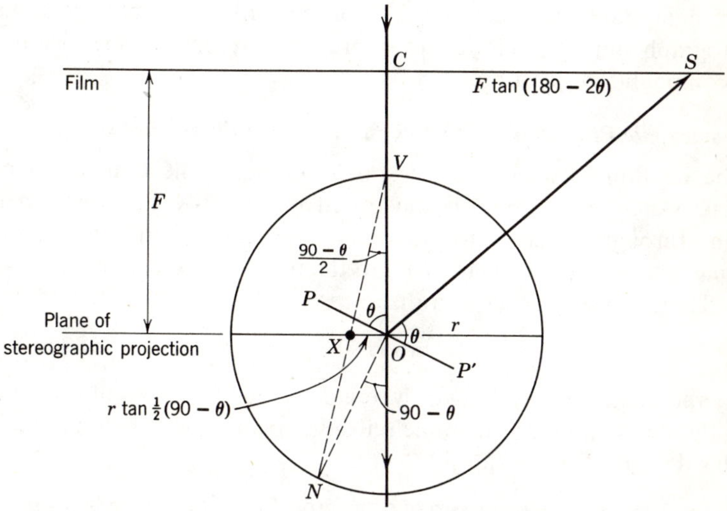

Fig. 7-7. Relation between a back-reflection Laue spot and the stereographic projection of the reflecting plane.

The projection of Laue photographs by the method just described is illustrated in more detail in Figs. 7-6*b* and *c*. Figure 7-6*b* is a tracing of a transmission Laue photograph of the cubic mineral spinel ($MgAl_2O_4$). The crystal was oriented in the beam with fourfold symmetry axes parallel to the N–S and W–E axes of the film. The third fourfold axis, which coincides with the x-ray beam direction, is clearly evident from the arrangement of spots about the center of the film. As stated in an earlier paragraph, the reflections from zones of planes lie on conic sections which pass through the direct beam spot on transmission photographs. Two well-defined zones of reflections, indicated by ellipses on the tracing, have been chosen to demonstrate the relation between the Laue photograph and its stereographic projection. The stereographic poles, corresponding to the two sets of Laue spots, fall on the projections of two great circles (Fig. 7-6*c*). The indices of most of the reflections in the two zones are immediately apparent when the projection is compared with the standard projection of cubic planes depicted in Fig. 6-8. The remaining reflections are usually readily indexed with the aid of a set of cubic angle tables, available in many crystallographic texts.

No one standard projection serves as a reference for all tetragonal or all hexagonal crystals. The interplanar angles of planes which are inclined to the c axes of these compounds depend on the c/a ratios of the particular crystals. Hence the positions of the poles of these general planes vary from compound to compound. Consequently, the projected Laue photograph must be referred to a projection prepared specifically for the compound in question.

Stereographic Projection of Back-Reflection Laue Photographs

The relation between a Laue spot on a back-reflection photograph and its stereographic pole is illustrated in Fig. 7-7. The diagram is a section through O, the center of the stereographic sphere. The section is normal to the reflecting plane PP', to the plane of projection, and to the film. The Laue spot S lies at a distance

$$F \tan (180 - 2\theta)$$

from the center of the film C where F is the crystal-to-film distance. X is the stereographic pole of the reflecting plane; the pole to projection-center distance OX is $r \tan \frac{1}{2}(90 - \theta)$.

BACK-REFLECTION LAUE STEREOGRAPHIC SCALE. Back-reflection Laue photographs may be projected with stereographic scales prepared by plotting values of $F \tan(180 - 2\theta)$ to the left of a central point and values or $r \tan \frac{1}{2}(90 - \theta)$, to the right (Fig. 7-8). Because of the finite size of the film, the 2θ range extends from 180 to about 120°. The stereographic poles of planes reflecting in this range lie close to the center of the primitive circle of the projection. This is in contrast to poles of front-reflecting planes ($2\theta = 0$ to 60°) which are near the circumference of the circle. Hence back-reflection projections are more crowded when plotted on the same scale and indexing is, in general, less reliable.

GRENINGER CHART. Figure 7-9 shows the reflection OS, from one plane of a zone defined by the axis OZ, intersecting a film at S. OZ lies in a plane that includes the x-ray beam CO and a film axis CY and is normal to the second film axis CX. ON is normal to the reflecting plane. The normals to other planes in the zone intersect the film on a line LL', parallel to CX. The reflections from these planes lie on a hyperbola HH'.

The position of S can be described by the film coordinates x and y. The orientation of the plane-normal ON is defined by the angular coordinates α and β, which enable one to plot the pole of ON on a stereographic

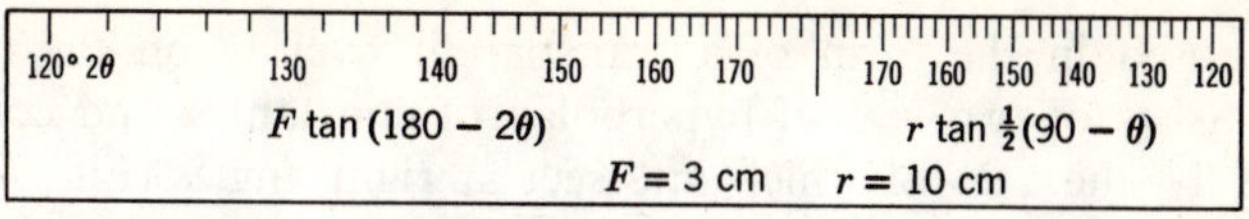

Fig. 7-8. Back-reflection Laue stereographic scale. $F = 3$ cm, $r = 10$ cm, 2/5 actual scale.

projection. Because the position of S depends on the orientation of ON, it is possible to express α and β in terms of x and y. From Fig. 7-9

$$x = SC \sin \mu \qquad y = SC \cos \mu$$

where SC is the Laue-spot-to-film distance. $SC = F \tan(180 - 2\theta)$, in which F is the crystal-to-film distance, and therefore

$$x = \tan(180 - 2\theta) \sin \mu$$

$$y = F \tan(180 - 2\theta) \cos \mu$$

The angle μ can be expressed as $\tan \mu = AN/AC$, but $AN = OA \tan \alpha$, $OA = F/\cos \beta$, and $AC = F \tan \beta$. Therefore

$$\tan \mu = \frac{F \tan \alpha}{F \cos \beta \tan \beta} = \frac{\tan \alpha}{\sin \beta}$$

These relations were used by Greninger to devise a chart which gives directly the α and β coordinates corresponding to a Laue spot, when it is

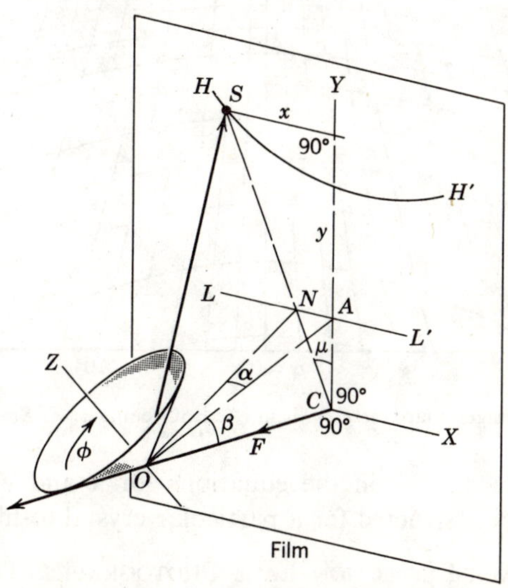

Fig. 7-9. Relation between the film coordinates x and y of a back-reflection Laue spot and the angular coordinates α and β of the plane-normal.

superimposed in the form of a transparent overlay on the film. The chart consists of two sets of hyperbolae of constant α and constant β, respectively, the axes of which intersect at right angles (Fig. 7-10). The hyperbolas are spaced at angular intervals of 2°. The presence of the

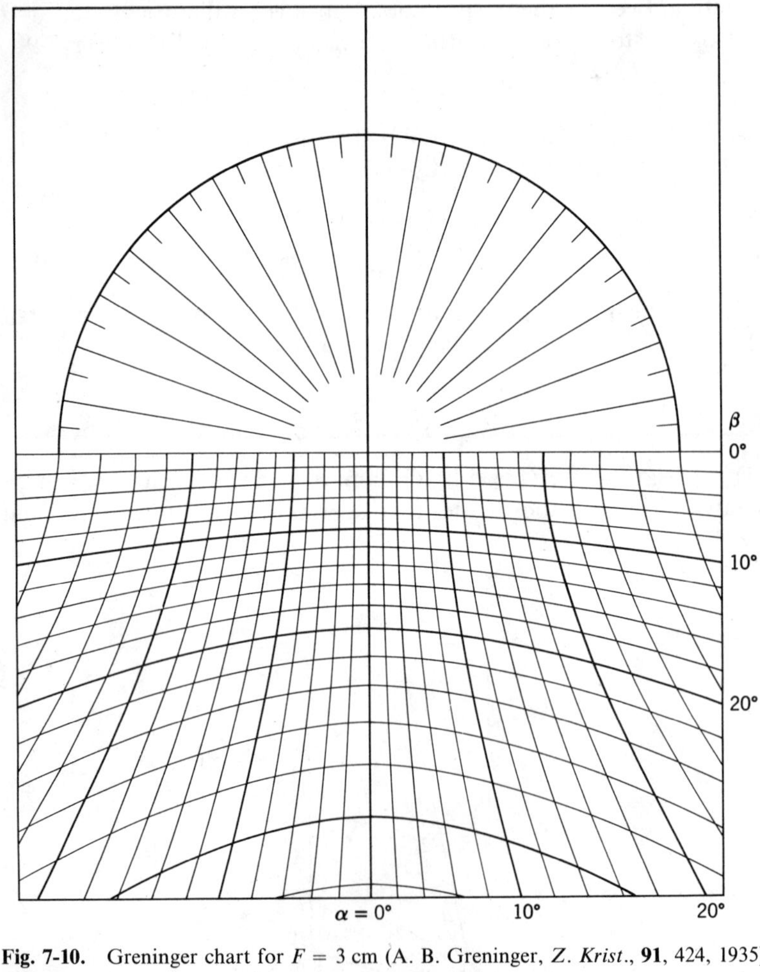

Fig. 7-10. Greninger chart for $F = 3$ cm (A. B. Greninger, *Z. Krist.*, **91**, 424, 1935).

instrumental constant F in the equations for x and y indicate that a chart has to be constructed for a particular crystal-to-film distance.

Indexing Back-Reflection Laue Photographs. The Laue photograph (Fig. 7-11) of an obviously misoriented aluminum crystal has been chosen to demonstrate the use of the back-reflection method to

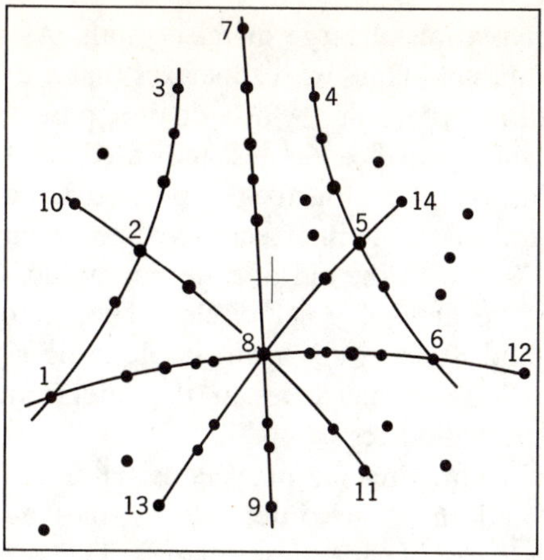

Fig. 7-11. Diagram of a back-reflection Laue photograph of a misoriented aluminum crystal. $F = 3$ cm. W radiation, 1/2 actual size.

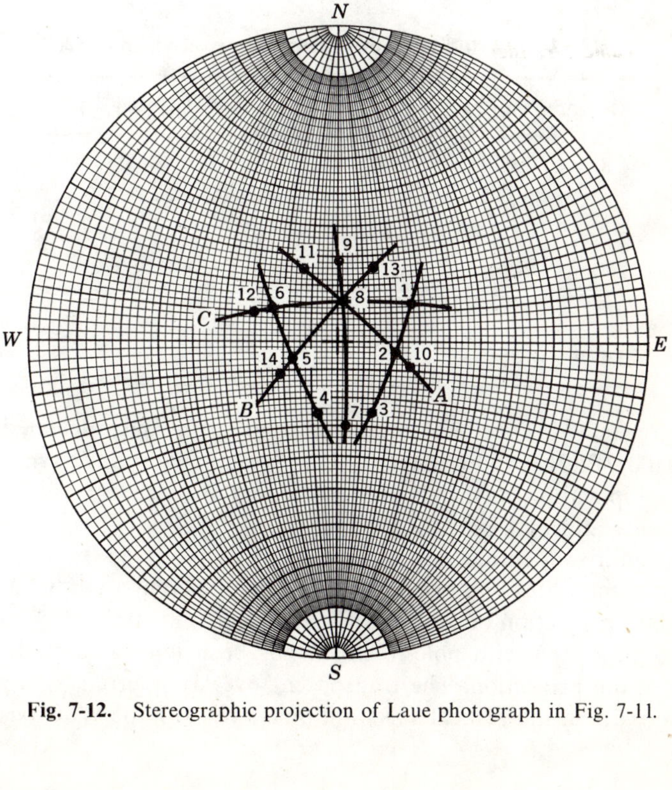

Fig. 7-12. Stereographic projection of Laue photograph in Fig. 7-11.

determine the orientation of large metal crystals. As stated earlier, the reflections from zones of planes with zone axes that are not at right angles to the beam outline hyperbolas which do not pass through the direct beam spot. Six well-defined zones are indicated on the tracing of the photograph in Fig. 7-11. The fourteen numbered Laue spots which outline these zones and mark the intersections of zones have been projected as shown in Fig. 7-12 with the aid of a stereographic scale. As expected, the poles lie on the arcs of great circles. The intersections of zones (poles 1, 2, 5, 6, and 8) generally have simple indices and are therefore relatively easy to index. Pole 8 lies at the intersection of four zones, suggesting very simple indices.

One method of identifying the poles consists of measuring the angles between pole 8 and the other intersection poles and comparing the values with an angle table for cubic planes. The projection shown in Fig. 7-12 is transferred to a transparent sheet, which is then placed as an overlay on a Wulff net of the same radius as the projection. The angle between pole 8 and another pole is measured by rotating the overlay until the two poles lie on a great circle of the net. The angle is then read directly from the net (Table 7-1).

Table 7-1. Identification of Principal Poles in Figure 7-12.

Measured Angles	Comparable Calculated Angles
$8 \wedge 1 = 25\frac{1}{2}°$ $8 \wedge 6 = 24\frac{1}{2}$	$001 \wedge 113 = 25° \; 14\frac{1}{2}'$
$8 \wedge 2 = 26\frac{1}{2}$ $8 \wedge 5 = 26\frac{1}{2}$	$001 \wedge 012 = 26 \;\; 33$

The agreement between the measured angles and the calculated cubic angles (001) $\wedge$ (013) and (001) $\wedge$ (012) is reasonably good. The method of crossmultiplication then indicates that zones A and B of Fig. 7-12 are [100] zones and that C is a [110] zone.

The tentative identification of the zones can be tested by rotating the entire projection into an orientation that can be compared directly with a standard projection of cubic planes, as in Fig. 6-8—that is, by reorienting the projection with pole 8, which has been indexed as 001, at the center of the projection. The transparent overlay referred to is rotated clockwise about the vertical axis until pole 8 rests on the W–E axis of the underlying Wulff net. Next, all the poles are moved from east to west along their respective small circles through an angular interval that brings pole 8 to the center of the projection. This is equivalent to rotating the poles from east to west about the *N–S* axis. The result of these two operations is shown in Fig. 7-13. Zones *A* and *B* are clearly at right angles to each other, and the positions of previously unidentified poles on these

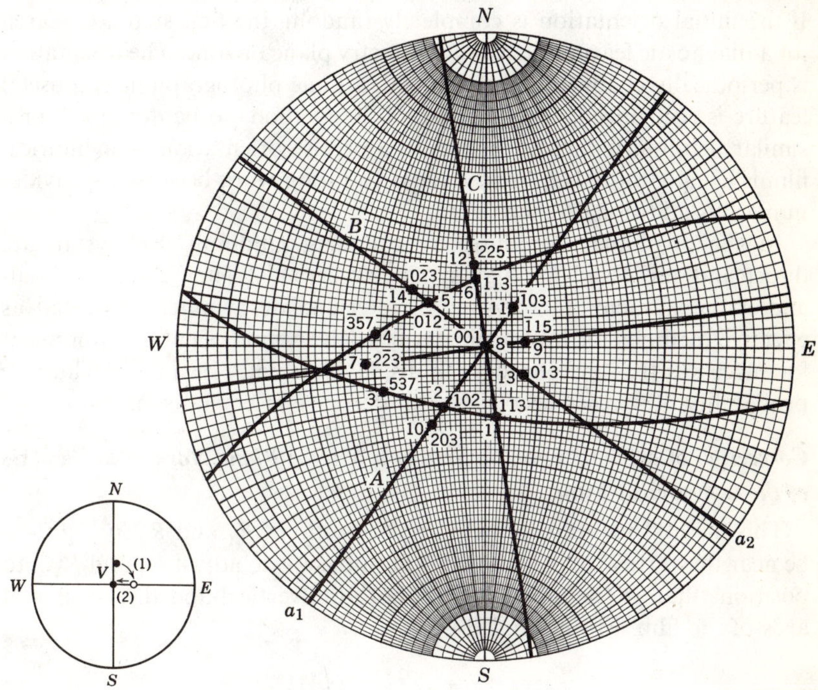

Fig. 7-13. Reorientation of projection in Fig. 7-12 to bring pole 8 into center.

zones (10, 11, 13, and 14) are compatible with the positions of simple-index *h*0*l* and 0*kl* cubic planes. Zone *C*, and a zone at right angles to it, occurs midway between zones *A* and *B*; the additional poles (1, 7, 9, and 12) are readily identified as simple *hhl* planes.

With some experience, the principal reflections 001, 111, and 101 can often be identified by an inspection of the symmetry displayed by the Laue photograph. A simplified procedure for confirming the identification consists of projecting this and a number of additional diagnostic Laue spots and immediately orienting the poles for a direct comparison with projections prepared with 001, 111, and 101 at the center of the projection.

Orienting a Crystal With the Laue Method

The Laue method is used extensively in some laboratories to orient a crystal for study with one of the powerful moving-crystal methods described in later chapters. A Laue photograph is sensitive to a small change in orientation. This property makes it possible to detect and correct a slight misalignment from a symmetrical position in the beam. If the initial orientation is completely random, the first step is a search for a diagnostic feature such as a symmetry plane or zone. The orientation is periodically changed and tested with a Laue photograph until a useful feature is recognized. One or more of the methods to be described, or a similar method, is then used to gain a precise orientation. A cylindrical film is usually more helpful in Laue orientation work because it provides numerous additional reflections.

In general, Laue procedures for orienting randomly set crystals are direct and successful for high-symmetry crystals. They assume a trial-and-error character for crystals of intermediate symmetry and fail as effective methods for low-symmetry crystals. A fundamental requirement for success is that the photograph display symmetry or a zone. Thus the probability of success decreases with decrease in crystal symmetry.

Correction of a Small Error in Setting a Mirror Plane Normal to the Axis of the Cylindrical Film

This method requires that a pair of reflections, R_1 and R_2 in Fig. 7-14 be near the vertical axis of the film. The object is to adjust R_1 and R_2 into positions that are equidistant from both the vertical and the horizontal axes of the film. From Fig. 7-14

$$2\phi + 2\varepsilon = \tan^{-1}\left(\frac{x_1}{r}\right)$$

$$2\phi - 2\varepsilon = \tan^{-1}\left(\frac{x_2}{r}\right)$$

where ϕ is the angle of incidence required to produce a pair of reflections symmetrical about the horizontal film axis. The difference between the two relations gives the angle ε by which the plane responsible for the reflections is mis-set

$$\varepsilon = \frac{1}{4}\left[\tan^{-1}\left(\frac{x_1}{r}\right) - \tan^{-1}\left(\frac{x_2}{r}\right)\right]$$

A second Laue photograph with the crystal rotated through 90° provides a check on the correction applied to the setting or indicates the nature

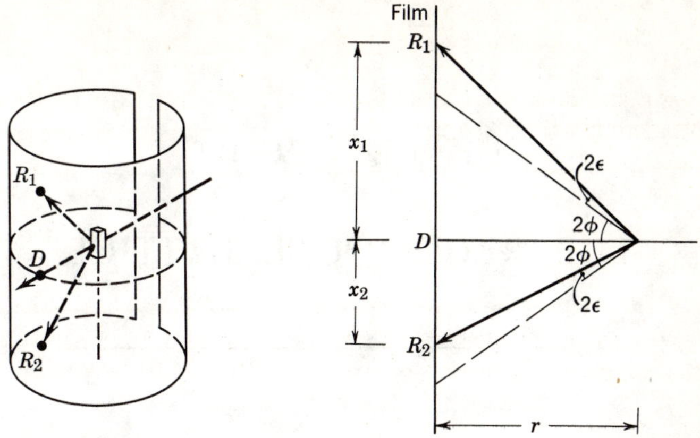

Fig. 7-14

and extent of any further adjustments. The method is applicable to both flat and cylindrical films.

Orientation of a Plane Which Reflects in the Equatorial Plane of the Film, Precisely Parallel to the Beam

The fact that the plane reflects in the equatorial plane implies that it is parallel to the axis of the cylindrical film. The object is to rotate the plane until its reflection (R of Fig. 7-15) coincides with the direct beam spot D. The angle of rotation is evidently θ. From Fig. 7-15

$$\frac{s}{2r\pi} = \frac{2\theta}{360}$$

and therefore $\theta = (90/r\pi)s$. If the radius of the film is $90/\pi = 28.65$ mm, 1 mm s measured on the film equals $1°\ \theta$.

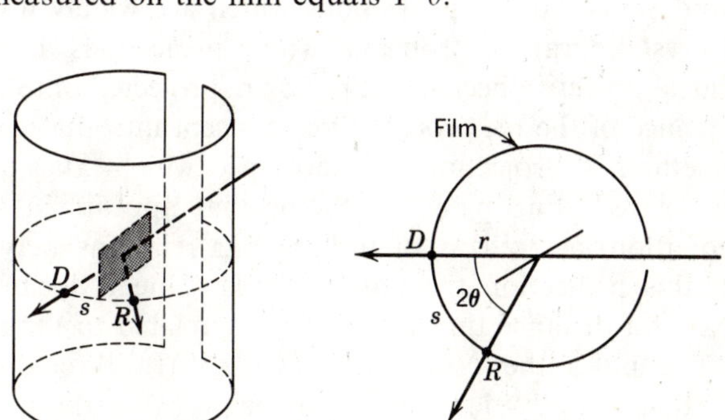

Fig. 7-15

CHAPTER EIGHT

THE RECIPROCAL LATTICE

In Chapter 3 the diffraction effects from a single crystal in the path of an x-ray beam were resolved into a series of diffraction cones, coaxial with the reference axes of the crystal. A diffraction direction was represented as the common intersection of three cones, each coaxial with a different axis. The direction was shown to be geometrically equivalent to the reflection of the incident beam from a lattice plane whose indices (referred to the same axes) correspond to the wavelength orders of the intersecting cones.

A cone is comprised of a restricted set of reflections. The first-order cone about the *c* axis, for example, is composed of reflections from the (*hk*1) planes; the second-order cone, of (*hk*2) planes, and so on. It is not easy to visualize the relative slopes of the reflecting planes directly, even for one cone, because the planes [such as the (111), (131), and (321) planes reflecting in the *hk*1 cone) have a complex angular relation to one another. A simplification occurs if the (two-dimensional) planes are replaced by their (one-dimensional) normals. This is the basis of the gnomonic and stereographic projections which are widely used for the solution of crystallographic problems. These projections, however, fall short of an ideal objective because all poles are projected onto one plane. As a consequence of the reciprocal lattice concept introduced by Ewald in 1921, a method of projection has developed which takes account of the spacings of reflecting planes (a factor that appears in the Bragg reflection equation) as well as of the angular relations between their normals. In this projection the pole of each plane is located on the plane-normal at a distance that is reciprocally related to the spacing of the plane. The poles then sort themselves into a three-dimensional lattice-array that is aptly described as a reciprocal lattice. The planes responsible for the reflections in a diffraction cone are represented by one level of the Ewald lattice, situated normal to the cone axis, and the series of diffraction cones associated with a crystal axis is related to a series of simple, regularly spaced lattice planes of poles normal to the axis. The poles (and therefore the reflecting planes) are indexed on the basis of a unit cell chosen from the reciprocal lattice.

The Reciprocal Lattice

The relation between the space or direct lattice and the reciprocal lattice is shown in Fig. 8-1. The direct lattice, identified by solid circles, is oriented in this diagram with the *c* axis normal to the page; hence all

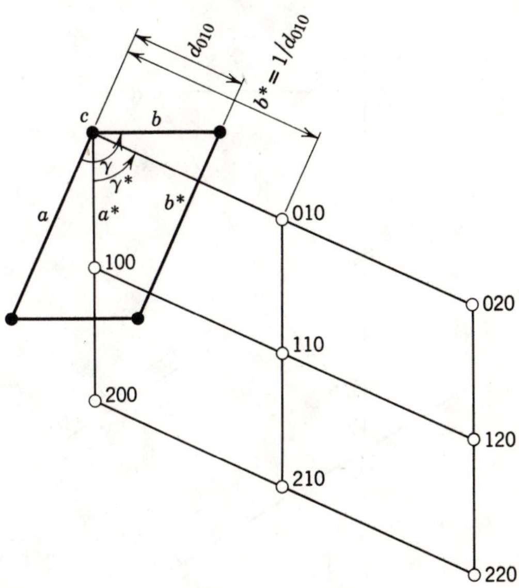

Fig. 8-1. The relation between the direct lattice (heavy lines, solid circles) and the reciprocal lattice.

($hk0$) planes are also normal to the page. Each plane of the direct lattice is represented by a point (open circle) that lies on the normal to the plane at a distance of $1/d$ (the reciprocal of the interplanar spacing) from the origin. These points constitute the reciprocal lattice. The two individuals of a direct-reciprocal pair of lattices have the same symmetry, but their shapes evidently bear an inverse relation to each other. Reciprocal-lattice constants are customarily designated by starred symbols such as a^*, β^*, and d^*_{010} in contrast to the corresponding unstarred symbols of the direct lattice.

It follows from the concept of the reciprocal lattice that reciprocal rows are normal to planes of the direct lattice. The periodicity (p^*) of a row can be expressed in terms of the direct lattice as

$$p^*_{hkl} = \frac{1}{d_{hkl}}$$

Three special cases of this relation are

$$a^* = \frac{1}{d_{100}} \qquad b^* = \frac{1}{d_{010}} \qquad c^* = \frac{1}{d_{001}}$$

where a^*, b^*, and c^* are the unit translations of the reciprocal lattice.

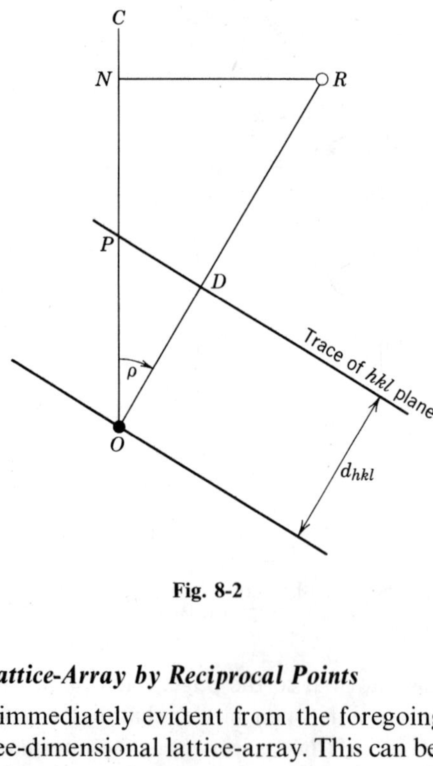

Fig. 8-2

Formation of a Lattice-Array by Reciprocal Points

It may not be immediately evident from the foregoing that reciprocal points form a three-dimensional lattice-array. This can be proved with the aid of Fig. 8-2 which shows the trace of an (hkl) plane of the direct lattice intersecting the c axis, OC, at P. The intercept of the plane on the c axis is $OP = c/l$. R is the corresponding reciprocal lattice point, and therefore $OR = 1/d_{hkl}$. OD equals the spacing, d_{hkl}, of the planes. RN is constructed normal to OC. From the triangle ODP

$$\cos \rho = \frac{OD}{OP} = \frac{d_{hkl} l}{c}$$

From triangle ORN

$$ON = OR\cos\rho = \frac{1}{d_{hkl}}\cos\rho$$

and therefore $ON = (1/c)l$. Because the value of $1/c$ is a constant for a particular lattice, reciprocal points with identical l indices have the same value of ON. In other words, the points lie in planes that are normal to the c axis and are spaced at intervals of $1/c$. The value of ON is zero when $l = 0$, hence the plane that contains the $hk0^*$ points—the $hk0^*$ level—passes through the origin of the direct lattice (Fig. 8-3). The $hk1^*$ level of points is situated at a distance $1/c$ from the origin measured along the c axis, the $hk2^*$ level is at $2/c$, and so on.

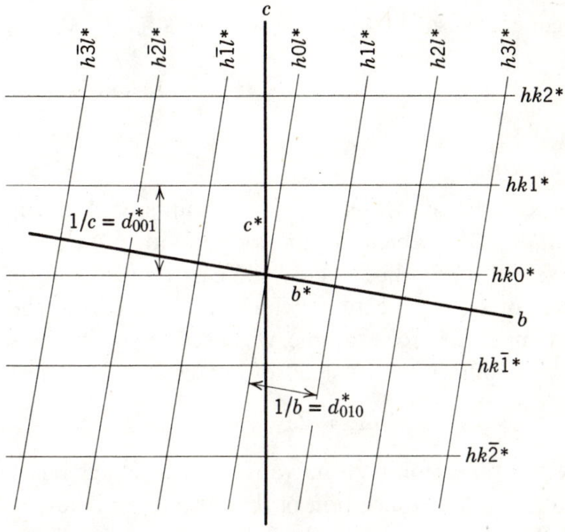

Fig. 8-3. The relation between reciprocal-lattice planes and direct-lattice rows.

It can be similarly shown that the $h0l^*$, $h1l^*$, and $h2l^*$ levels are normal to the b axis and spaced at intervals of $1/b$—also that the $0kl^*$, $1kl^*$, and $2kl^*$ levels are normal to the a axis and spaced at intervals of $1/a$.

The reciprocal point 123^* occurs in three levels: $1hl^*$, $h2l^*$, and $hk3^*$. Consequently, it occurs at the intersection of these levels. Similarly, each reciprocal point is common to three levels. Therefore reciprocal points occur only at the level intersections and, because the levels are regularly spaced, the points form a three-dimensional lattice-array.

Relation Between Reciprocal-Lattice Planes and Direct-Lattice Rows

The preceding section and Fig. 8-3 have already demonstrated a second fundamental property of the reciprocal lattice—namely, that reciprocal planes are normal to direct-lattice rows. The successive levels, which are normal to the c axis (Fig. 8-3) and include the reciprocal points $hk0^*$, $hk1^*$, and $hk2^*$, are evidently successive $(001)^*$ planes of the reciprocal lattice with spacing $d^*_{001} = 1/c$. Similarly, levels that comprise the $h0l^*$, $h1l^*$ and $h2l^*$ reciprocal points are successive $(0k0)^*$ reciprocal planes with spacing $d^*_{010} = 1/b$. Hence

$$c = \frac{l}{d^*_{001}} \qquad b = \frac{l}{d^*_{010}} \qquad a = \frac{l}{d^*_{100}}$$

and in the general case

$$p_{hkl} = \frac{l}{d^*_{hkl}}$$

The Centered Lattices

The reciprocal lattice of Fig. 8-1 shows points with multiple indices—200*, 020*, and 220*—which do not correspond to unique planes of the direct lattice. Multiple indices were used earlier in this book to describe various orders of x-ray reflections which can occur from the same set of planes. The symbol 200, for example, was used to signify the second-order reflection from (100) planes. Instead of expressing the reflection as

$$2\lambda = 2d_{100} \sin \theta$$

the order of the reflection (i.e. the value of n in Bragg's law) was incorporated into the index description of the reflecting planes. Following this scheme, the Bragg equation has the form

$$\lambda = 2d_{200} \sin \theta$$

for this reflection. Because n is given a value of unity, the practice is equivalent to treating the reflection as a first-order reflection. The symbol (200) can then be used to represent a fictitious set of planes in the direct lattice, parallel to the (100) planes but with the spacing $d_{200} = \frac{1}{2}d_{100}$. The reciprocal-lattice point 200*, which lies a distance of $1/d_{200} = 2/d_{100}$ from the origin, represents this set of planes.

It is interesting now to examine the reciprocals of the centered lattices. Lattice centering systematically halves or thirds the spacings of certain planes, which doubles or triples the distances $1/d_{hkl}$ from the origin to the corresponding reciprocal points. C centering, as explained in Chapter 4, halves the spacings of all planes for which $h+k$ is odd and therefore doubles the distances to the first reciprocal point along normals to these planes. Such points have multiple indices: 200*, 020*, 220*, 202*, and so on. The arrangement is C-centered (Fig. 8-4b); the missing points (compared with the arrangement generated by a direct *primitive* lattice of the same shape and size, Fig. 8-4*a*) have indices for which $h+k$ is odd. The reciprocal of a body-centered lattice is all-face-centered and the reciprocal of an all-face-centered lattice is body-centered (Fig. 8-4*c*,*d*). In both cases, as with the end-centered lattice, the indices of the reciprocal points conform to the condition for reflection by planes of the direct lattice. In general, then, each reciprocal point represents a possible reflection from

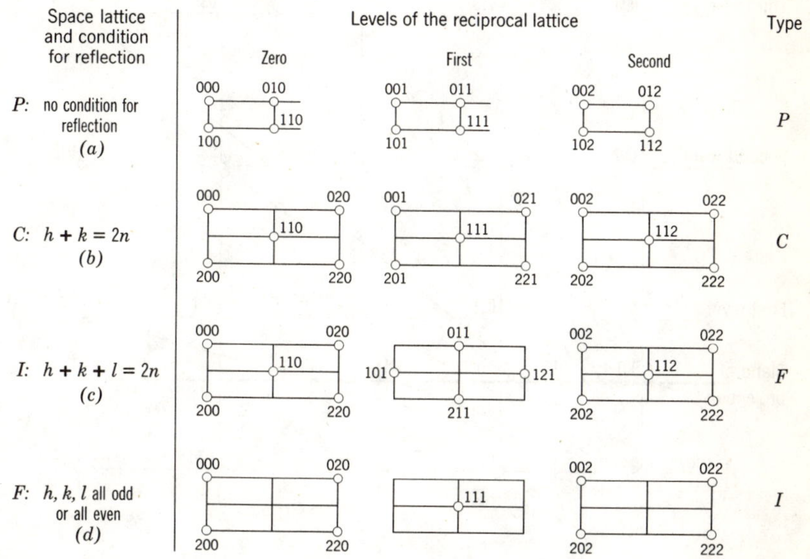

Fig. 8-4. The reciprocals of the space lattices.

the direct lattice. Naturally, it is the crystal structure that determines the character of the x-ray diffraction spectrum; hence the spectrum predicted by the lattice type is modified by the presence of space-group symmetry elements and by the structure amplitude factor.

Relation Between the Reciprocal Lattice and the Gnomonic Projection

A gnomonic projection is constructed by drawing normals to crystal faces (which are parallel to direct-lattice planes) from a common point. Therefore the gnomonic pole and the corresponding reciprocal-lattice point lie on the same normal (Fig. 8-5). Whereas the reciprocal points occur in levels, all gnomonic poles lie in one plane—the plane of projection. The gnomonic projection is a kind of collapsed reciprocal lattice.

The problem of the morphological crystallographer is to deduce the shape (he cannot deduce the size) of the reciprocal unit cell from this array of poles. It is solved if the first-level poles can be distinguished. Figure 8-5 demonstrates in one dimension that the poles of first-level faces form a plane-lattice array in which the distances of the nodes in any row make an arithmetic series when measured from any node in that row. This level is generally more fully represented by face-poles than any higher level, and therefore the first-level plane lattice should account for

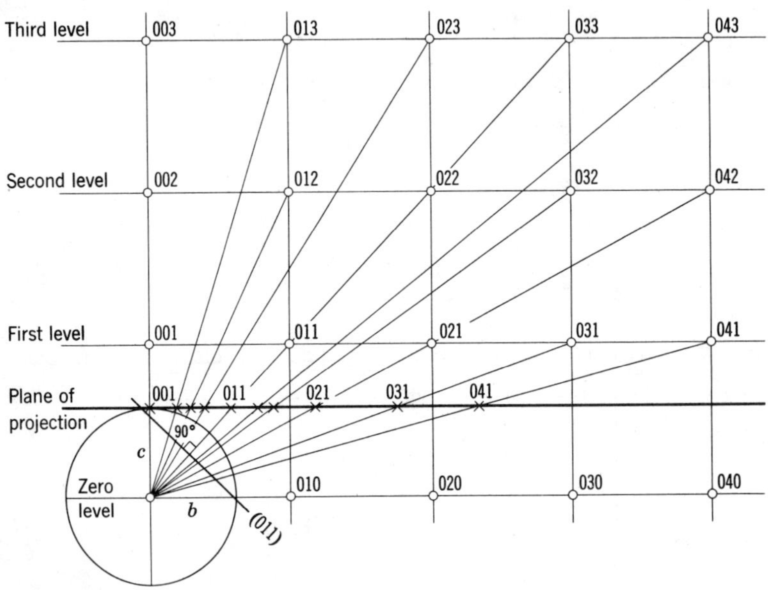

Fig. 8-5. Relation between the reciprocal lattice and the gnomonic projection.

most of the poles on the gnomonic projection. All poles not at the nodes of this plane lattice represent upper level points; a pole lying halfway between two first-level poles (e.g., 012 and 032 in Fig. 8-5) represents a second-level point and a pole one third or two thirds of the way between two first-level poles represents a third-level point. Once the first-level points have been found, axial directions can be chosen to outline the cell, and thereafter the poles are easily indexed.

The Reflection of X-Rays in Terms of the Reciprocal Lattice

The Sphere of Reflection

The reciprocal lattice is precisely related, geometrically, to the direct lattice and must therefore lend itself to an equally precise expression of the concept of x-ray reflection.

Figure 8-6*a* shows a set of planes of the direct lattice with spacing d, inclined to the beam at the Bragg angle θ for these planes. PQ is constructed $1/d$ in length and normal to the planes. Hence Q may be taken as the corresponding reciprocal point and P, as the origin of the lattice. AQ is constructed normal to PQ and therefore parallel to the trace of the reflecting planes. Then the angle PAQ equals θ and

$$AP = \frac{PQ}{\sin\theta} = \frac{1}{d\sin\theta}$$

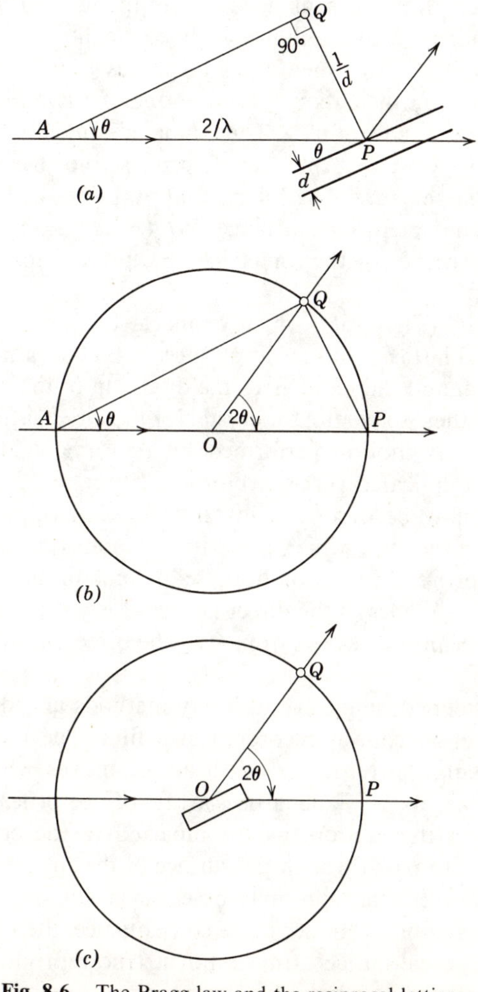

Fig. 8-6. The Bragg law and the reciprocal lattice.

The Bragg law states that

$$\frac{1}{d_{nh\cdot nk\cdot nl}\sin\theta} = \frac{2}{\lambda}$$

when a set of planes is reflecting x-rays. Therefore $AP = 2/\lambda$. Note that the order of the reflection (n of the Bragg law) is here embodied in the value of d.

Because the angle AQP is a right angle, the points A, Q, and P lie on the circumference of a circle with diameter $AP = 2/\lambda$ (Fig. 8-6*b*). Each set of planes that reflects in the plane of the page leads to the construction of the same circle. Reflecting planes that are inclined to the page result in circles which are correspondingly inclined to the page but share the same diameter AP. Thus, when a set of planes is reflecting x-rays, its reciprocal point is on the circumference of one of a family of circles with a common diameter of length $2/\lambda$. The diameter coincides with the beam. This is equivalent to saying that the reciprocal point lies on the surface of a sphere of diameter $2/\lambda$. The sphere is appropriately called the sphere of reflection. The reciprocal distance PQ ($= 1/d$) cannot exceed AP —consequently, planes of the direct lattice with spacings less than $\lambda/2$ cannot reflect.

The radius OQ (Fig. 8-6*b*), which connects the center of the sphere with a reciprocal lattice point on its surface, makes an angle 2θ with the undeviated beam and therefore gives the direction of the reflected beam. It is convenient, then, to picture the crystal (or the direct lattice) as situated at O (Fig. 8-6*c*). A motion performed by the crystal at this point is simultaneously duplicated by the reciprocal lattice at P. The process of reflection may now be described in terms of the reciprocal lattice as follows. At the instant at which a reciprocal point passes through the surface of the sphere of reflection because of some motion of the crystal, the corresponding planes of the direct lattice satisfy the Bragg condition, and a reflected beam moves out from O in the direction of the reciprocal point.

The various refined, single-crystal x-ray methods are identical in that only one diffraction cone is recorded on a film. The reciprocal-lattice concept represents the reflecting direct-lattice planes which contribute to any one cone, as a plane-lattice array of reciprocal points. The methods differ in the motion that is imparted to the crystal to cause reciprocal points to pass through the surface of the sphere and "reflect"; furthermore, the recording film may be either a flat sheet or it may be bent into the shape of a cylinder. As a consequence, the film representation of the reciprocal-lattice array is not a true reproduction in some methods, but rather an orderly distortion or projection of the array. In either case the crystallographer deals with a photographic representation of the reciprocal lattice, which is magnified by two factors:

(1) the crystal-to-film distance (the factor F in the de Jong–Bouman and precession methods and r in the rotation and Weissenberg methods);
(2) the wavelength λ of the incident radiation.

When the effect of the crystal-to-film distance is maintained constant, a reciprocal period on the film is proportional to λ/d. It is convenient, then, in deriving the instrumental settings for photographing levels of the

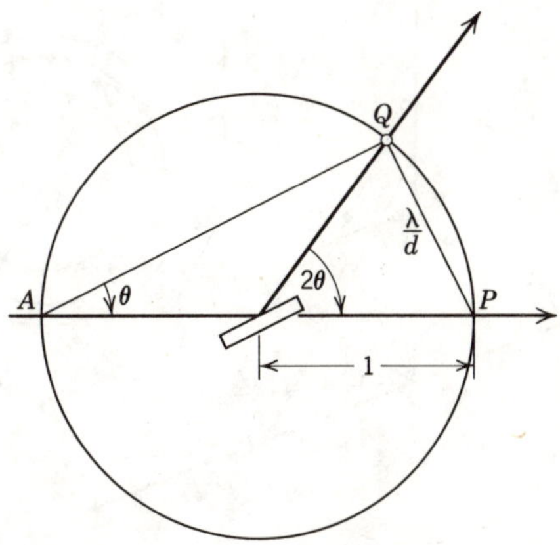

Fig. 8-7. The sphere of reflection of unit radius.

reciprocal lattice and deducing cell constants from the photographs, to work with a reciprocal lattice in which the period PQ is given a value of λ/d instead of $1/d$ (Fig. 8-7). Under these circumstances

$$AP = \frac{PQ}{\sin\theta} = \frac{\lambda}{d\sin\theta}$$

The Bragg law states that the relation $\lambda/(d\sin\theta) = 2$ prevails during reflection, and therefore $AP = 2$ and the sphere has unit radius. It will become evident in the following pages that the unit radius feature of the modified sphere simplifies the description of diffraction methods. In this version of the sphere the radius and reciprocal lattice vectors (λ/d) are dimensionless quantities. Nevertheless, it is a common practice to refer to the quantities in terms of reciprocal lattice units (rlu).

Diffraction from Planes of Reciprocal Points

Figure 8-8 shows the sphere of reflection with OC, representing the c axis or [001] row of a crystal, oriented at an angle $\bar{\mu}$ to the beam. The

$hk0^*$, $hk1^*$, $hk\bar{1}^*$, etc., levels of reciprocal points are normal to this row. They intersect the sphere in circles with diameters PQ, AB, DE, etc. The $hk0^*$ level includes the origin P of the reciprocal lattice. The Bragg condition is satisfied for any $hk0^*$ point, for example, which happens to be on the surface of the sphere—more particularly in this case on the circumference of the circle PQ. (The corresponding circles for the $hk1^*$

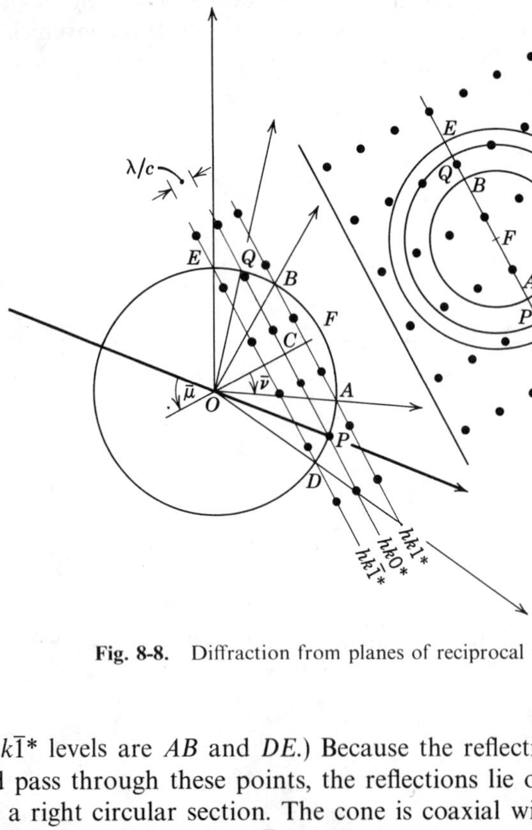

Fig. 8-8. Diffraction from planes of reciprocal points.

and $hk\bar{1}^*$ levels are AB and DE.) Because the reflections emanate from O and pass through these points, the reflections lie on a cone of which PQ is a right circular section. The cone is coaxial with OC. Reflections corresponding to $hk1^*$ and $hk\bar{1}^*$ points lie on coaxial cones with different half-apex angles. The half-apex angle of the zero level is equal to $\bar{\mu}$, but the angle for an n level depends on both $\bar{\mu}$ and the spacing of the reciprocal planes. Thus for the $hk\bar{1}^*$ level

$$\cos \bar{\nu} = \frac{OF}{OA} = OC + \frac{\lambda}{c}$$

$$OC = OP \cos \bar{\mu} = \cos \bar{\mu}$$

and consequently

$$\cos \bar{\nu} = \cos \bar{\mu} + \frac{\lambda}{c}$$

It can be shown that for the $hk\bar{1}^*$ level

$$\cos \bar{\nu}' = \cos \bar{\mu} - \frac{\lambda}{c}$$

When rearranged into the form

$$c(\cos \bar{\nu} - \cos \bar{\mu}) = 1\lambda$$

$$c(\cos \bar{\nu}' - \cos \bar{\mu}) = -1\lambda$$

the expressions for diffraction from the two first-level reciprocal planes can be recognized as the Laue equations for first-order diffraction cones associated with the *c* axis of a crystal.

The reciprocal planes depicted in Fig. 8-8 will give rise to few diffracted beams, unless the planes are given some motion that will cause a representative number of points to pass through the surface of the sphere. A description of single-crystal x-ray methods is, in large part, a description of the methods used to generate diffraction cones and record the effects. The methods are described in some detail in the succeeding chapters.

CHAPTER NINE

THE BUERGER PRECESSION METHOD

De Jong and Bouman described in 1938 the first method for recording the reciprocal lattice in undistorted form. The lattice is recorded level-by-level with their method, and, if the instrumental settings are arranged to magnify each level by the same factor, it is a straightforward matter to visualize the reciprocal lattice and its symmetry, to choose the unit cell and measure its dimensions, and to deduce the space-group symmetry. De Jong and Bouman called their instrument a retigraph—literally, an instrument for drawing a net or lattice. The crystal and the film rotate during the exposure; hence the instrument is known as a rotation-retigraph.

Figure 9-1 illustrates the de Jong-Bouman method for recording an $hk0^*$ level of reciprocal points. A circular area of the level, of diameter PQ, is tilted into the sphere of reflection by setting the plane-normal (the c axis of the crystal) at an angle ($\bar{\mu}$) to the beam. The crystal is rotated about the c axis during the exposure, which causes the level to rotate about its normal at P. Then reciprocal points from a circular area of the level of radius PQ pass through the surface of the sphere and "reflect"; the reflections lie on the surface of a cone, coaxial with the c axis and have a half-apex angle $\bar{\mu}$. The undeviated portion of the beam is a line in the cone. The reflections are recorded to give an undistorted photograph of the level on a flat film by causing the film to duplicate simultaneously the motion of the level with reference to the x-ray beam. This is accomplished by rotating the film about its normal at P' in the path of the beam, the motion being synchronized with that of the crystal. A lattice point at Q produces a reflection at Q', the distance $P'Q'$ on the film representing in magnified form the reciprocal-lattice period PQ. As additional lattice points are rotated onto the surface of the sphere, their positions with respect to Q and P will be faithfully represented on the film with respect to Q' and P', because the rotations of the film and

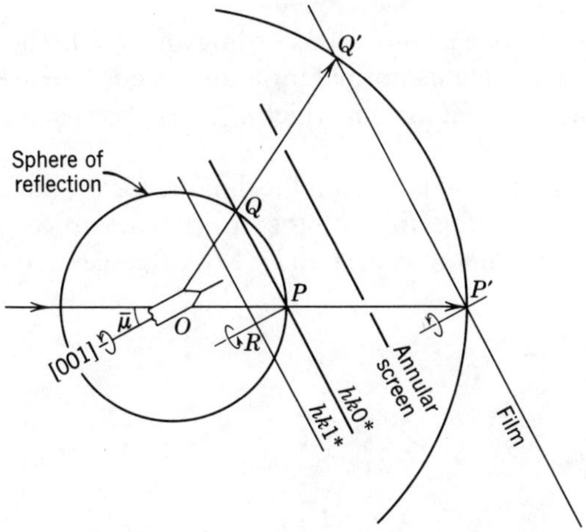

Fig. 9-1. De Jong–Bouman method of recording the zero level.

reciprocal level are synchronized and the crystal-to-film distance remains unchanged. Thus the geometrical pattern of lattice points in the level is accurately reproduced on the film. A screen with an annular opening of appropriate radius is inserted between crystal and film to permit only the zero-level reflections to reach the film.

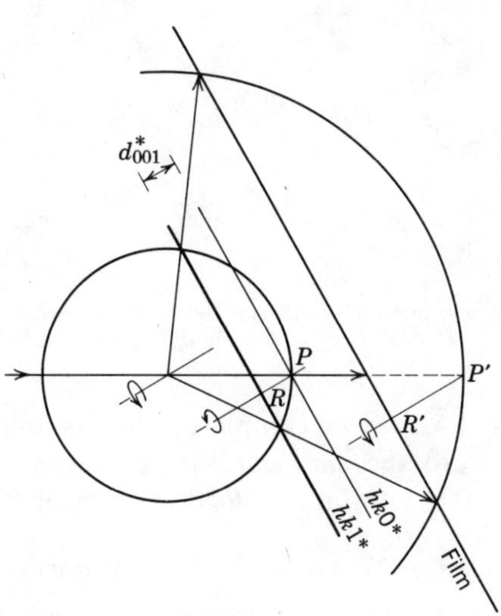

Fig. 9-2. The general method of recording an *n* level with a de Jong–Bouman apparatus.

Figure 9-2 shows one method of recording an n level. The film is moved toward the crystal along its normal for a distance $P'R' = PR \cdot (OP'/OP) = d^*_{001} \times$ (magnification factor), in which d^*_{001} is the spacing between the zero and the n-level plane.

The precession method is a second technique for producing symmetry-true, undistorted photographs of the reciprocal lattice, level-by-level, on flat films. The method was devised by M. J. Buerger and first described

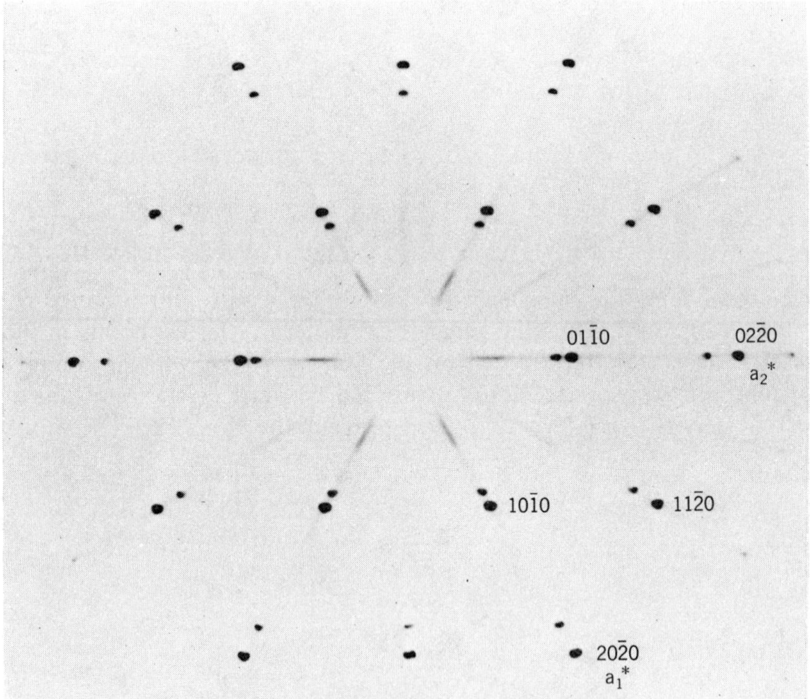

Fig. 9-3. Zero-level photograph of quartz with $c[0001]$ as precession axis; $\bar{\mu} = 25°$, $r_s = 15$ mm, $s = 32.5$ mm, unfiltered Cu radiation.

in a short note in 1942.[1] More complete accounts followed in 1944[2] and 1964.[3] In contrast to the simple rotation movement of a de Jong–Bouman crystal, the crystal in this method is given an unusual motion

[1] M. J. Buerger, *X-Ray Crystallography*, John Wiley and Sons, New York (1942), pp. 210–211.

[2] M. J. Buerger, *The Photography of the Reciprocal Lattice*, ASXRED Monograph No. 1 (1944).

[3] M. J. Buerger, *The Precession Method*, John Wiley and Sons, New York (1964).

in the path of the x-ray beam, which has the desirable effect of producing well-shaped reflection spots (Fig. 9-3) in comparison with de Jong–Bouman spots. It is superior to the powerful Weissenberg method, described later, for indexing reflections, especially when the crystal has a very small reciprocal cell; mainly for this reason, it ranks as one of the most useful single-crystal techniques.

Motion of the Crystal

The motion of the crystal is illustrated in Fig. 9-4. The crystal is held in a yoke which permits simultaneous oscillation about a horizontal, *H*, and a vertical, *V*, axis. The crystal is oriented on the device with a chosen axis normal to the horizontal axis. The direction of the x-ray beam is normal to the vertical axis.

An oscillation of the horizontal axis causes the crystal to oscillate in a vertical plane. The motion is controlled to produce an oscillation which is symmetrical with respect to the x-ray beam. Similarly, a controlled motion of the vertical axis causes the crystal to oscillate in a horizontal plane in a manner which is also symmetrical with reference to the beam. When the two oscillations are combined and when they are of the same magnitude (say, $2\bar{\mu}$ degrees), but 90° out of phase as shown in the figure, the crystal axis sweeps out a cone which is coaxial with the beam, and has a half-apex angle $\bar{\mu}$. This means that throughout its motion, the crystal axis maintains an angle $\bar{\mu}$ to the x-ray beam.

The movement of the crystal axis about the beam is not a pure precession motion. The term precession implies that the axis rotates while it sweeps out the cone. Actually, the oscillations confine rotation of the crystal to axes which are mutually perpendicular to the precession axis.

Motion of the Zero-Level Plane

The motion imparted to reciprocal planes which are normal to the crystal axis is best visualized by considering first a zero-level plane. This plane (*MN* of Fig. 9-5*a*,*b*) has its center fixed at *P*, the origin of the reciprocal lattice. It makes an angle $(90 - \bar{\mu}) = \mu$ with the beam and maintains this angle during precession by tilting about directions through *P*, which are parallel to the two oscillation axes of the yoke.

The reciprocal plane intersects the sphere of reflection in a circle of diameter $PQ = PR$. The diameter of this circle is determined by the precession angle and remains unchanged during precession, although the circle rolls about *P*, traversing a circular area of the reciprocal plane of radius *PQ*. Reflection occurs when the circumference of the circle passes through a lattice point (that is, when a lattice point passes through

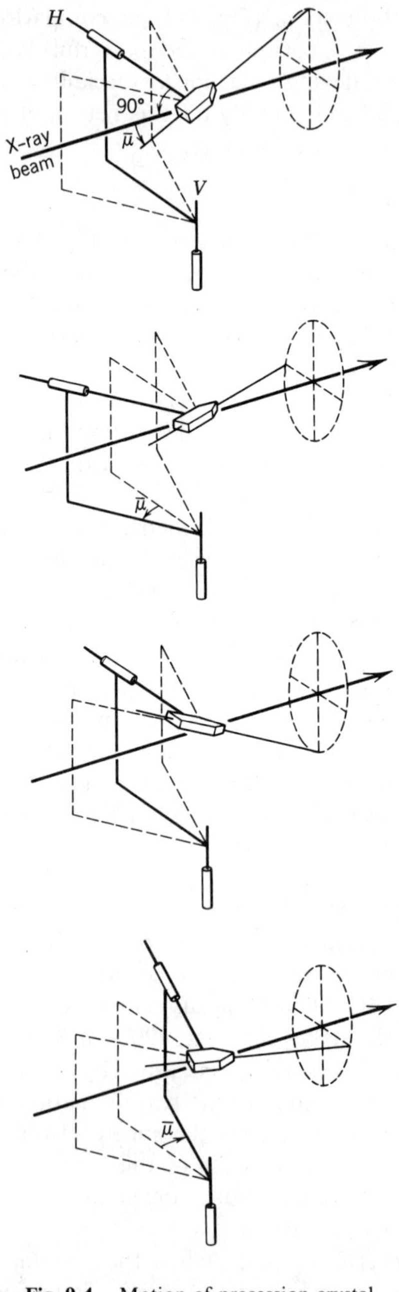

Fig. 9-4. Motion of precession crystal.

the surface of the sphere). The circular area, of radius PQ, represents that portion of the reciprocal plane which will diffract at a particular $\bar{\mu}$ angle. It is obviously important that this be a representative portion of the plane.

It is now possible to visualize the pattern of zero-level reflection directions. They emanate from O and travel through lattice points which

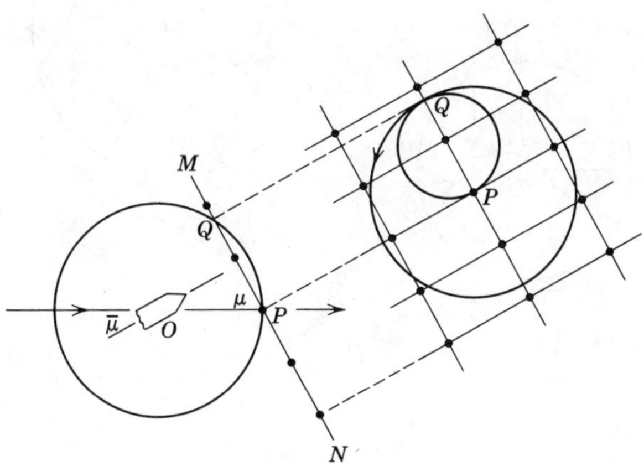

(a)

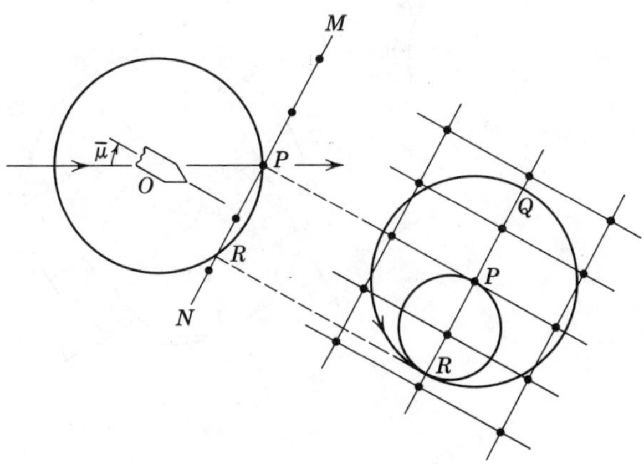

(b)

Fig. 9-5. Motion of zero level through reflection sphere.

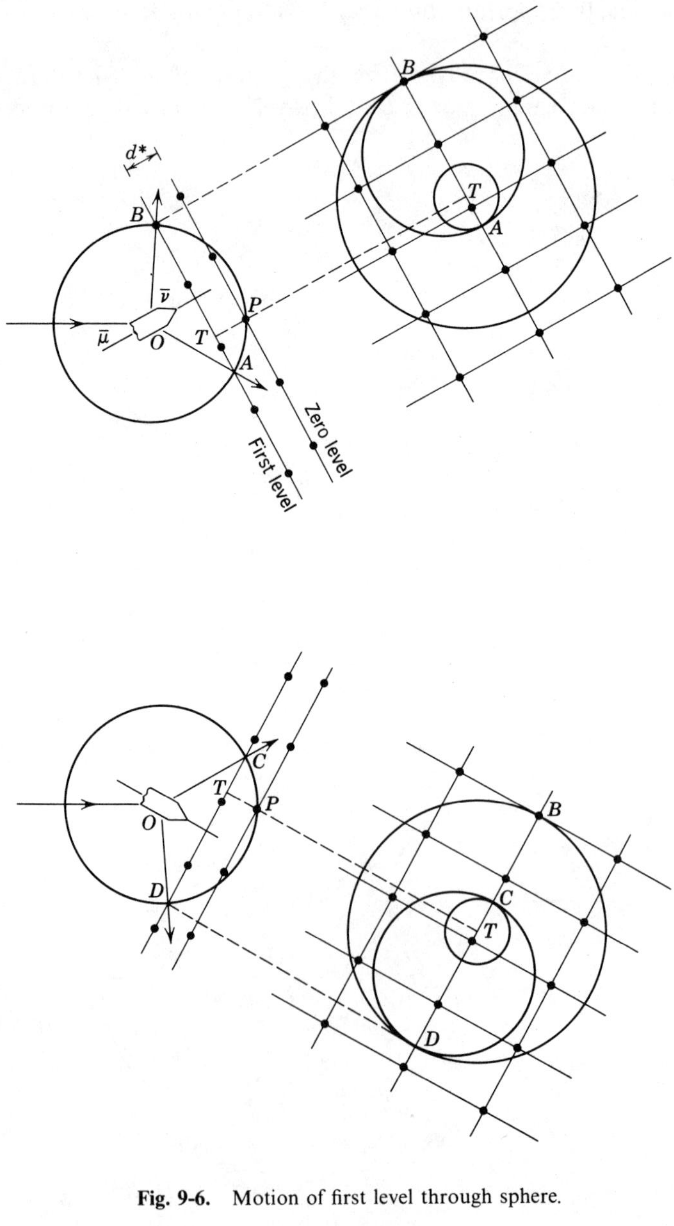

Fig. 9-6. Motion of first level through sphere.

are intercepted by the circle of intersection. The directions therefore lie on the surface of a cone which rolls about *OP*, the x-ray beam direction. The cone has a half-apex angle $\bar{\mu}$, and is coaxial with the crystal axis.

Upper-Level Reciprocal Planes

The first level of the reciprocal lattice (taken in a direction toward the crystal) intersects the sphere of reflection in a circle of diameter $AB = CD$ (Fig. 9-6*a,b*). During a complete precession of the crystal, a circular area of the plane, of diameter *BD*, migrates through the sphere.

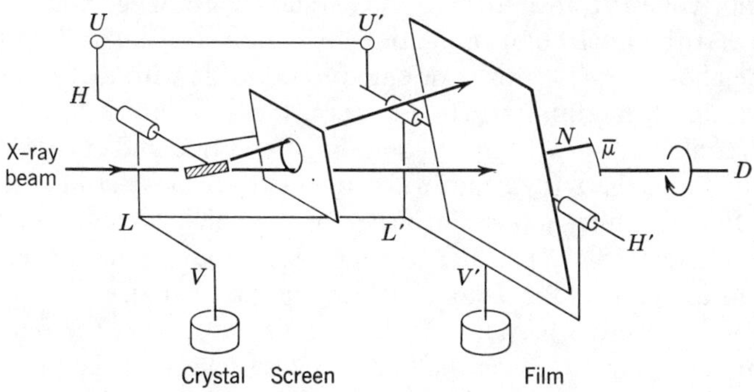

Fig. 9-7. Buerger precession instrument.

A small central portion, of radius $TA = TC$, does not leave the sphere throughout the motion of the plane, and as a result planes of the direct lattice represented by reciprocal points in this area do not reflect. A consideration of the diagrams shows that the radius of this central zone increases with increasing spacing, d^*, between the first and zero level, and with decreasing $\bar{\mu}$.

The reflections corresponding to this level lie on the surface of a cone of half-apex angle $\bar{\nu}$. This cone, like the zero-level cone, is coaxial with the precession axis.

The Precession Instrument

A reciprocal plane may be recorded in undistorted form, as in the de Jong–Bouman method, if the film and plane are caused to perform simultaneously the same motion with reference to the x-ray beam direction. This is accomplished in the precession instrument as shown diagrammatically in Fig. 9-7.

The film is held in a yoke with oscillation axes H' and V', similar to the crystal yoke portrayed in Fig. 9-4. A shaft *N*, normal to the film at the film center, is connected with a circular arm that rotates about a horizontal driveshaft D. The angle $\bar{\mu}$, that N makes with the driveshaft is adjustable from 0 to about 30°. The film can be moved forward along its shaft out of the plane of the yoke axes. The x-ray beam is directed at the intersection of the yoke axes and along the driveshaft. When the driveshaft is rotated, the film-normal N, precesses about the x-ray beam forcing the film to oscillate about the H' and V' axes at a constant angle $(90 - \bar{\mu}) = \mu$ to the beam.

The motion of the chosen crystal axis, oscillating in its yoke, is synchronized with the motion of the film-normal. The vertical oscillation of the film yoke is transmitted to the crystal by a linkage connecting L and L', and the horizontal oscillation by a universal linkage at U and U'. If the chosen crystal axis is oriented to coincide with the beam when the $\bar{\mu}$ setting of the film-normal is at zero, the axis and the film-normal remain parallel when $\bar{\mu}$ is given some angular value and precess simultaneously about the x-ray beam when the instrument is set in motion. It follows that the film and reciprocal planes, which are normal to the precession axis, remain parallel to each other and execute an identical motion in the path of the x-ray beam during precession.

A screen is inserted between crystal and film to permit the reflections from only one plane to reach the film. Because reflections lie on the surfaces of cones, which are coaxial with the precession axis, the screen must have an annular opening, whose axis coincides with and duplicates the motion of the precessing axis. The latter condition is met by attaching the screen with an arm s to the horizontal axis H, which carries the crystal. With the instrument set at zero $\bar{\mu}$, the beam strikes the center of the screen. Provision is made for altering the crystal to screen distance, and for substituting screens with different annular openings.

Photographing Reciprocal Levels

The geometrical relations for recording the zero level are shown in Fig. 9-8. A reciprocal lattice point Q on the surface of the sphere reflects to Q' on the film. The film distance $P'Q'$ is related to the reciprocal plane distance PQ as

$$\frac{P'Q'}{PQ} = \frac{OP'}{OP} = F$$

where F is the crystal-to-film distance, and therefore

$$P'Q' = F \cdot PQ$$

Consequently, F is the magnification factor of the record.

The arrangement for recording the first level is shown in Fig. 9-9. When the film is in position to record the zero level, its center is at P', a

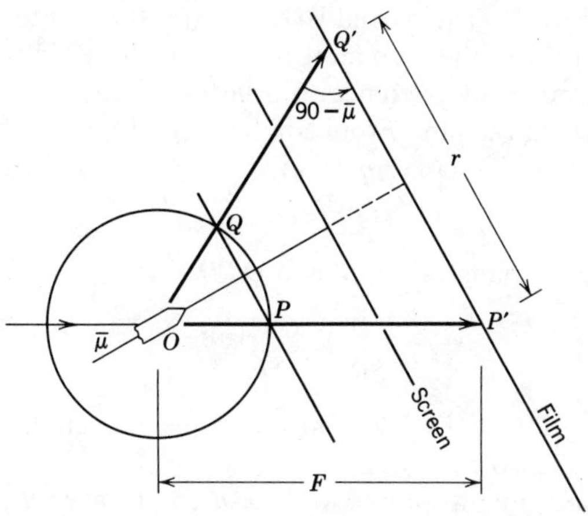

Fig. 9-8. Arrangement for photographing zero level.

distance F from the crystal. The center of that portion of the first level which moves through the sphere is at T, a distance $PT = d^*$ along the normal to the plane through P. If the scales of the zero- and first-level records are to be identical, the film must be moved along its normal a

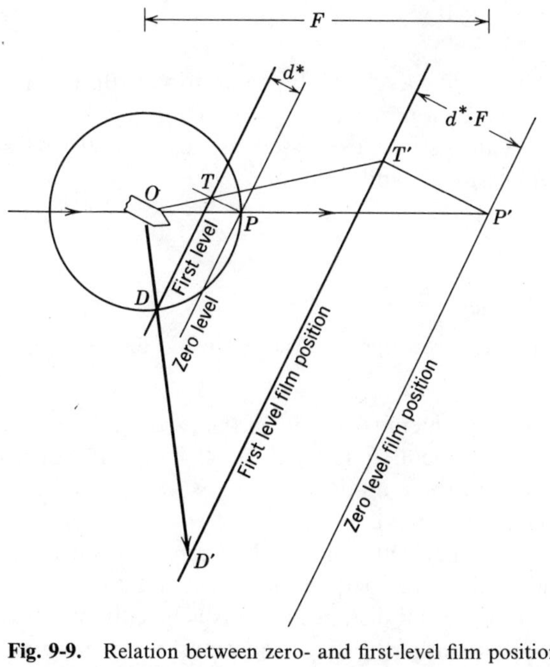

Fig. 9-9. Relation between zero- and first-level film positions.

distance $F \cdot d^*$ to T'. (The second level requires the film to be displaced a distance $2d^* \cdot F$ from the zero-level position.) The validity of this statement can be shown by reference to a lattice point D on the first level, reflecting to D' on the film. From similar triangles $OP'T'$ and OPT

$$\frac{OT'}{OT} = \frac{OP'}{OP} = F$$

and from similar triangles $OT'D'$ and OTD

$$\frac{T'D'}{TD} = \frac{OT'}{OT} = F$$

and therefore $T'D' = F \cdot TD$. Thus the magnification factor of the first-level film is also F.

Although the film is moved a measured distance $F \cdot d^*$ along its normal towards the crystal, the oscillation axes continue to act through P' (Fig. 9-9), and therefore the film duplicates the motion of the reciprocal level which is controlled by the oscillation axes acting through P.

Reciprocal Level Screen Setting

The radius r_s and setting s of the annular opening is given by

$$\frac{s}{r_s} = \cot \bar{\nu}$$

where $\bar{\nu}$ is the half-apex angle of the cone (Fig. 9-10). For the zero level $\bar{\nu} = \bar{\mu}$, and the expression becomes $s = r_s \cot \bar{\mu}$.

The angle $\bar{\nu}$ can also be expressed as $\cos \bar{\nu} = OE/OD = OC - d^*$. But $OC = OP \cos \bar{\mu} = \bar{\mu}$ and therefore

$$\cos \bar{\nu} = \bar{\mu} - d^*$$

Instrumental Settings

Four instrumental settings must be considered for each upper-level film:

1. The film displacement ($F \cdot d^*$) from the zero-level position. In recent models of the instrument, F is fixed at 60 mm, and consequently the setting is $60 \times d^*$ in these models.
2. The precession angle, $\bar{\mu}$.
3. The radius of the reciprocal-level screen, r_s. Four sizes, with radii 15, 20, 25, and 30 mm are adequate for most situations.
4. The screen-to-crystal distance, s. This is usually the last adjustment made, and consequently the two equations for $\bar{\nu}$ may be profitably

combined to give this value in terms of the other variables

$$s = r_s \cot \cos^{-1}(\cos \bar{\mu} - d^*)$$

A nomogram (Fig. 9-11) designed by Adams, Evans, and Tilden[4] provides a graphical solution of this expression and permits a rapid selection of values for each of the settings.

*The Determination of d**

The determination of d^* (the spacing of reciprocal planes normal to the precession axis) is a necessary prelude to recording n levels of the

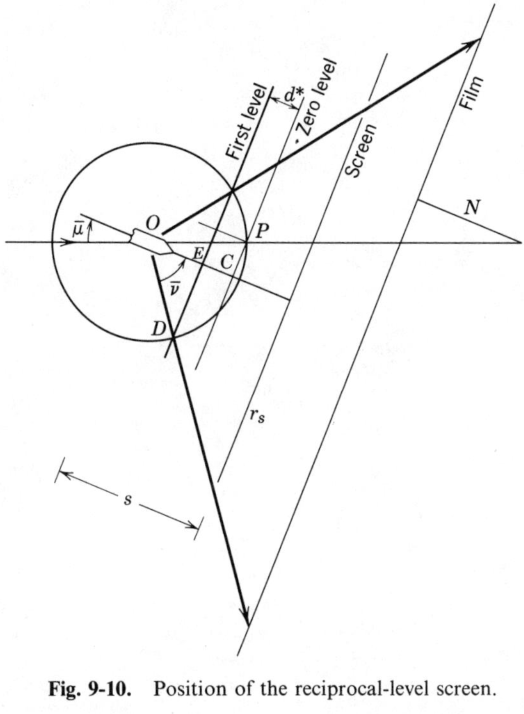

Fig. 9-10. Position of the reciprocal-level screen.

lattice. This constant is needed for calculating both the film displacement $F \cdot d^*$ and the reciprocal-level screen setting s. The constant is given by $d^* = \lambda/p$, where p is the period in the direct lattice, parallel to the precession axis. The period can be determined by the rotation method described elsewhere in this book, but this requires that the crystal be remounted with the precession axis parallel to the axis of the crystal

[4] D. P. Adams and H. T. Evans, Jr., *Rev. Sci. Instr.*, **20**, 150 (1949); H. T. Evans, Jr., S. G. Tildon, and D. P. Adams, *ibid.*, 155.

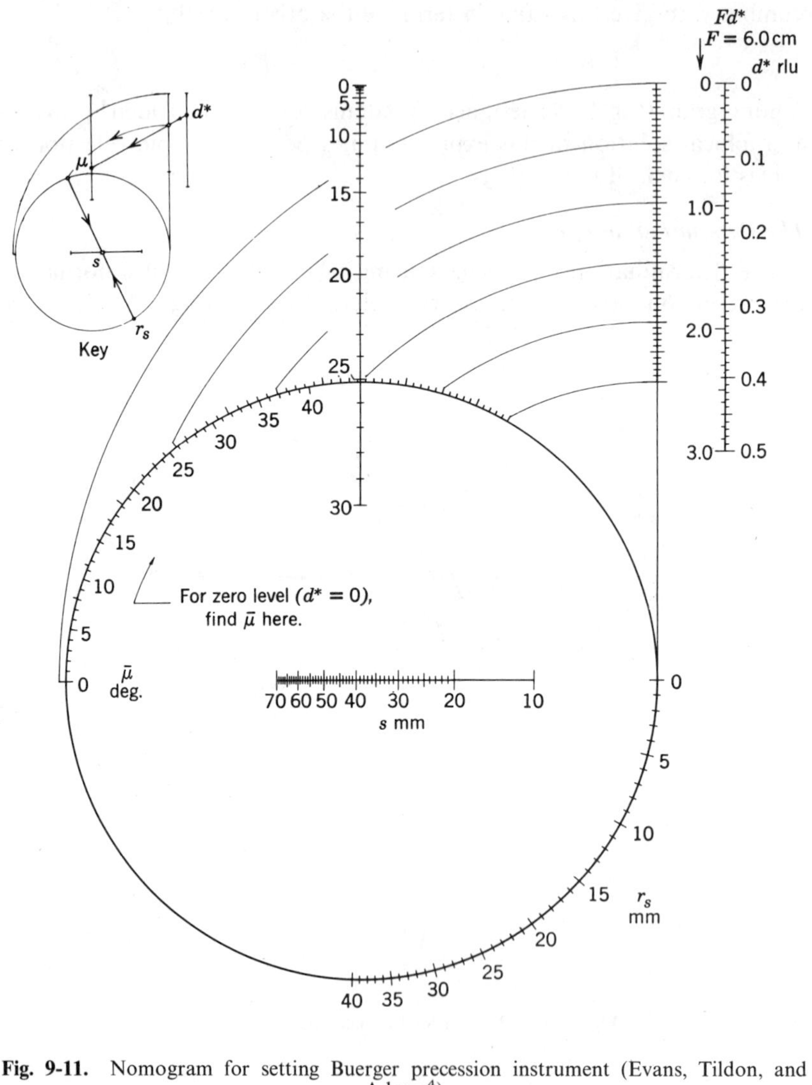

Fig. 9-11. Nomogram for setting Buerger precession instrument (Evans, Tildon, and Adams[4]).

holder. The procedure is troublesome if the crystal is very small or lacks faces suitable for orienting about this new direction on the goniometer. A value of d^*, sufficiently accurate for recording n-levels, can be determined with the precession instrument without altering the crystal setting.

A useful relation connection d^* with $\bar{\mu}$ and $\bar{\nu}$, the half-apex angle of the first level reflection cone

$$d^* = \cos \bar{\mu} - \cos \bar{\nu}$$

was derived in another form in the section dealing with the reciprocal-level screen setting. This gives the desired constant, provided that a method can be found for measuring $\bar{\nu}$. Buerger devised the *cone axis* photograph for this purpose. The motion of the reciprocal-level screen is designed to keep the precession axis (OC of Fig. 9-10) directed at its center

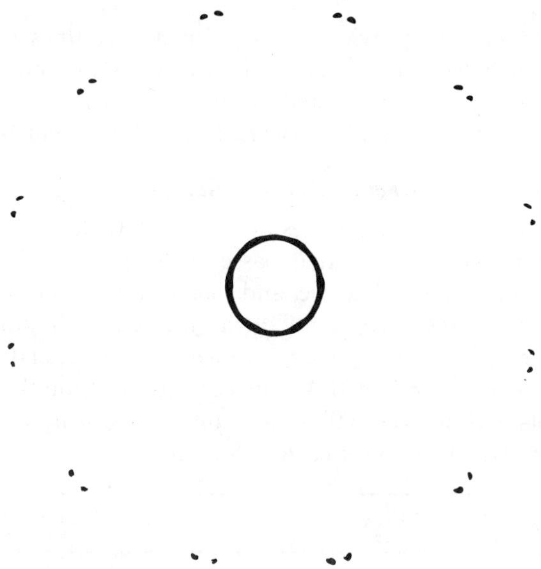

Fig. 9-12. Cone axis photograph of quartz with c[0001] as precession axis; $\bar{\mu} = 10°$, $s = 35$ mm, Ni-filtered Cu radiation. The measured radius of first-level circle = 36.0 mm leading to a value of 5.37 Å (compared to 5.40 Å as determined by precision methods) for p_{0001}.

and normal to the screen. Because the diffraction cones are coaxial with the axis, they are intercepted as concentric circles on the screen (except that ordinarily one selected cone is passed through the annular opening of the screen by choosing an appropriate s setting). A cone axis photograph is a record of the circles (Fig. 9-12), prepared by inserting a sheet of film in the reciprocal-level screen holder in place of the screen. From Fig. 9-10

$$\frac{\text{radius of first level circle } r_s}{\text{crystal-to-film distance along } OC,\ s} = \tan \bar{\nu}$$

and therefore

$$d^* = \cos \bar{\mu} - \cos \tan^{-1}\frac{r_s}{s}$$

The value of d^* can also be determined graphically with the aid of the nomogram in Fig. 9-11 and a cone-axis photograph.[5] The nomogram is a geometrical representation of the relation

$$s = r_s \cot \cos^{-1}(\cos \bar{\mu} - d^*)$$

It is programmed to provide the instrumental settings s and r_s for a selected $\bar{\mu}$ angle, when d^* is known. In a cone-axis photograph, s and $\bar{\mu}$ are known and r_s can be measured—only d^* is unknown. Hence by using the nomogram in reverse a graphical solution for d^* can be obtained.

The Effect of the Wavelength of Incident Radiation

The precession angle determines the radius PQ (Fig. 9-5) of the circular area of the zero-level plane that passes through the sphere. However, the scale of the reciprocal lattice, and therefore its extent in the circular area, depends on the wavelength of the incident radiation. The $(hk0)^*$ reciprocal level of the tetragonal mineral anatase (TiO_2), with cell dimensions $a = 3.73$, $c = 9.37$ Å, will serve to illustrate this. Reciprocal-lattice points with indices $h00$ occur at intervals of $\lambda/d_{100} = \lambda/3.73$ from the origin P. The interval for each of four commonly used radiations is:

		Cr	Fe	Cu	Mo
λ	=	2.29	1.93	1.54	0.71
$\lambda/3.73$	=	0.61	0.52	0.41	0.19

Mo radiation results in the smallest interval, hence the greatest number of lattice points in the circular area. The extent of the $(hk0^*)$ reciprocal plane is depicted in Fig. 9-13 for three kinds of radiation when the circular area has unit radius (that is, when $\bar{\mu} = 30°$, because at this angle $PQ = OP = 1$). Five lattice points, with indices (100) to (500), lie within the circle when Mo radiation is used, compared with only one with Cr (and Fe) and two with Cu radiation. The figure emphasizes the need for short-wavelength radiation in precession studies of compounds with small unit cells. Conversely, the precession record of very large cells (very small reciprocal cells) is crowded when short wavelengths are employed; a Cr target is useful for dispersing the diffraction effects from large organic cells to easily deciphered, proportioned cells.

[5] Privately communicated by Dr. Howard T. Evans, Jr.

The wavelength of the incident radiation is also a factor in recording upper levels of the reciprocal lattice. Recent models of the instrument have a fixed crystal-to-film distance (F) of 60 mm. The film adjustment ($F \cdot d^*$) is limited to about 30 mm when the $\bar{\mu}$ angle is 15° or less, due to interference between the film holder and other components of the apparatus. As the angle is increased, this distance becomes progressively more restricted, decreasing to about 20 mm when $\bar{\mu}$ is 30°. This means that an upper-level reciprocal plane can be recorded only if it lies within $F \cdot d^*/60 = 30/60 = 0.50$ rlu of the zero-level plane at small precession angles, and within $20/60 = 0.33$ rlu at $\bar{\mu} = 30°$. The significance of this, in working with small direct cells, can be illustrated with the monoclinic

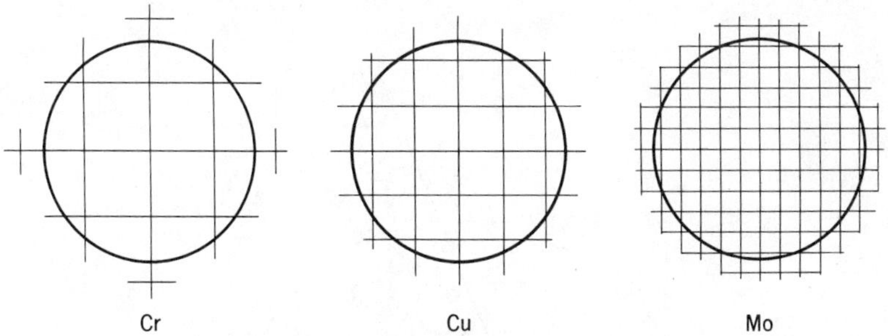

Fig. 9-13. The extent of the $(hk0)^*$ record of anatase (TiO_2) at $\bar{\mu} = 30°$, for several kinds of radiation.

compound cupric bromide ($CuBr_2$), which has the cell constants $a = 7.14$, $b = 3.46$, $c = 7.02$ Å, $\beta = 110°\ 05'$. The most favorable precession axis is the b axis because the zero level, $(h0l)^*$, permits a direct measurement of the β angle. The first level, $(h1l)^*$, is separated from this level by a distance $d^*_{010} = \lambda/3.46$. The value of this constant for four radiations is as follows:

	Cr	Fe	Cu	Mo
$d^*_{010} = \lambda/3.46 =$	0.66	0.56	0.44	0.21

The values show that the first level is beyond the reach of the instrument with Cr and Fe radiation, but it can be recorded with Cu radiation provided that a moderate $\bar{\mu}$ angle is used. The constant is sufficiently small, with Mo radiation, so that a film record of the second level, which is separated from the zero level by the spacing $2d^*_{010} = 0.42$ rlu, is possible. Records of one, or at most two levels, in addition to the zero level are usually sufficient to establish the symmetry and dimensions of the unit cell. The precession method can provide these records, but short-wavelength radiation is necessary when the period in the direct lattice along the precession axis is small.

Orientation Photographs

A valuable feature of the precession instrument is that a misalignment of the reciprocal plane, with respect to the film, can be detected and readily corrected.

The photograph of a properly oriented zero-level reciprocal plane shows the lattice net in a circular area which represents, in magnified

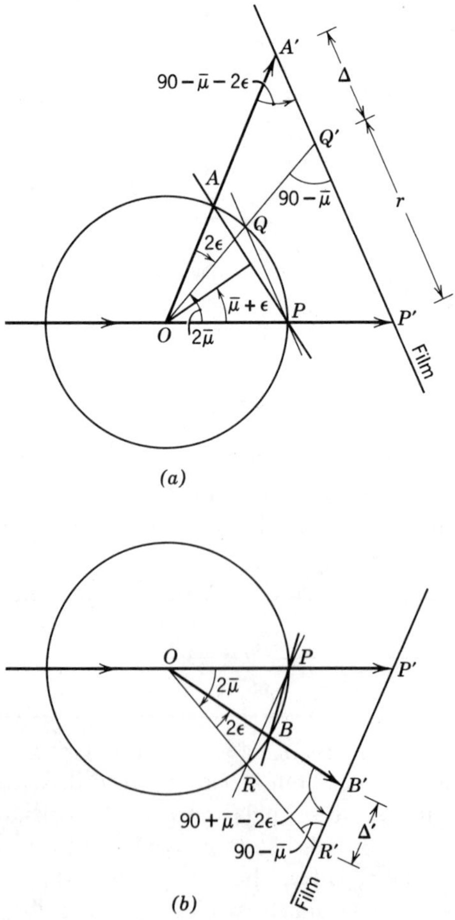

Fig. 9-14. Misalignment of the reciprocal plane.

form, that portion of the plane which passes through the sphere of reflection. The circular area is usually clearly defined if unfiltered x-radiation has been used, because the white radiation in the beam produces Laue trails from the center through the lattice points to the edge of the reflection circle (Fig. 9-3). The radius of the circle r, depends on the angle $\bar{\mu}$ and the crystal-to-film distance F. From Fig. 9-8

$$r = 2F \sin \bar{\mu}$$

If the precession axis has inadvertently been set at an angle $\bar{\mu} + \varepsilon$ to the x-ray beam (Fig. 9-14*a*), the zero level is not in the position PQ, but is instead situated at PA. The reflections now extend to A' on the upper half of the film instead of to Q'. The distance $Q'A' = \Delta$ is a measure of the misalignment of the precession axis and is easily expressed in terms of $\bar{\mu}$ and the angular alignment error ε. In the triangle $OQ'A'$

$$\frac{Q'A'}{OQ'} = \frac{\Delta}{F} = \frac{\sin 2\varepsilon}{\sin(90 - \bar{\mu} - 2\varepsilon)} = \frac{\sin 2\varepsilon}{\cos(\bar{\mu} + 2\varepsilon)}$$

and

$$\Delta = \frac{F \sin 2\varepsilon}{\cos(\bar{\mu} + 2\varepsilon)}$$

Thus the Laue trails will extend a maximum distance of

$$r + \Delta = F\left[2 \sin \bar{\mu} + \frac{\sin 2\varepsilon}{\cos(\bar{\mu} + 2\varepsilon)}\right]$$

from the center of the film at P'.

Figure 9-14*b* shows the situation after the crystal has precessed through 180°. The precession axis now makes an angle $\bar{\mu} - \varepsilon$ with the beam. The reciprocal plane is in the position PB, and the reflections extend to B' instead of R'. The distance $R'B' = \Delta'$ is also a measure of the misalignment of the crystal. In the triangle $OR'B'$

$$\frac{R'B'}{OR'} = \frac{\Delta'}{F} = \frac{\sin 2\varepsilon}{\sin(90 + \bar{\mu} - 2\varepsilon)} = \frac{\sin 2\varepsilon}{\cos(\bar{\mu} - 2\varepsilon)}$$

and

$$\Delta' = \frac{F \sin 2\varepsilon}{\cos(\bar{\mu} - 2\varepsilon)}$$

The distance from the film center to the edge of the reflection field is then

$$r - \Delta' = F\left[2 \sin \bar{\mu} - \frac{\sin 2\varepsilon}{\cos(\bar{\mu} - 2\varepsilon)}\right]$$

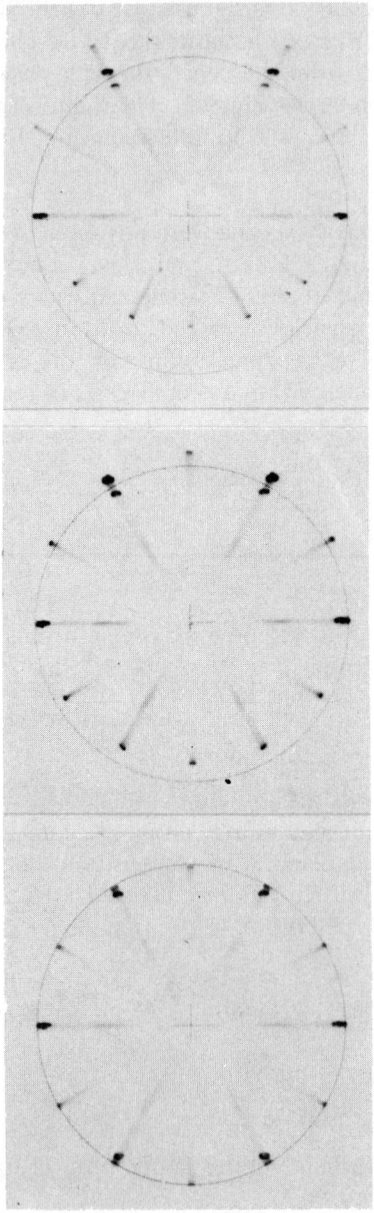

Fig. 9-15. Orientation photographs at $\bar{\mu} = 10°$ and ε, respectively, 0°, 1°, and 3° in setting the H or dial axis.

A misalignment of the precession axis, and its direction, can be detected by describing on the film, a circle of radius $2F \sin \bar{\mu}$ centered at the intersection of the Laue trails (Fig. 9-15). The angular alignment error ε may be determined by measuring either $(r + \Delta)$ or $(r - \Delta')$; the first value can be estimated more accurately, especially for large angular errors.

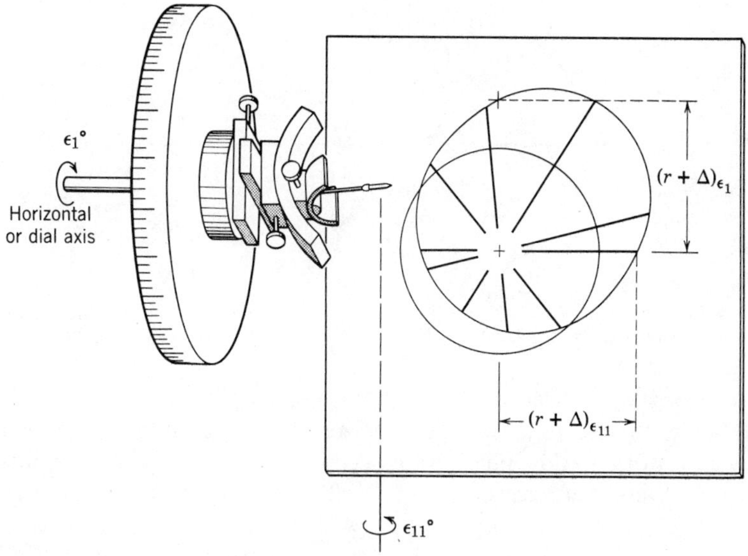

Fig. 9-16. A misalignment that requires a correction of both the *H* and *V* settings.

A precession photograph, in which the Laue trails have both a vertical and horizontal displacement (Fig. 9-16), indicates that the crystal requires adjustment about both oscillation axes of the instrument. The orientation error has two components on the film:

1. $(r + \Delta)_{\varepsilon_1}$ measured along the vertical axis of the film requires a rotation of ε_1 about the horizontal or dial axis of the precession instrument.

2. $(r + \Delta)_{\varepsilon_{11}}$, measured along the horizontal film axis represents an angular error ε_{11} with respect to the vertical oscillation axis. This correction is effected by an adjustment of the arc settings on the crystal holder.

The Preparation of Orientation Photographs

In low-symmetry crystals, upper levels of the reciprocal lattice may be offset with respect to the zero level (as in Fig. 9-6), in which case the

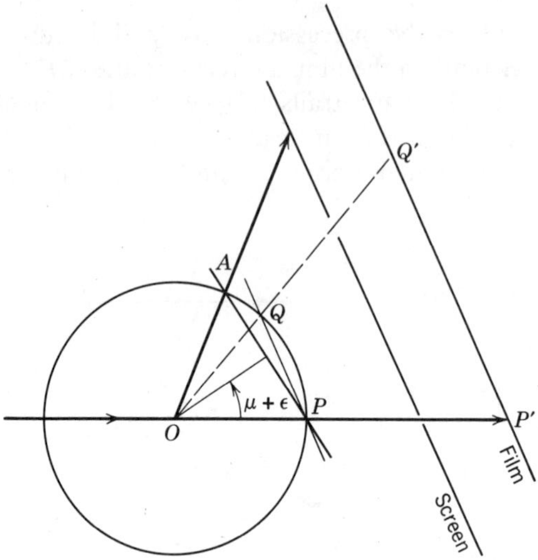

Fig. 9-17. The reflection OA, which results from the misalignment of the zero-level plane PQ, is intercepted on the screen, unless the error ε is very small.

record of these reflections will not be centered at the film center. It is advisable, therefore, to permit only zero-level reflections to reach the film in an orientation photograph. A reciprocal-level screen cannot be used for this purpose because it will also screen those zero-level

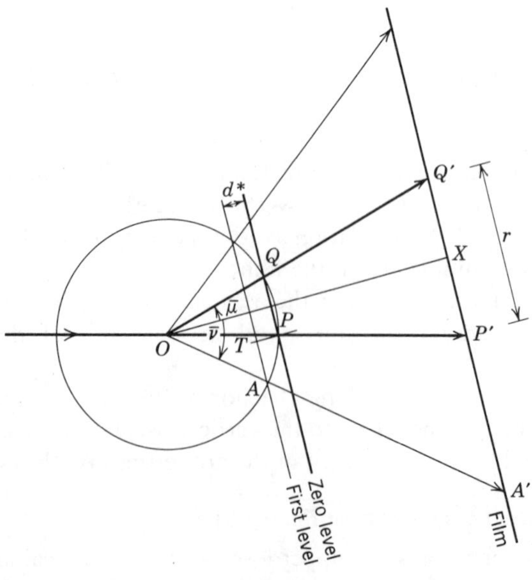

Fig. 9-18

reflections which result from the misalignment of the reciprocal plane (Fig. 9-17), unless the error is very small.

A circular area at the center of the first level (of radius $AT = CT$ in Fig. 9-6) does not pass through the surface of the sphere during the precession motion. Consequently, planes represented by reciprocal points in this region do not reflect, and the film shows a blank area at its center, the size of which increases with decreasing $\bar{\mu}$ and increasing d^*. If $\bar{\mu}$ is chosen sufficiently small, the reflection field of the zero level will lie entirely within this blank zone and no level screen is necessary. The geometrical relations are shown in Fig. 9-18. OX coincides with the precession axis and is, therefore, normal to the film. The diameter of the

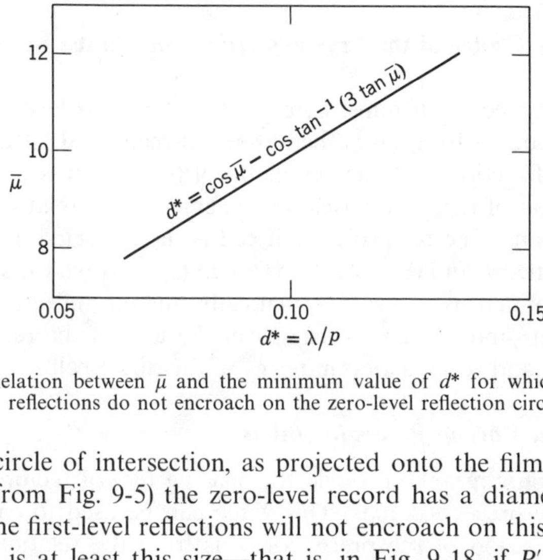

Fig. 9-19. Relation between $\bar{\mu}$ and the minimum value of d^* for which the first level reflections do not encroach on the zero-level reflection circle.

zero-level circle of intersection, as projected onto the film, is $P'Q'$, and therefore (from Fig. 9-5) the zero-level record has a diameter $2P'Q'$ on the film. The first-level reflections will not encroach on this record if the blank spot is at least this size—that is, in Fig. 9-18, if $P'A' > P'Q'$, or stated another way, if $XA' > 3XP'$. The relation between the minimum value of $\bar{\nu}$, the half-apex angle of the first level cone, and $\bar{\mu}$ is therefore

$$\frac{\tan\bar{\nu}}{\tan\bar{\mu}} = \frac{XA'/OX}{XQ'/OX} = \frac{XA'}{XQ'} = \frac{3}{1}$$

and therefore the minimum value of $\bar{\nu}$ is $\tan^{-1}(3\tan\bar{\mu})$.

The angle $\bar{\nu}$ depends on both $\bar{\mu}$ and d^* according to the relation

$$d^* = \cos\bar{\mu} - \cos\bar{\nu}.$$

derived previously. The minimum value of d^*, at which the first level reflections will just clear the zero-level record, is given by

$$d^* = \cos\bar{\mu} - \cos\tan^{-1}(3\tan\bar{\mu})$$

Figure 9-19 demonstrates this relation for $\bar{\mu}$ values ranging from 8 to 12°. The graph indicates that for a precession angle of 10°, the minimum value of d^* that gives satisfactory clearance between the zero- and upper-level reflections, is 0.1 rlu. This corresponds to a maximum value for the direct-lattice period p, along the precession axis of 7.1 Å for Mo radiation ($p = \lambda/d^* = 0.710/0.1$), 15.4 Å for Cu radiation, and 22.9 Å for Cr radiation. The angle is therefore satisfactory for most cases. Furthermore, the scale of the film record is sufficiently large to detect errors of 5′, yet the film will accommodate errors of more than 5°. Figure 9-20 (page 273) gives for this angle the angular error (ε) plotted against the linear displacement of the reflection circle ($r + \Delta$).

The Choice of the Crystal's Orientation in the Beam

A series of precession films, consisting of the zero level and one or more levels parallel to it, gives the pattern of reciprocal lattice points in space. The diffraction or Laue symmetry of the crystal is deduced from a consideration of the geometrical arrangements and relative intensities of the reflections. The reciprocal unit cell is then selected in conformity with the symmetry, and the reflections are indexed. Next, the space-group type is determined from the systematically missing reflections. Finally, the dimensions and angular constants of the unit of the reciprocal cell are measured and recalculated in terms of the direct cell.

Function of the Various Precession Films

Cone-axis photographs provide the spacing (d^*) of reciprocal planes normal to the precession axis. This value can be used to calculate the direct-lattice period of the precession axis from the relation $p = \lambda/d^*$. However, the reflections which outline cone-axis circles tend to be broad, and the circles are often poorly defined. For these reasons, cone-axis photographs are not suitable for the accurate estimate of lattice constants. Their main function is to provide d^* to calculate two of the instrumental settings for recording n-levels, namely, the film displacement $F \cdot d^*$, and the screen setting $s = [r_s \cot \cos^{-1}(\cos \bar{\mu} - d^*)]$.

The amount of "shear" or displacement between films of successive levels of crystals with oblique cells may be used to calculate certain angles and periods in the reciprocal cell. However, the values depend on measuring very small distances on the film and consequently lack precision.

The most accurate method of determining reciprocal-lattice dimensions and angles with the precession instrument is to photograph levels

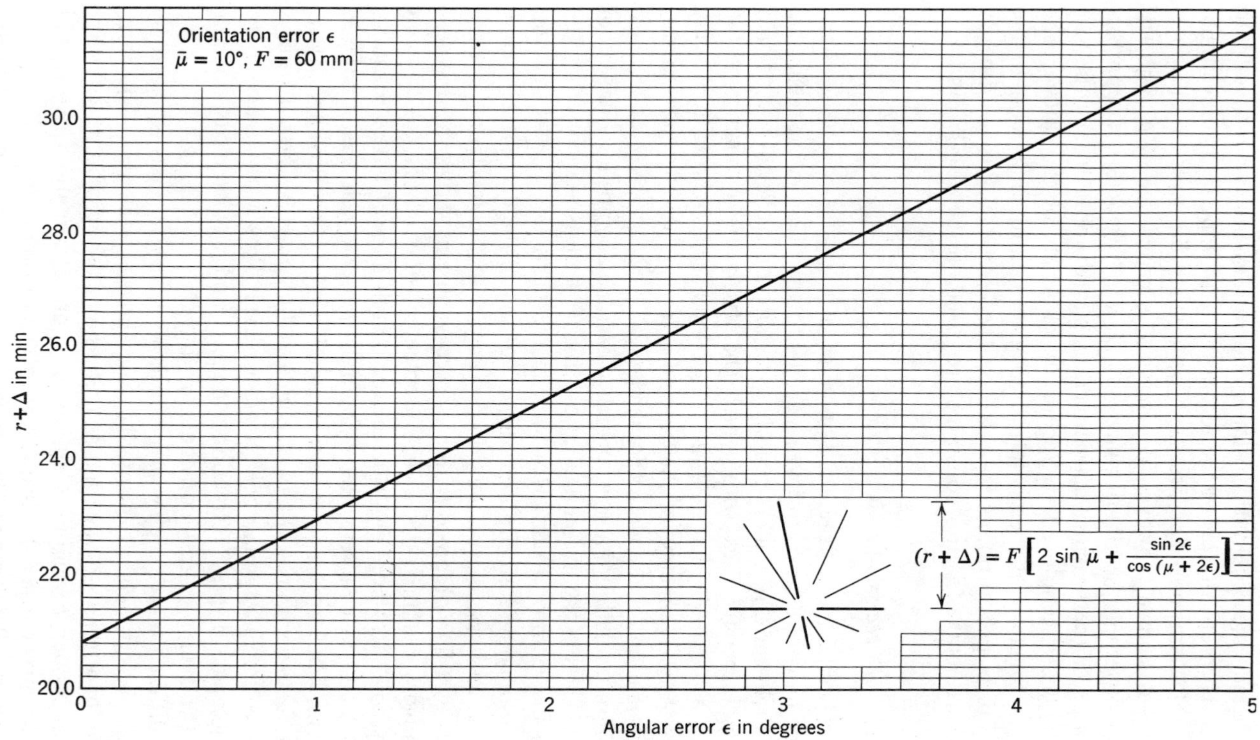

Fig. 9-20. Relation between the displacement of Laue trails $(r + \Delta)$ and the angular error of misalignment ε when $\bar{\mu} = 10°$.

which include the required constants and to measure them directly on the films. Theoretically, either the zero-level or an upper-level film should serve for this purpose. The precision with which the scale of an upper-level film is known, however, depends on a knowledge of d^* and the accuracy of setting the film displacement $(F \cdot d^*)$. In practice, an approximate value of the spacing, determined as stated above from cone-axis photographs, is employed to record upper levels of the lattice. The important function of these films is to give a third dimension to the pattern of reciprocal points recorded on the zero-level film—ordinarily all measurements of lattice constants are made on the zero-level film.

The Choice of Precession Axis

Theoretically, the diffraction symmetry of a crystal can be deduced from the films of a series of levels parallel to any plane in the reciprocal lattice. Practically, it is more readily ascertained if the levels are normal to crystallographic axes with which the *diagnostic* symmetry of the crystal is associated—that is, if the precession axes are chosen to display the diagnostic symmetry on the films of a series. Table 9-1 summarizes this information for the eleven Laue classes.

The zero level film of a series which shows the diagnostic symmetry of the Laue class of a cubic crystal also provides a measurement of a^*, the edge of the reciprocal unit. However, in the remaining crystal systems, the corresponding film supplies only a portion of the data which define the shape and size of the unit. For example, in the tetragonal system the zero-level film obtained with c as precession axis shows the diagnostic symmetry but provides a measurement of only a^* (Tables 9-1 and 9-2); a second zero-level film is needed to establish the c^* lattice constant. It is advisable to choose a precession axis that will yield easily indexed reflections. Thus an a axis in a tetragonal crystal is an obvious choice, because the reflections will have the indices $(0kl)$ or $(h0l)$. The situation is analogous for hexagonal crystals; zero-level films with a and c as precession axes display diffraction symmetry and provide measurements for the two reciprocal cell constants, a^* and c^*.

The diffraction symmetry of orthorhombic crystals is displayed by films of levels which are normal to the three crystal axes. The zero-level film of a series—say, with a as precession axis—provides two reciprocal dimensions, b^* and c^*. Consequently, precession about two crystal axes is necessary to fix the size of the reciprocal unit.

Reciprocal cells of monoclinic crystals are defined by four constants: a^*, b^*, c^*, and β^*. Films of levels normal to the b axis are most useful—they show the characteristic twofold symmetry associated with this

Table 9-1. Symmetry Exhibited by Precession Films of Selected Reciprocal Levels

Crystal System	Laue Class	Precession Axis	Symmetry	
			Zero Level	Upper Levels
Cubic	$\frac{4}{m}\bar{3}\frac{2}{m}$	a_1, a_2, or a_3	$4\,m\,m$	$4\,m\,m$
	$\frac{2}{m}\bar{3}$	a_1, a_2, or a_3	$2\,m\,m$	$2\,m\,m$
Tetragonal	$\frac{4}{m}\frac{2}{m}\frac{2}{m}$	c	$4\,m\,m$	$4\,m\,m$
	$\frac{4}{m}$	c	4	4
Hexagonal	$\frac{6}{m}\frac{2}{m}\frac{2}{m}$	c	$6\,m\,m$	$6\,m\,m$
	$\frac{6}{m}$	c	6	6
	$\bar{3}\frac{2}{m}$	c	$6\,m\,m$	$3\,m$
	$\bar{3}$	c	6	3
Orthorhombic	$\frac{2}{m}\frac{2}{m}\frac{2}{m}$	a, b, or c	$2\,m\,m$	$2\,m\,m$
Monoclinic	$\frac{2}{m}$	b	2	2
Triclinic	$\bar{1}$	a, b, or c	2	1

axis, and the zero-level film, which is composed of $h0l$ reflections, permits the measurement of a^*, c^*, and β^*. The b^* dimension is given by the zero-level record prepared with either the a or c axis as precession axis.

Triclinic crystals require the determination of six constants: a^*, b^*, c^*, α^*, β^*, and γ^*. Each film of a zero level which is normal to a crystal axis—such as the a axis—provides the length of two translations and the angle between them—b^*, c^*, and α^*. Films of three zero levels, prepared with each of the crystal axes as precession axes, are necessary to provide direct measurements of the constants.

Table 9-2. Most Favorable Crystal Orientation for Determination of Reciprocal Constants

Symmetry System	Precession Axis	Indices of Reflections on Zero-Level Film	Reciprocal Elements from Zero Level	Most Favorable Plane Normal to Dial Axis
Cubic	a	$(0kl)$	a^*	$\{100\}$
Tetragonal	c	$(hk0)$	a^*	$\{100\}$
	a	$(0kl)$	a^* c^*	
Hexagonal	c	$(hk\bar{i}0)$	a^*	$\{10\bar{1}0\}$
	a	$(h0\bar{h}l)$	a^* c^*	
Orthorhombic	a	$(0kl)$	b^* c^*	(100) or
	b	$(h0l)$	a^* c^*	(010) or
	c	$(hk0)$	a^* b^*	(001)
Monoclinic	a	$(0kl)$	b^* c^*	(100) or
	b	$(h0l)$	a^* c^* β^*	(001)
	c	$(hk0)$	a^* b^*	
Triclinic	a	$(0kl)$	b^* c^* α^*	(100) or
	b	$(h0l)$	a^* c^* β^*	(010) or
	c	$(hk0)$	a^* b^* γ^*	(001)

Relation Between the Axis of Crystal Holder and Precession Axes

The preceding section has established that unless a crystal has cubic symmetry, it must be presented in two or even three different orientations in the x-ray beam, if the diffraction symmetry is to be displayed to best advantage and the lattice constants measured on zero-level films. By properly orienting a crystal with respect to the horizontal or dial axis of the instrument (H of Fig. 9-7), one mount can be made to serve for two zero-level films. This advantage is a consequence of the precession axis being normal to the dial axis. If the crystal plane that includes two selected precession axes is oriented normal to the dial axis, first one and then the other axis can be oriented for precession by a rotation of the dial. For example, if the crystal is mounted so that its (010) plane is normal to the dial axis, the a and c crystal axes are also normal to it and can be brought into coincidence with the beam for precession by a rotation of the dial. An examination of Table 9-2 shows that, except for triclinic crystals, two properly chosen precession axes can provide films for direct measurement of all reciprocal constants. In the triclinic case, the zero-level films from a crystal that has been mounted with the (010) plane normal to the dial axis give the following data:

Precession Axis	Reciprocal Elements
a	b^* c^* α^*
c	a^* b^* γ^*

The missing reciprocal-lattice element is β^*. Now the difference in reading on the dial of the instrument for the two zero-level settings is equal to the angle between the a and c axes, which is the angle β (or $180° - \beta$). The two triclinic relations (Appendix 6)

$$V^* = a^*b^*c^*[1 - \cos^2 \alpha^* - \cos^2 \beta^* - \cos^2 \gamma^* + 2 \cos \alpha^* \cos \beta^* \cos \gamma^*]^{1/2}$$

$$V^* = a^*b^*c^* \sin \alpha^* \sin \beta \sin \gamma^*$$

can be combined to give

$$\sin \alpha^* \sin \beta \sin \gamma^* = [1 - \cos^2 \alpha^* - \cos^2 \beta^* - \cos^2 \gamma^* + 2 \cos \alpha^* \cos \beta^* \cos \gamma^*]^{1/2}$$

from which the unknown reciprocal element, β^*, can be calculated.

Orienting a Crystal for Precession with Two Axes

Crystals are easily oriented on the two-circle goniometer for precession with two axes, if the requisite faces are present and give good optical reflections. However, if the faces are so unfavorable as to permit only a rough orientation, the final adjustment can be made on the precession instrument.

When the plane which contains the precession axes (say, the $ac = (010)$ plane) is precisely oriented normal to the axis of the crystal holder, a reciprocal axis (the b^* axis) is parallel to it and manifests itself as a horizontal row on the film. The true horizontal of the instrument is accurately marked by two holes drilled in the back of the film holder, which can be opened to admit light and to produce fiducial spots on the film. A misalignment of the plane gives the row an angular displacement with respect to the horizontal (Fig. 9-21). The correction is made in two steps. The crystal holder is clamped with one of its arc adjustments parallel to the film. After exposure, the angular error is measured on the film and corrected by adjusting the setting of the arc. The procedure is then repeated with the other arc parallel to the film. The reciprocal

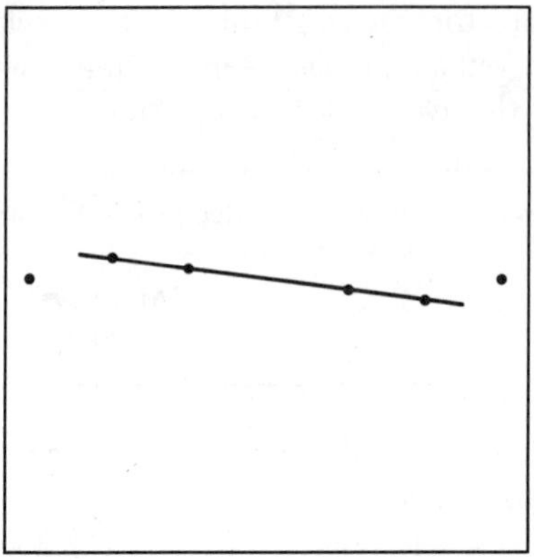

Fig. 9-21. Fiducial spots for aligning a reciprocal row with the *H* or dial axis of the instrument.

planes, $(0kl)^*$ and $(hk0)^*$, normal respectively to the a and c axes of the crystal, are now parallel to the axis of the crystal holder. The planes can therefore be arranged parallel to the film for precession, each in turn, by rotating the crystal about this axis—the dial axis of the instrument, the precise setting being determined by means of orientation photographs.

Measurement of Zero-Level Films

A Measuring Instrument

Measurements on zero-level precession films are conveniently made with the instrument depicted in Fig. 9-22. The film is fastened to a square of glass (1), fitted into a circle of metal (2) that is free to rotate in the metal frame (3). The circle is graduated in degrees, and a vernier (4) permits estimating a setting to two minutes of arc. A cursor (5), with a fine vertical hairline rides on a horizontal millimeter bar-scale (6), which is attached to the lower part of the frame. A setting can be estimated on this scale to within 0.05 mm with the aid of a vernier.

The Determination of Lattice Constants

The angle between two reciprocal-lattice rows may be measured with the instrument by setting the hairline on one row and then rotating the circle and shifting the cursor until the hairline rests on the other row. The difference in the readings for the two positions of the graduated circle is equal to the angle.

The circle also serves to rotate rows into a position parallel to the hairline, so that distances between the rows may be measured on the horizontal bar-scale. If the measurement is between adjacent rows, and

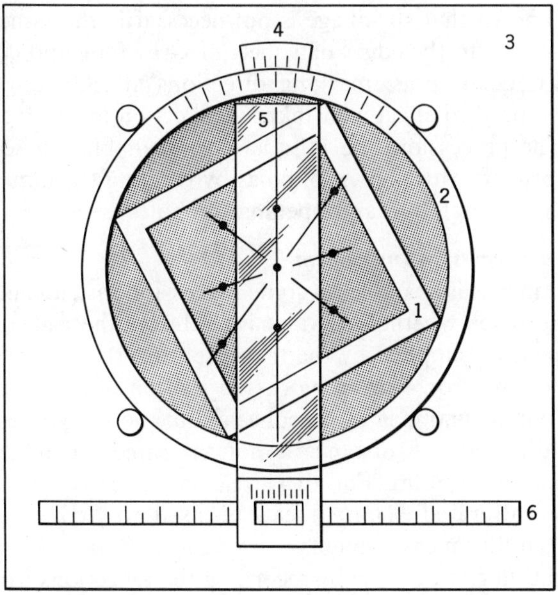

Fig. 9-22. A measuring device for precession films.

successive levels superimpose without offsets, the distance s (in mm) is numerically equal to the reciprocal spacing magnified by a factor F, namely Fd^* (in rlu). Then $d^* = s/F$. Reciprocal spacings (in rlu) are related to periods of the direct lattice (in Å units) as

$$p_{hkl} = \frac{\lambda}{d^*_{hkl}}$$

Correction for Film Shrinkage

The fiducial spots, printed by opening the two holes in the back of the film holder, mark the dial-axis (or horizontal) direction on the film. The distance s_F between the holes can be accurately determined, for example, by direct measurement with a travelling microscope. If $s_{F(\text{meas})}$ is the measured distance between the fiducial spots, the ratio

$$\frac{s_F}{s_{F(\text{meas})}}$$

is a factor which can be applied to film measurements to correct for film shrinkage. The corrected value for the reciprocal spacing is therefore

$$d^* = \frac{s}{F} \cdot \frac{s_F}{s_{F(\text{meas})}}$$

It is well known that shrinkage is not necessarily the same in the two directions parallel to the edges of a sheet of x-ray film, and consequently two sets of fiducial spots, marking directions at right angles to each other, are required to effect a shrinkage correction for all measurements on a film. The precession film is square, thus enabling a second set of spots to be printed with one set of holes by rotating the film or the back of the holder, through 90°, and opening the holes a second time.

Calibrating a Precession Instrument

The F distance is adjustable on some precession instruments and fixed on others. However, whether the value is read from the scale of an adjustable instrument or supplied by the manufacturer, the distance is subject to a construction error. Furthermore, the motion of the precession axis and/or the film normal may be eccentric about the x-ray beam direction, and then the *effective* F distance is not the same for all directions of measurement on the film. For these reasons, a precession instrument needs to be calibrated before it can be used to derive accurate cell constants from film measurements.

The calibration can be done by recording the reflections from a crystal whose cell constants are very accurately known and by calculating the F distances for each film measurement from the relation

$$d^* = \frac{\lambda}{p} = \frac{s}{F} \cdot \frac{s_F}{s_{F(\text{meas})}}$$

Table 9-3. Determination of F for Three Directions of Measurement on an $(hk\bar{i}0)^*$—Level Precession Film of Quartz

Direction of Measurement on Film	Film Measurement s	$s(s_F/s_{F(\text{meas})}) = 4F \cdot d^*_{10\bar{1}0} = 4F(\lambda_{\text{Cu}K\beta}/a_{\text{quartz}})^a$	F(calc)
N–S	68.20 mm	68.30 mm	60.25 mm
NE–SW[b]	68.50	68.60	60.51
NW–SE[b]	68.50	68.60	60.51

s_F, the distance between the fiducial holes in the camera back = 122.17 mm, $s_{F(\text{meas})}$, the measured distance between the fiducial spots = 122.00 mm; $\lambda_{\text{Cu}K\beta}$ = 1.39217 Å; a_{quartz} = 4.9125 Å [based on a recent determination, by H. D. Keith, *Am. Mineralogist*, **40**, 530 (1955), rounded here to five figures].
[b] At 60° to the N–S direction.

Clear, well-crystallized quartz (naturally occurring SiO_2 with symmetry 32) is useful for this purpose. The cell constants of this material differ from locality to locality, but the variation amounts to less than 1 part in

10,000. Because the accuracy that can ordinarily be expected from the precession instrument is less than this, the material is acceptable as a standard. In any case, the cell constants of quartz from a particular locality can be accurately determined on the finely ground powder by the Debye–Scherrer method.

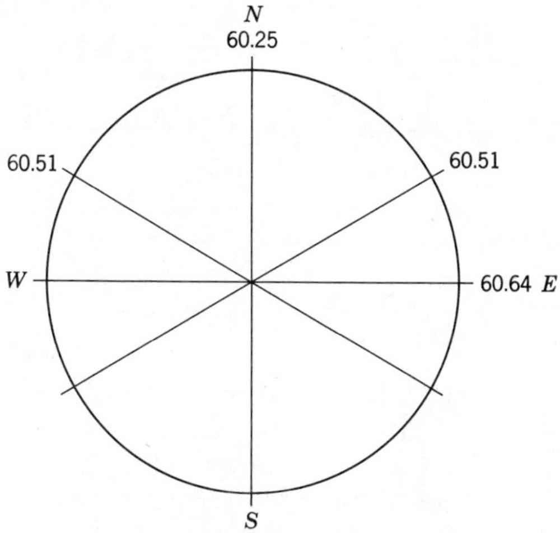

Fig. 9-23. Variation of F with direction on the film from one precession instrument.

Figure 9-3 shows a reproduction of the $(hk\bar{i}0)^*$-level film obtained from a quartz crystal measuring 0.2 mm in cross section. The film allows three determinations of F, at angular intervals of 60°, from measurements of the photographic representation of the spacing, $d^*_{10\bar{1}0}$. Table 9-3 gives the results (column 2) of measuring a distance s, equal to four times this "film" spacing, between rows of the appropriate β reflections (rather than the less sharp α reflections) on the film. After correcting for film shrinkage (column 3), this quantity is equal to $4F \cdot d^*_{10\bar{1}0} = 4F(\lambda_{\text{CuK}\beta}/a_{\text{quartz}})$. Column 4 shows that the F values depart markedly from 60.0 mm in this instrument and differ noticeably for different directions on the film. A plot of F against film direction (Fig. 9-23) outlines an ellipse whose principal axes have the values: N–S = 60.25 and W–E = 60.64 (by extrapolation).

CHAPTER TEN

THE ROTATION AND OSCILLATION METHODS

In the methods described in this chapter the crystal is mounted with a selected axis of the direct lattice parallel to the axis of the crystal holder (in the precession method, a selected direct-lattice axis is normal to this direction). In the rotation method the crystal is rotated continuously about this axis during the exposure. If the angular range is restricted, so that less than the full spectrum of reflections is generated, the record is described as an oscillation, rather than a rotation, photograph.

Rotation and oscillation photographs present on one film the reflections from a number of levels of the reciprocal lattice, clearly separated from each other. Thus one index of each reflection is available by inspection. They provide a simple and rapid way of measuring the period of the (direct-lattice) rotation axis. For these reasons the methods were formerly much used for the complete analysis of a crystal. Although superseded for this purpose, they remain convenient and valuable methods for measuring a direct-lattice period—especially the oscillation method, because it gives a usable photograph in a shortened exposure time.

The methods continue to have another important function. A photograph (preferably oscillation because of exposure time) is invariably prepared as a preliminary to a Weissenberg study: it serves to test the precision of the crystal's orientation and provides the instrumental settings for recording upper levels with the Weissenberg method.

Oscillation photographs are more easily indexed than their rotation counterparts—fewer reflections are recorded on the film, and consequently the possibility of overlapping reflections and general complexity of the photograph is reduced.

Rotation Photographs

The Method

A beam of characteristic x-rays is directed at the chosen axis, usually at an angle of 90° because this simplifies the interpretation of the photograph. Figure 10-2 shows the relation between the x-ray beam direction, the rotating crystal axis through O at the center of the sphere of reflection, and the orientation and motion of reciprocal planes normal to the rotation axis. In this case the rotation axis is $c[001]$. The undeviated beam emerges from the sphere at P, and consequently this point marks the origin of the reciprocal lattice. The reciprocal planes which are normal to the c axis and consist respectively of $hk0^*$, $hk1^*$, and $hk2^*, \ldots$ points, simultaneously rotate about a parallel axis through P. The planes are spaced at intervals of $d^*_{001} = \lambda/c$.

During the rotation, $hk0^*$ reciprocal points pass through the surface of the sphere on the circumference of a circle whose trace is AP. The

Fig. 10-1. A single-crystal rotation/oscillation goniometer. (Courtesy Unicam Instruments Ltd., Cambridge, England.)

corresponding $hk0$ reflections emanate from O and lie in the plane of this circle—that is, on the surface of a flat cone. The $(hk1)^*$ and $(hk\bar{1})^*$ reciprocal planes cut the sphere in circles whose traces are respectively BC and $B'C'$. The corresponding $hk1$ and $hk\bar{1}$ reflections are lines in symmetrical Laue cones, coaxial with the rotation axis, and characterized

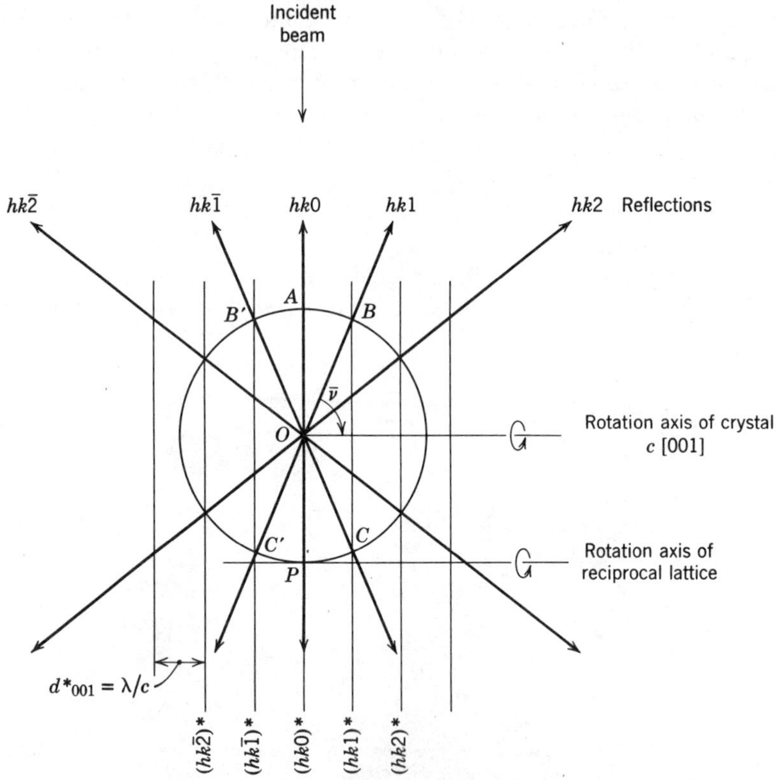

Fig. 10-2. A section through the sphere of reflection showing reciprocal planes and associated Laue cones for a crystal rotating about the c[001] axis.

by the half-apex angle $\bar{\nu}$. Similarly, the $(hk2)^*$ and $(hk\bar{2})^*$, and other reciprocal planes which happen to cut the sphere, result in pairs of symmetrical Laue cones, also coaxial with the rotation axis, but with smaller half-apex angles.

Recording the Reflections

The reflections are usually recorded on a cylindrical film whose axis coincides with the rotation axis. The beam is generally admitted between

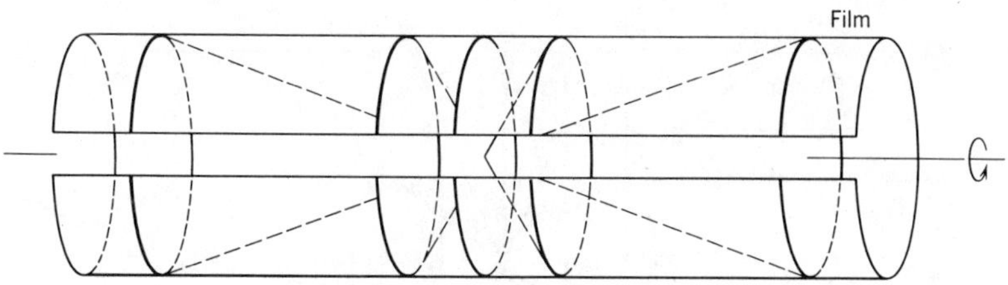

Fig. 10-3. Diffraction cones intersecting a coaxial, cylindrical film.

the edges of the film (Fig. 10-3). The cones intersect the film in a set of circles, and when the film is unrolled the reflections which outline the various circles appear in straight lines (Fig. 10-4). These rows of reflections are known as *layer lines*, because they are related to layers or levels in the reciprocal lattice. The layer lines are the most striking and useful feature of rotation photographs.

Determination of the Direct-Lattice Period in the Rotation Axis

Figure 10-5 shows a section through the cylindrical film, parallel to and including the rotation axis. The half-apex angle $\bar{\nu}$ of a diffraction cone depends on the spacing XY of the reciprocal planes. If the rotation

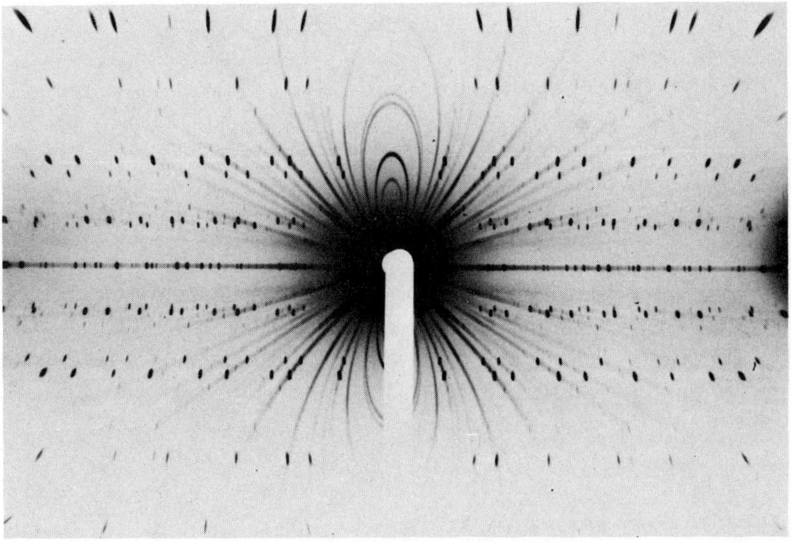

Fig. 10-4. Normal beam rotation photograph of quartz (SiO_2); rotation axis $c[0001]$; unfiltered Cu-radiation, film radius 28.65 mm, about 7/10 actual size.

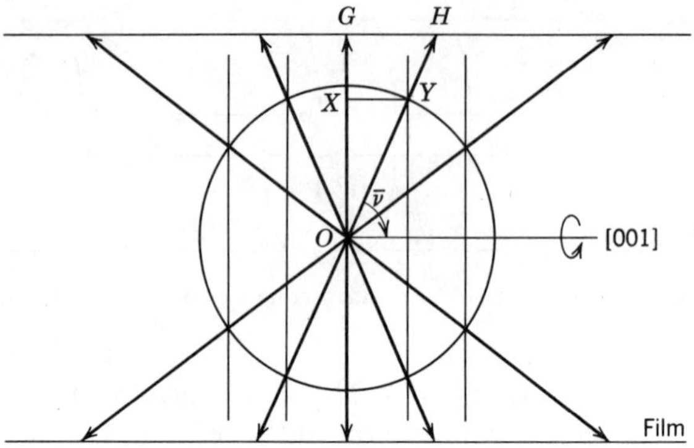

Fig. 10-5. Relation between a film measurement and the cone angle $\bar{\nu}$ on a rotation photograph.

axis is $c[001]$, $XY = d^*_{001}$. Now $d^*_{001} = \lambda/c$ and thus, in terms of the direct lattice, the angle $\bar{\nu}$ depends on the wavelength of the incident radiation and on the period in the rotation axis. The relation for the first-order diffraction cone is given by the triangle OXY

$$\cos \bar{\nu} = \frac{XY}{OY} = \frac{\lambda}{c}$$

because OY is a radius of the unit sphere. Then

$$c = \frac{\lambda}{\cos \bar{\nu}}$$

For the nth order cone

$$\cos \bar{\nu}_n = n \cdot XY = n\frac{\lambda}{c}$$

and

$$c = n\frac{\lambda}{\cos \bar{\nu}_n}$$

The angle $\bar{\nu}$ is readily determined from measurements on the film. Triangle OGH is similar to triangle OXY. OG is the radius r of the camera; GH is the distance y between the zero and first layer lines as measured on the film. Then

$$\cot \bar{\nu} = \frac{GH}{OG} = \frac{y}{r}$$

and

$$c = \frac{\lambda}{\cos \cot^{-1}(y/r)}$$

For the nth layer this relation becomes

$$c = n \frac{\lambda}{\cos \cot^{-1}(y_n/r)}$$

where y_n is half the distance between the pair of n layer lines.

The β component of the beam produces a similar set of layer lines but with smaller y values. Usually this radiation is not filtered from the beam, unless the two sets of reflections are difficult to distinguish. In fact, the presence of β reflections can be an advantage. The α layer lines are actually $\alpha_1\alpha_2$ doublets which, at best, are only imperfectly resolved. As a result, the reflection spots tend to be elongated. The most intense portion of each layer line corresponds reasonably well with the weighted average wavelength $(2\alpha_1 + \alpha_2)/3$, and this value is customarily used in the calculation of the rotation period. The β component of the beam is essentially monochromatic, and consequently the reflection spots are sharper and may lead to more accurate estimations of the rotation period.

The rotation method is a rapid and reasonably accurate method for determining the direct-lattice period which coincides with the rotation axis of the crystal. In the past, unit-cell dimensions were often based entirely on rotation photographs, prepared by remounting and rotating a crystal successively about a requisite number of crystallographic axes, and, in the case of monoclinic and triclinic crystals, by measuring the interaxial angles on an optical goniometer. It is found not infrequently that such determinations are in error when they are repeated with one of the newer methods, such as the precession or Weissenberg methods which give an unambiguous determination of the size and shape of the unit cell from one crystal mount. The error can usually be traced to the difficulty of recognizing the requisite axes on small, imperfectly developed crystals. The early work on the orthorhombic mineral low-chalcocite (Cu_2S) illustrates this. The cell dimensions were originally given as

$$a = 11.8,\ b = 27.2,\ c = 22.7\ \text{Å}$$

largely on the basis of rotation studies. Later, Weissenberg films showed that the [102] axis had inadvertently been selected for rotation in place of the c[001] axis. The correct dimensions proved to be

$$a = 11.90,\ b = 27.28,\ c = 13.41\ \text{Å}$$

Table 10-1 gives the results of measuring a film (reproduced in Fig. 10-4) obtained by rotating a small (0.2 mm in cross section) clear crystal of quartz about its c[0001] axis. The calculations lead to a value of 5.41 Å for the c period with a probable accuracy of ± 0.01 Å, based on an

Table 10-1. Measurement of a Rotation Film of Quartz (SiO_2)

Rotation axis—c[0001]; Cu radiation ($\lambda_{K\alpha} = 1.5418$, $\lambda_{K\beta} = 1.3922$ Å); camera radius, $r = 28.65$ mm.

Layer Line =	1α	2α	3β	3α
$2y$ (measured mm)	17.05	39.80	69.75	94.65
y	8.53	19.90	34.88	47.33
$y/r = \cot \bar{v}$	0.298	0.6946	1.217	1.652
$\cos \bar{v}$	0.285	0.5705	0.7727	0.8555
$\dfrac{\lambda}{\cos \bar{v}} = \dfrac{c}{n}$	5.40	2.703	1.802	1.802
c	5.40	5.406	5.406	5.406

$c = 5.41$

estimated error of ± 0.05 mm in measuring y. The value for this period is acknowledged to be 5.404(5) $\pm$ 0.0002 Å on the basis of precise determinations in several laboratories by using the powder film method. The arrangement of the film in the standard rotation camera does not provide the means for determining the amount of film shrinkage due to darkroom processing. It is unlikely, however, that the greatest error in the rotation method lies in measuring the distance y on the film. To illustrate—another crystal of quartz, which gave markedly longer spots, led to a value of 5.42 Å due to uncertainties in setting the hairline of the measuring device. In general, only exceptional rotation photographs give an accuracy of one part in 500.

Indexing Rotation Photographs

It was shown in the preceding sections that reciprocal planes situated normal to the rotation axis manifest themselves as layer lines on rotation photographs and that the distances between these layer lines are measures of the periodicity along the rotation axis. A layer line is composed of the reflections that are produced when the points in a rotating reciprocal plane pass through the surface of the reflection sphere. The relative positions of the reflection spots in the layer line evidently depend on some detail of the reciprocal plane. This relation is examined in the following paragraphs; mathematical equations are developed which give the cylindrical coordinates, ζ and ξ, of any reciprocal point in terms of the film coordinates, x and y, of its reflection on the rotation photograph. It would appear, then, that having measured the film coordinates of a reflection, these equations might be used to index the reflection—that is, to identify the responsible reciprocal point in the lattice. The degree of certainty with which this can be accomplished for all reflections on the photograph will be considered.

Each reciprocal point (such as (011) of Fig. 10-6) can be referred to two cylindrical coordinates—ζ, measured parallel to, and ξ, measured normal to the rotation axis which passes through the origin of the reciprocal lattice. These coordinates are shown for a first-level reciprocal point Q

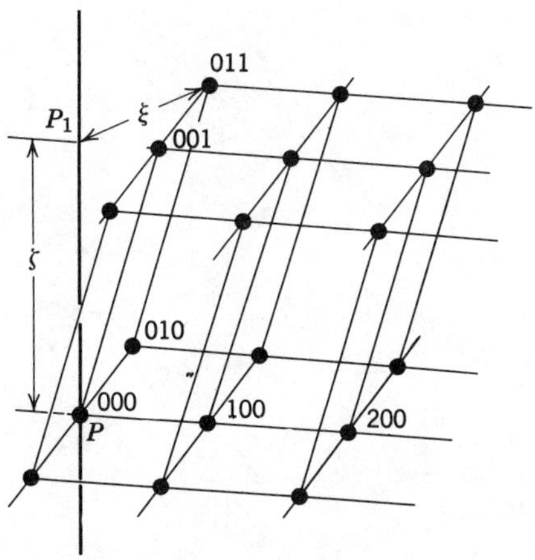

Fig. 10-6. The cylindrical coordinates of a reciprocal point referred to the rotation axis PP' and the origin.

on the connected sketches of Fig. 10-7. Figure 10-7*a* is a longitudinal section through the rotation camera; it includes the rotation axis of the crystal and the point Q which happens to be situated on the surface of the reflection sphere. Figure 10-7*b* is a section in the plane of the first level; the rotation axis of the crystal OO_1 is normal to the page in this view. The reciprocal lattice rotates about an axis through P_1, which lies a distance OO_1 directly above the point at which the undeviated beam emerges from the sphere.

The point Q results in a reflection at R whose position can be defined by the film coordinates x and y with zero values at the film center; x is measured along the equator and y along the meridian of the film, as illustrated in the unrolled view of the film in Fig. 10-7*c*. From Fig. 10-7*a*, $\zeta = OQ \cos \rho$ and $\cot \rho = y/r$, where OQ is a radius of the unit sphere and r is the radius of the camera. Consequently

$$\zeta = \cos \cot^{-1} y/r \tag{1}$$

It is evident from Fig. 10-7*a* that all reciprocal points in any one level normal to the rotation axis have the same ζ coordinate, and therefore the reflections have the same value of y. In other words, the reflections are aligned in rows on a cylindrical film: these are the layer lines mentioned previously.

The cylindrical coordinate ξ is the side P_1Q of the oblique triangle O_1P_1Q of Fig. 10-7*b*. In this triangle, O_1P_1 is a radius of the unit sphere,

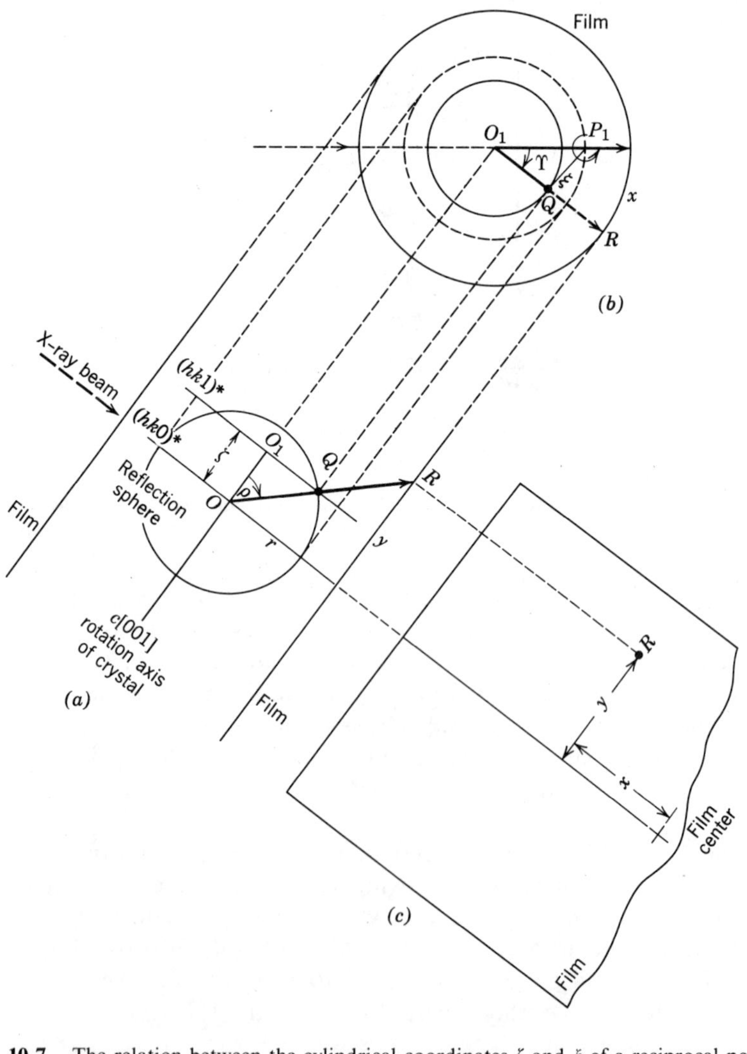

Fig. 10-7. The relation between the cylindrical coordinates ζ and ξ of a reciprocal point Q and the film coordinates x and y of its reflection spot R.

and $O_1Q = \sin\rho = \sin\cot^{-1}(y/r)$. The angle included by the latter two sides, Υ, has a value of x/r radians or $(x/r)(180/\pi)$ degrees. From the law of cosines[1]

$$\xi = \left[1 + \left(\sin\cot^{-1}\left(\frac{y}{r}\right)\right)^2 - 2\sin\cot^{-1}\left(\frac{y}{r}\right)\cos\left(\frac{x}{r}\cdot\frac{180}{\pi}\right)\right]^{1/2} \tag{2}$$

To eliminate the labor of calculating the cylindrical coordinates from these relations, Bernal devised charts which present the solutions for (1) and (2) in graph form. Figure 10-8 shows such a chart for a cylindrical film of radius 28.65 mm. The vertical lines on the chart are lines of constant ζ at intervals of 0.05 reciprocal lattice units (rlu) measured along the horizontal. The curves represent constant values of ξ for a range of ζ values. In practice, the chart is used as a transparent overlay on the illuminated rotation film, permitting the observer to read the values of ζ and ξ directly for each reflection.

The following example illustrates one method of indexing a rotation photograph. Let us assume that a series of rotation films have provided the dimensions of the reciprocal lattice and that we wish to index the film obtained by rotating the crystal about its $c[001]$ axis. The reflections in the zero-order layer line have indices $hk0$, in the first-order layer line $hk1$, and so on. Thus the l index for each reflection is obtained by inspection of the order of the layer line, and it is only necessary to identify the h and k indices for each reflection. A sufficient portion of the zero level of the reciprocal lattice is drawn on a suitable scale (Fig. 10-9). The ξ values, read from the zero layer line with the aid of the Bernal chart, are plotted to scale on a radial line PT. These values are the perpendicular distances to the reciprocal points from the rotation axis which passes through the origin at P. Each reflection can now be identified with its reciprocal point and indexed, by describing an arc from the plotted ξ position on PT through the lattice until it intersects a reciprocal point. The procedure is then repeated for the other layer lines on the photograph.

The example illustrates a weakness of the method. The ξ values for 200 and 030 are identical and there is no way of deciding whether the reflection is caused by one or both of the planes represented by the points. This is also true for the points 230 and 140 which exhibit a difference in ξ that is too small to be resolved by measurement on the film. Reciprocal lattices with very small periods are more difficult to index, because they offer more chances for points to have similar ξ distances.

[1] $a = [b^2 + c^2 - 2bc\cos A]^{1/2}$

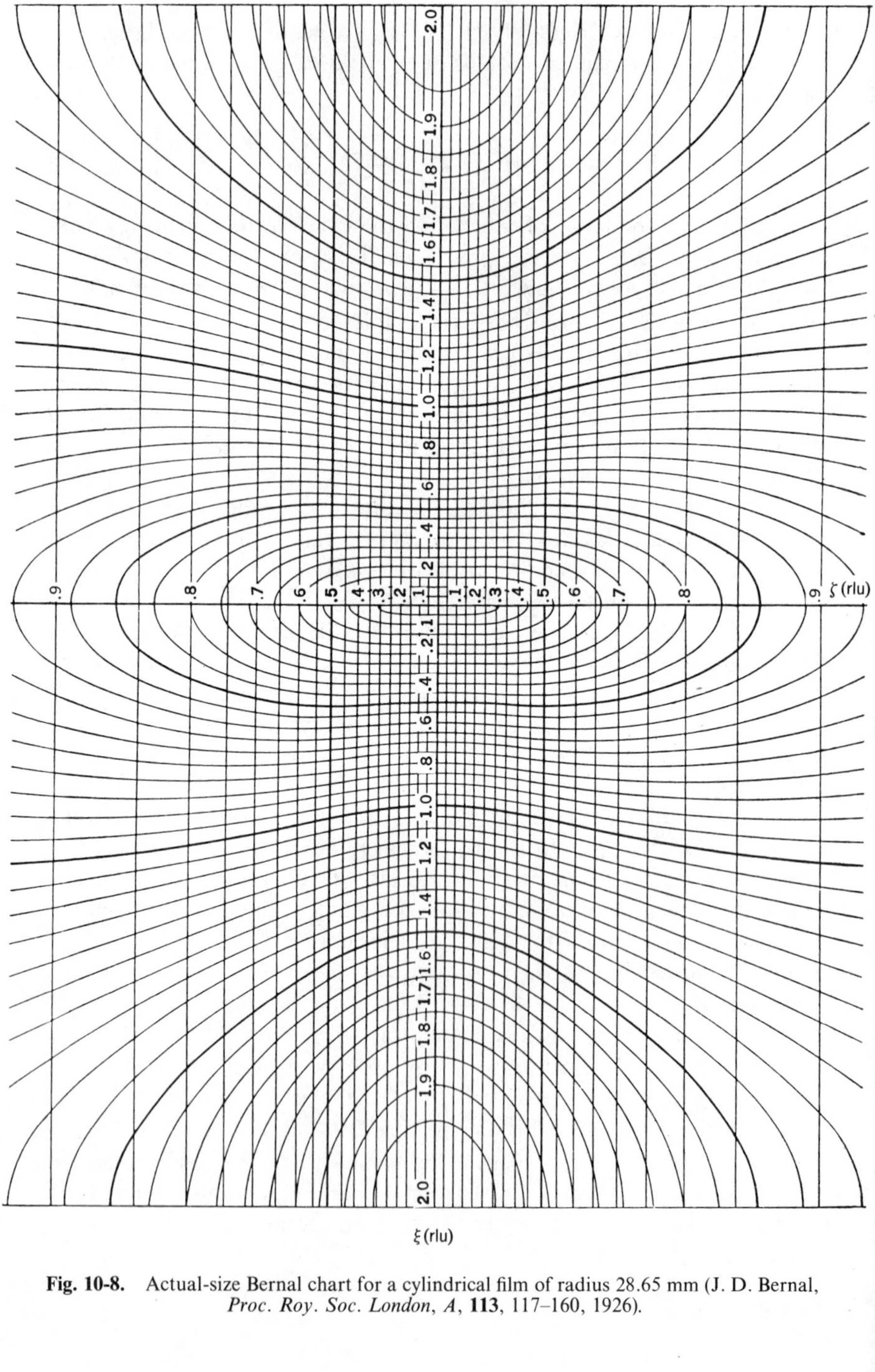

Fig. 10-8. Actual-size Bernal chart for a cylindrical film of radius 28.65 mm (J. D. Bernal, *Proc. Roy. Soc. London*, *A*, **113**, 117–160, 1926).

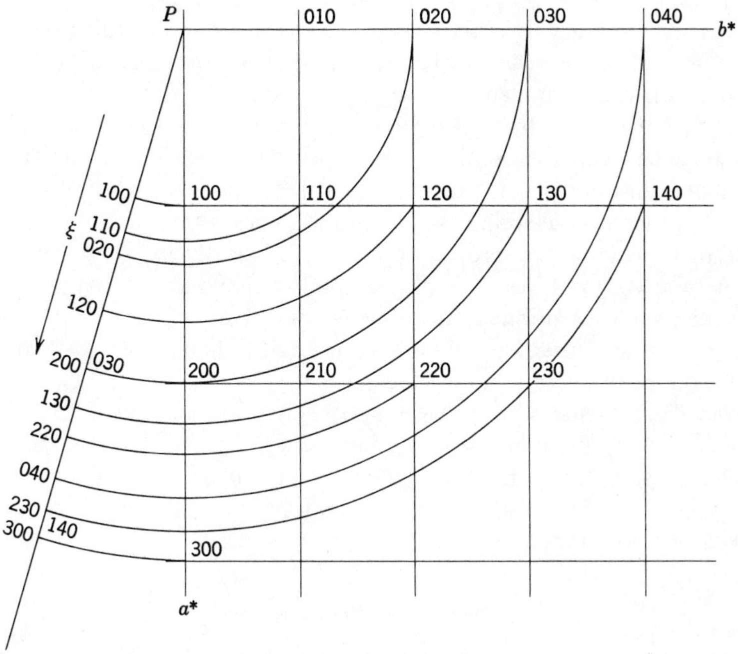

Fig. 10-9. A graphical method for indexing reflections on a rotation photograph.

Summary of the Rotation Method

The rotation method furnishes the shape and size of the unit cell of an unknown crystal, provided that the crystal can be oriented to rotate about the requisite directions. If the symmetry is cubic, the three cell edges are the same, and a single rotation photograph about any one of the axes leads to the size of the cell. Tetragonal and hexagonal crystals require two photographs, whereas three photographs are needed for orthorhombic, monoclinic, and triclinic crystals provided, in the latter two crystal systems, that the interaxial angles have been found by optical measurements on the external faces of crystals. A knowledge of the symmetry is necessary to decide on the directions about which a crystal must be rotated to determine the unit cell. Unfortunately, the rotation method provides only indirect information about the symmetry —for example, if two of three photographs obtained about three mutually perpendicular axes are identical, the crystal is tetragonal. For this reason, the rotation method plays a supplementary role in any laboratory in which the determination of symmetry is an important part of the research program.

A rotation photograph can give only a limited amount of the data which are necessary to identify a specific reciprocal point with a reflection (that is, index the reflection). The order of a layer line is evident by inspection, which immediately furnishes one of the three Miller indices of each reflection. The x film coordinate provides ξ (the perpendicular distance between the reciprocal point and the rotation axis). The coordinate does not provide the position of this point in relation to other points in the level. Herein lies a second weakness of the rotation method. The method fails as a procedure for unequivocally indexing the reflections of any crystal, because a layer line is a one-dimensional pattern of reflections representing a two-dimensional array of reciprocal points. The pattern can provide only one coordinate, whereas two are required to fix the position of a point in its level.

The rotation method has been superseded, for the complete analysis of a crystal, by the more powerful precession and Weissenberg methods, which portray the diffraction symmetry of the crystal, lead to an unambiguous choice of the unit cell, and permit each reflection to be unequivocally indexed.

Oscillation Photographs

A number of advantages can be gained by oscillating a crystal over a small angular range, instead of rotating it through 180°. An oscillation of 10–15°, for example, is usually adequate to outline the layer lines and enable one to measure the period of the rotation axis—and a photograph of usable intensity can be obtained in a fraction of the time required for the corresponding rotation photograph.

Some of the difficulties which accompany the indexing of rotation photographs were referred to previously and were illustrated in Fig. 10-9. One of the uncertainties in the particular case of this figure arose because the calculated reflection positions for the reciprocal points 230 and 140 practically coincide, making it impossible to decide whether the reflection recorded on the film is caused by one or both of the two kinds of planes represented by these points. The difficulty can be resolved by oscillating the crystal over a limited range, such that only one kind of plane is given the opportunity to satisfy the Bragg condition. In terms of the reciprocal lattice, this means that only one of the two kinds of reciprocal points is permitted to pass through the surface of the reflection sphere. The reciprocal plane of Fig. 10-9 is shown again in Fig. 10-10*a* but in greater detail, giving the eight reciprocal points which have a bearing on the reflection. Two of the reciprocal points—$2\bar{3}0$ and 230—are within the sphere. All eight points pass through the sphere when the crystal is rotated continuously, as in the rotation method. However, if the rotation is restricted to an oscillation of 15°, only one of the eight points—namely 230 or $2\bar{3}0$, depending on whether the initial oscillation is to the right or left—is permitted to pass through the surface of the sphere, and it is possible to determine whether the plane represented by this point contributes to the reflection observed on the rotation photograph. Figure 10-10*b* shows the situation after the completion of an anticlockwise motion. The point 230 is now outside the sphere. If no

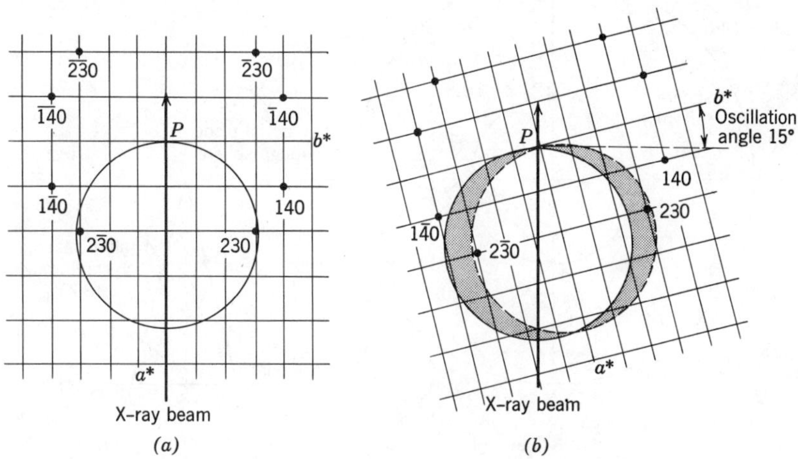

Fig. 10-10. Oscillation method for distinguishing between the 140 and 230 reflections of Fig. 10-9. The shaded area in (*b*) represents the portion of the zero level that has passed through the surface of the sphere.

reflection has occurred, it may be taken that the 140 planes are solely responsible for the reflection. However, if the oscillation is accompanied by a reflection, it is necessary to determine whether the 140 family of planes also contributes. This can be done by choosing a new oscillation range, which sends one or more 140-type points through the surface of the sphere, while restricting the movement of 230 points.

The reciprocal lattice must be known to index an oscillation photograph. If unknown, the lattice must first be determined, presumably from a series of rotation photographs. Oscillation must be restricted to a small angular range, usually 10 to 15°, in order to isolate, for reflection, portions of the lattice of which the interpretation is questionable on rotation photographs. Therefore, a number of oscillation photographs is needed for each layer, the number depending on the complexity of the lattice and its symmetry.

Orienting a Crystal for Rotation about a Direct-Lattice Row

A crystal which is approximately adjusted on a set of arcs for rotation about a direct-lattice row, can be accurately oriented with the help of two 15° oscillation exposures, differing by 180° in azimuthal angle, but recorded on the same film. Two procedures will be described. In the first, the arcs are set at 45° to the beam at the center of the oscillation range; in the second, the arcs are respectively parallel and normal to the beam in this position. The missetting errors can be determined, and the corrections to the arc settings can be made in an ordinary rotation camera. The procedures are a valuable adjunct to the powerful Weissenberg method described in the next chapter because the Weissenberg crystal, too, rotates about a chosen direct-lattice row. Most Weissenberg instruments can be modified to perform the oscillation, if the original design does not allow it, and then it is possible to test and correct the orientation of a crystal in the course of an extended Weissenberg study without the need to transfer it to another instrument.

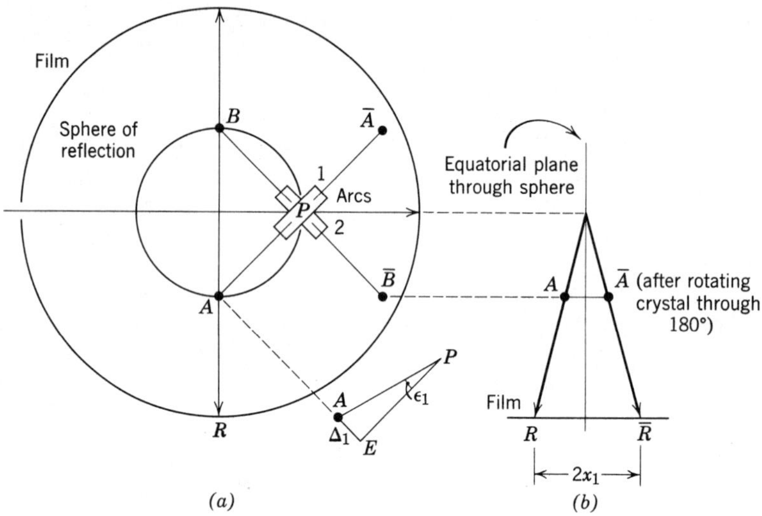

Fig. 10-11. Double-oscillation method of orienting a misset crystal with the arcs set in the 45° position: (*a*) position of arcs in middle of oscillation, (*b*) reflections R and $\bar{R}$ corresponding to A and $\bar{A}$ (not to scale).

Arcs Set in the 45° Position. Suppose that two reflecting, zero-layer reciprocal points A and B (Fig. 10-11) lie in the planes of the arcs and, therefore, give rise to reflections in the $\theta = 45°$ region of the film. If the crystal is maladjusted, neither point in general is in the equatorial plane of the reflection sphere. Suppose that A, in the present case, lies above the plane and is misset by an angle ε_1. Similarly, B is misset, either above or below the equatorial plane, by an angle ε_2. If the angle of missetting is small, Δ_1 the displacement of A is practically equal to $PE \cdot \varepsilon_1$ (in which ε_1 is in radians). Then, because the equatorial circle has unit radius

$$\Delta_1 \sim \sqrt{2}\varepsilon_1$$

It follows that x_1, the displacement of the reflection R, from the center line of the film is

$$x_1 \sim \sqrt{2}r\varepsilon_1$$

for a film of radius r. When the crystal is rotated through 180° in preparation for the second oscillation, the reciprocal point $\bar{A}$ is brought into position to diffract. Because this point now lies as far below the equatorial plane as A was above it, its reflection $\bar{R}$ is as far below the center line of the film as R is above it. The distance between the reflections R and $\bar{R}$ is therefore $2x_1$, and

$$2x_1 \sim 2\sqrt{2}r\varepsilon_1$$

The correction ε_1, which must be applied to arc 1 to bring A and $\bar{A}$ onto the plane, is

$$\varepsilon_1 = \frac{x_1}{\sqrt{2}r} \text{ (radians)} = \frac{x_1}{\sqrt{2}r}\frac{180}{\pi} \text{ (degrees)}$$

For a camera of standard radius ($90/\pi$ mm), the latter relation simplifies to

$$\varepsilon_1{}^\circ = \sqrt{2}x_1$$

If one of the oscillations is given, say, four times the exposure of the other, reflections R and $\bar{R}$ are readily distinguished and the direction of adjustment of the arc is immediately apparent from an inspection of the film.

The correction for the other arc is given by the relation

$$\varepsilon_2{}^\circ = \sqrt{2}x_2$$

in which x_2 is half the distance between corresponding reflections in the 45° region on the other half of the film.

The method assumes that the arc adjustments to bring A and B into the equatorial plane are independent. The assumption is invalid, but at small missetting angles the adjustments are practically independent. A

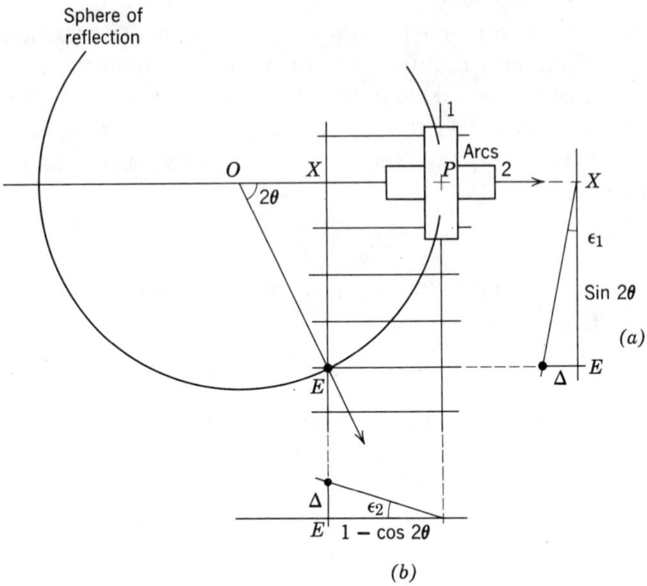

Fig. 10-12. The misset error of a reciprocal point resolved into two errors: ε_1 in arc 1 and ε_2 in arc 2.

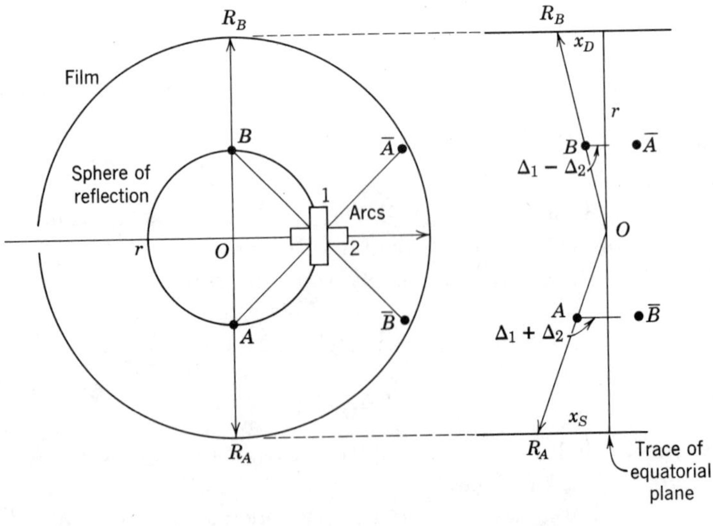

Fig. 10-13. Double-oscillation method of orienting a misset crystal with the arcs set respectively parallel and normal to the beam.

second double oscillation photograph provides a test of the adjustments. This will usually show the need for an additional correction if the crystal was originally misset by more than 5°.

The precision of the method is sensitive to the size of the reflection spots and is limited by the ability to detect the separation of spots at small missetting angles and to measure the distance between them. In favorable cases, a maladjustment of as little as 20′ can be observed and corrected.

ARCS SET RESPECTIVELY PARALLEL AND NORMAL TO THE BEAM. The displacement Δ of a zero-layer reciprocal point from its ideal position at E on the equatorial plane has, in general, two components Δ_1 and Δ_2, due to two errors: ε_1 in arc 1 and ε_2 in arc 2. If the missetting error is confined to arc 1, then according to Fig. 10-12 $\Delta = \Delta_1 \sim EX \cdot \varepsilon_1$ (ε_1 in radians) if the error is small. Because $EX = OE \cdot \sin 2\theta = \sin 2\theta$,

$$\Delta_1 \sim \sin 2\theta \cdot \varepsilon_1$$

Similarly, if the maladjustment is due entirely to an error in arc 2, $\Delta = \Delta_2 \sim XP \cdot \varepsilon_2$. But $XP = (1 - OX) = (1 - \cos 2\theta)$ and therefore

$$\Delta_2 \sim (1 - \cos 2\theta)\varepsilon_2$$

Δ_1 is evidently a maximum and Δ_2 a half-maximum at $\theta = 45°$. At this Bragg angle the total displacement of a zero-level point, misset by both arcs, is due either to the sum of the errors or to the difference of errors in the arcs. If the displacement of a point A (in Fig. 10-13) is due to the sum of the errors, the displacement of a point B in the diametrically opposite region of the sphere is due to the difference in errors. The displacements of the reflections, R_A and R_B due to these points, from the center line of the film, are respectively

$$x_S = r(\Delta_1 + \Delta_2)$$

$$x_D = r(\Delta_1 - \Delta_2)$$

in which x_S and x_D are the displacements due respectively to the sum and the difference of errors, and r is the film radius. After substituting the values of Δ_1 and Δ_2 derived above, these relations become

$$x_S = r[\sin 90° \cdot \varepsilon_1 + (1 - \cos 90°)\varepsilon_2] = r(\varepsilon_1 + \varepsilon_2)$$

$$x_D = r[\sin 90° \cdot \varepsilon_1 - (1 - \cos 90°)\varepsilon_2] = r(\varepsilon_1 - \varepsilon_2)$$

and therefore

$$\varepsilon_1 = \frac{(x_S + x_D)}{2r}$$

$$\varepsilon_2 = \frac{(x_S - x_D)}{2r}$$

For a film of standard radius ($90/\pi$ mm), the value of the ε_1 missetting angle, in degrees, is

$$\varepsilon_1{}^\circ = \frac{(x_S + x_D)}{2(90/\pi)}(180/\pi) = x_S + x_D$$

Similarly

$$\varepsilon_2{}^\circ = x_S - x_D$$

It is found in practice that the accuracy of the method is improved by an arrangement which allows the double distances $2x_S$ and $2x_D$ to be measured on the film. Reflections in the $\theta = 45°$ region of the film, corresponding to R_A and R_B, are recorded with a 15° oscillation, in which one of the arcs is parallel to the beam at the center of the range. The crystal holder is then rotated through 180°, and a second oscillation photograph is superimposed on the first. Reflections corresponding to points $\bar{A}$ and $\bar{B}$ are recorded in this latter position and will lie exactly opposite (with reference to the theoretical position of the zero-layer line) to reflections R_A and R_B. If one oscillation is given, say, four times the exposure of the other, the directional sense of the correction is known by inspection.

CHAPTER ELEVEN

THE EQUI-INCLINATION WEISSENBERG METHOD

The rotation method fails for the complete analysis of a crystal, for a number of reasons:

1. It does not provide direct evidence of the symmetry of the crystal. The symmetry either must be inferred by comparing photographs taken about several rotation axes or obtained directly by resorting to a supplementary single-crystal method.
2. Unless the symmetry is cubic, the crystal must be remounted and rotated about two or more axes to establish the cell constants. This means that the requisite axes must be recognizable on the crystal.
3. The reflections cannot be indexed without ambiguity because each (two-dimensional) reciprocal level is represented by a layer line which is a one-dimensional pattern of reflections. The reflections can be indexed with oscillation photographs, by successively oscillating selected segments of the lattice through the sphere of reflection.

A successful single-crystal x-ray method should lead to an unambiguous reconstruction of the reciprocal lattice from measurement of photographs obtained with only a single mount of the crystal. This is achieved with the de Jong–Bouman and Buerger-precession methods by recording successive levels as two-dimensional patterns of reflections on separate films. It can also be accomplished with a refined version of the rotation apparatus in which

(1) only one selected Laue cone is allowed to reach the film, and
(2) the film is given a controlled translating motion so that the reflections which comprise a layer line are no longer confined to a line, but are drawn across the film to form a decipherable pattern.

This is the basis of the three Weissenberg methods.

The Apparatus

The Weissenberg methods are similar to the rotation method in that a selected axis of the crystal (a direct-lattice row) is oriented to coincide with the axis of the cylindrical film. When the crystal is oscillated in the beam, it gives rise to Laue cones which are coaxial with the rotation axis; these cones produce the layer lines on rotation photographs. Thus a cylinder with an equatorial slit forms an effective *layer line screen* when

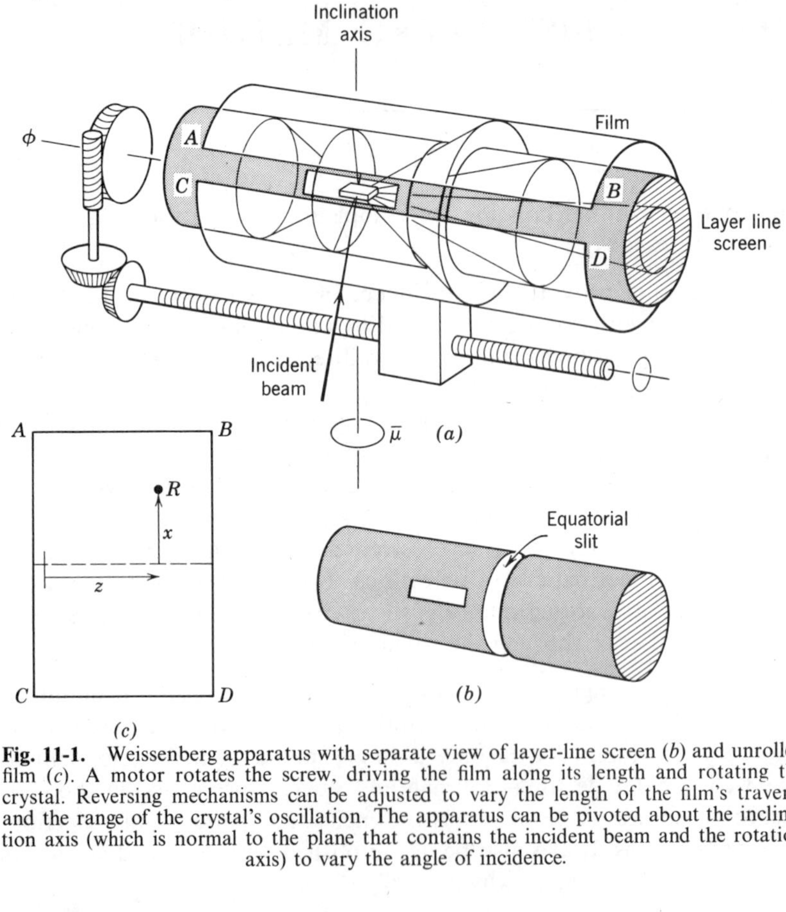

Fig. 11-1. Weissenberg apparatus with separate view of layer-line screen (*b*) and unrolled film (*c*). A motor rotates the screw, driving the film along its length and rotating the crystal. Reversing mechanisms can be adjusted to vary the length of the film's traverse and the range of the crystal's oscillation. The apparatus can be pivoted about the inclination axis (which is normal to the plane that contains the incident beam and the rotation axis) to vary the angle of incidence.

inserted between crystal and film and positioned to permit only one cone to reach the film (Fig. 11-1).

The film is translated along its axis, back and forth across the path of the x-ray beam, the motion being synchronized with the oscillation of the crystal. An oscillating range of at least 200° is built into the instrument. As a result of translating the film, the reflections are spread across the film and form a two-dimensional pattern of reflections in contrast to the one-dimensional layer line produced with the rotation method (Fig. 11-2). Each reflection now has two film coordinates:

1. x, measured along the circumference of the film (Fig. 11-1*c*) is identical with the x coordinate on rotation films and gives, by calculation, the ξ cylindrical coordinate of the reciprocal point in its level (Fig. 10-6).

2. z (Fig. 11-1*c*), measured parallel to the axis of the film, derives from the translating motion of the film. The measurement of this coordinate has an angular interpretation, because the rotation of the crystal is synchronized with the translation of the film. It will be shown that z, in combination with x, helps to fix the position of a reciprocal point relative to other points in the level.

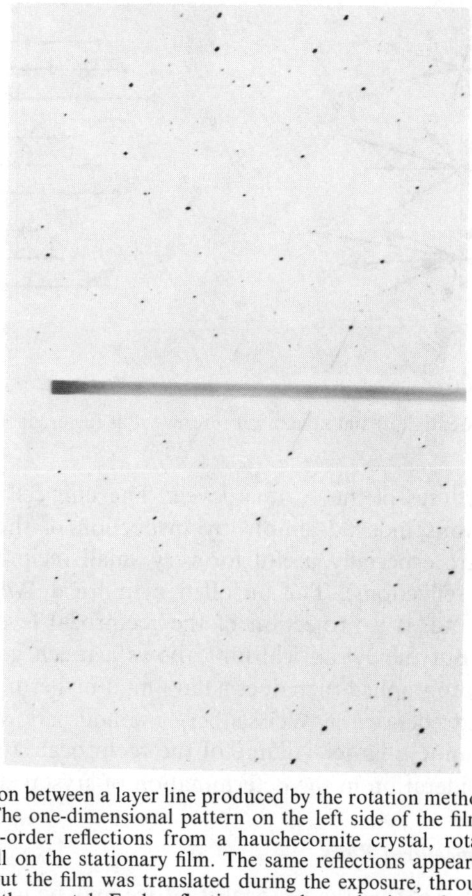

Fig. 11-2. Relation between a layer line produced by the rotation method and its Weissenberg projection. The one-dimensional pattern on the left side of the film was prepared by allowing the zero-order reflections from a hauchecornite crystal, rotating continuously about [001], to fall on the stationary film. The same reflections appear on the right-hand side of the film, but the film was translated during the exposure, through 1 mm for each 2° of rotation of the crystal. Each reflection now has a horizontal as well as a vertical coordinate.

The zero level is invariably recorded with the x-ray beam directed at right angles to the rotation axis. With this arrangement, the reciprocal levels under investigation are parallel to the beam. Upper-level photographs are prepared in several ways. The beam is not parallel to the reciprocal planes in the preferred methods, and to meet this requirement, the instrument (Fig. 11-1) has an inclination axis which allows the reciprocal planes to be inclined at any angle up to 45° to the beam.

The De Jong–Bouman and precession methods record successive reciprocal levels on separate, flat films which are caused to duplicate the motion of the level being photographed. The result is a set of undistorted,

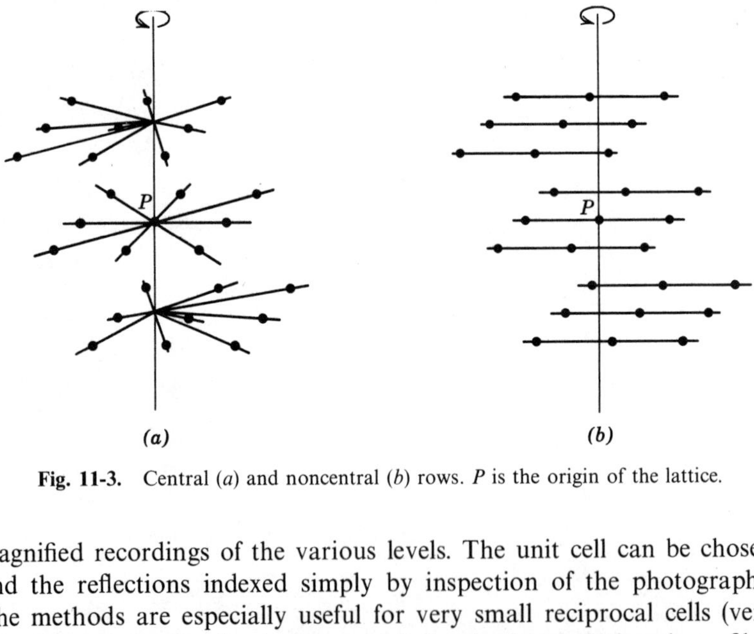

Fig. 11-3. Central (*a*) and noncentral (*b*) rows. *P* is the origin of the lattice.

magnified recordings of the various levels. The unit cell can be chosen and the reflections indexed simply by inspection of the photographs. The methods are especially useful for very small reciprocal cells (very closely spaced reflections). The unrolled cylindrical Weissenberg film, on the other hand, is a projection of the reciprocal level and requires interpretation. But the cylindrical film allows a much greater extent of a level to be photographed than does a flat film. Furthermore, more levels can be recorded; hence the Weissenberg methods allow the crystallographer to examine a larger volume of the reciprocal lattice. This is an important consideration in the determination of crystal structures.

Central and Noncentral Rows

It has been shown that the reciprocal lattice may be usefully described as a series of levels which are normal to and spaced at equal intervals along the rotation axis that passes through the origin of the lattice. The task of visualizing the diffraction effects on a Weissenberg film when a level rotates through the sphere of reflection, is greatly simplified if the level is resolved into simple components, such as rows. Two methods of subdividing a level offer themselves:

1. Subdividing into rows of reciprocal points which intersect at the rotation axis. Such rows are known as *central rows* in the terminology of x-ray crystallography. Figure 11-3*a* shows a special property of the zero level—its central rows always intersect at the origin of the lattice.
2. Subdividing into sets of parallel, equally spaced rows (Fig. 11-3*b*). One row of each set on the zero level is evidently always a central row. The remainder are described as *noncentral rows* because they do not intersect the rotation axis. Upper levels may show central rows but this depends on the symmetry of the crystal.

Both concepts manifest themselves on Weissenberg films and are useful in interpreting the reflections. The diffraction effects which are characteristic of each concept will now be considered.

Diffraction in Terms of Central Rows

Weissenberg Projection of a Central Line in the Zero Level

Figure 11-4 shows, in several cross sections through a Weissenberg camera, a line PL in the zero level rotating through the sphere of reflection. This is a central line, because it passes through the origin of the reciprocal lattice at P. The unrolled films (not to scale) are on the right.

The diagrams show that a central line in the zero level continuously has the potential to develop one reflection as it rotates through the circle of reflection—except when it is normal to the x-ray beam. This exception is depicted in Fig. 11-4*a*. Assume that at this instant the film has reached a position, in translating along its track, such that the direct beam strikes it at P'. Figure 11-4*b* shows the line after it has rotated through an angle ω. The film has translated a distance $P'P''$, and the undeviated beam now strikes it at P''. The line intersects the circle of reflection at Q. Consequently, OQ is a potential reflection direction, and a reflection R is actually developed if a reciprocal point is situated at Q. The azimuth angle of the reflection Υ bears the simple relation

$$\Upsilon = 2\omega \tag{1}$$

to the angle of rotation ω. The relation holds for every position of the line PL, as it rotates through the sphere. The angles Υ and ω may be described as the reflection coordinates of the reciprocal point Q.

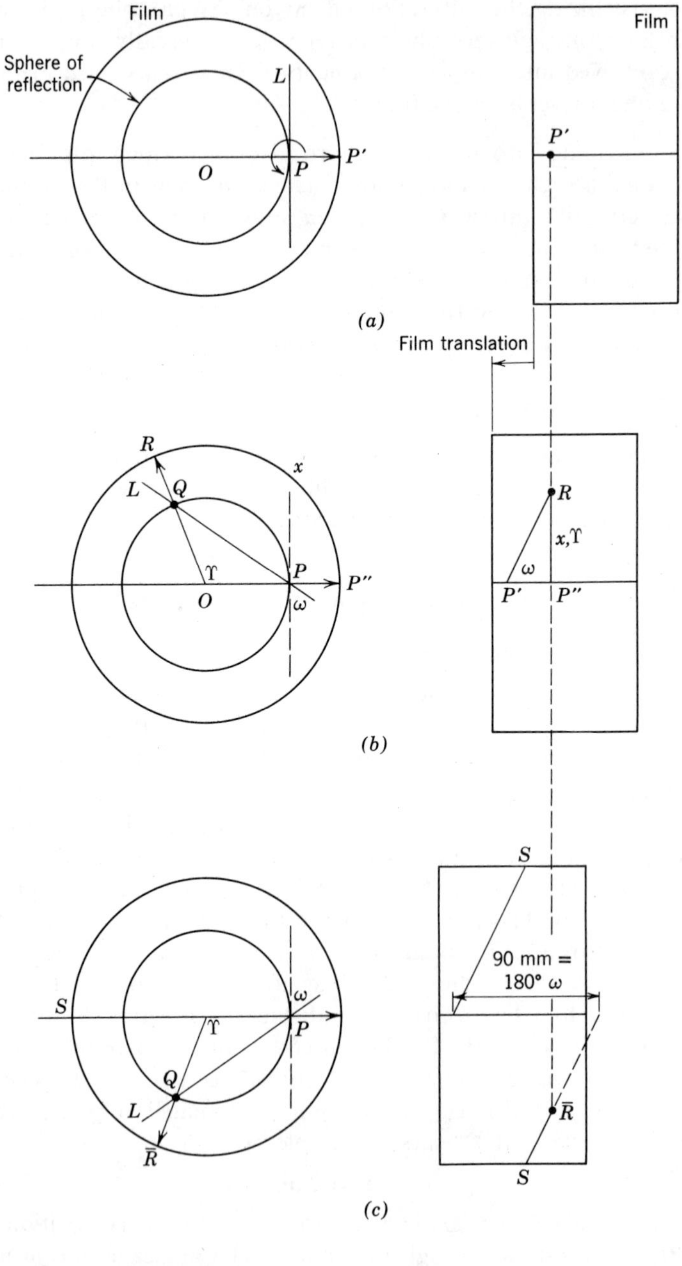

Fig. 11-4. The Weissenberg projection of a central line PL in the zero level.

The reflection directions transfer to the lower half of the film as the reciprocal line passes through the primary beam at S (Fig. 11-4c), and a new lattice line projection is begun on this portion of the film. The azimuth angle Υ decreases from 180° to 0, the trace of the line returning to the upper half of the film when the value of Υ reaches 0°.

The height ($P''R = x$ of Fig. 11-4b) of the reflection R above the center line of the film is related to Υ as

$$\frac{x}{\Upsilon} = \frac{\text{circumference of film}}{360} = \frac{2\pi r_f}{360}$$

where r_f is the radius of the film. This relation shows that the value of x depends only on Υ and the film radius. In designing a Weissenberg instrument it is desirable to choose a film radius that is small enough to give good intensities in a reasonable exposure time, yet large enough to house the layer line screen and crystal holder. A radius of $90/\pi$ or 28.6 mm (equal to the radius of the small "standard size" powder camera) is a good compromise. Furthermore, the relation between x and Υ reduces to

$$\frac{x}{\Upsilon} = \frac{2\pi}{360} \cdot \frac{90}{\pi} = \frac{1}{2}$$

for this radius and then

$$1 \text{ mm measured along } x \text{ represents } 2° \ \Upsilon \text{ or } 1° \ \frac{\Upsilon}{2} \tag{2}$$

A comparison of Figs. 11-4b and 8-7 shows that if Q is a point in the zero level

$$\Upsilon = 2\theta \tag{3}$$

and consequently for this level

$$1 \text{ mm } x = 1° \ \theta \tag{4}$$

This means that zero-level film measurements of x (in mm) are numerically equal to θ (in degrees). According to Bragg's law and the reciprocal-lattice concept

$$\theta = \sin^{-1}\left(\frac{\lambda}{2d}\right) = \sin^{-1}\left(\frac{p^*}{2}\right)$$

because $\lambda/d = p^*$. Thus the relation allows the rapid conversion of film measurements to direct-lattice spacings d (especially if spacing tables are available) or reciprocal-lattice periods p^*.

The translation distance $P'P''$ is obviously independent of the film radius, but is related to the angle of rotation ω and the film-translation/

crystal-rotation constant of the instrument. It is convenient to be able to measure the Υ and ω coordinates of a point with the same scale. This can be arranged by designing the instrument to translate the film 1 mm for each 2° of crystal rotation. Then 1 mm measured along $P'P''$ represents 2° ω. Weissenberg photographs which have the same angular value in the vertical and horizontal directions are described by Buerger[1] as having an undistorted scale.

The slope of the line $P'R$, in terms of reflection coordinates, is

$$\frac{\Upsilon}{\omega} = \frac{2\omega}{\omega} = 2$$

The angle which the line makes with the horizontal is therefore $\tan^{-1} 2 = 63° 26'$ on photographs of undistorted scale. The $90/\pi$ mm film radius and the 1 mm/2° translation/rotation constant are generally accepted and have the status of standards for ordinary Weissenberg work. The descriptions that follow and the calculations that they include are based on these values.

Weissenberg Projection of the Angular Relations Between Central-Lattice Rows in the Zero Level

As the crystal rotates in the path of the beam, each central row in the level rotates in turn through the sphere and generates reflections which lie on a lattice line projection. The complete Weissenberg film consists of numerous lattice row projections, each composed of an arrangement of reflections which is characteristic of the reciprocal row it represents (Fig. 11-2). The relative positions of the lattice row projections depend on the angular relations between the corresponding lattice rows. This is illustrated in Fig. 11-5 for two rows I and II, separated by an angle $(\omega_1 - \omega_2)$. Lattice line projection I is initiated on the upper half of the film at the instant at which its reciprocal row passes a position NN', normal to the x-ray beam. The crystal rotates through $(\omega_1 - \omega_2)$ degrees before the projection of the second line is begun on the upper half of the film. In the interval, the film translates $(\omega_1 - \omega_2)/2$ mm if the translation/rotation constant of the Weissenberg instrument is 1 mm/2°. Therefore, the horizontal distance between the lattice row projections is also $(\omega_1 - \omega_2)/2$ mm.

Thus the angles between lattice rows are related in a simple manner to the horizontal distances between their corresponding lattice row projections. In practice these distances are usually measured from an arbitrarily chosen zero position along the center line of the film, and are designated as z coordinates (Fig. 11-5*b*). Under this scheme, all of the

[1] M. J. Buerger, *X-Ray Crystallography*, John Wiley and Sons, New York (1942).

reflections in a lattice row projection have the same z coordinate. The z coordinates are converted to angles according to the relation

$$1 \text{ mm } z = 2° \phi \tag{5}$$

for an instrument with a translation/rotation constant of 1 mm/2°. The arbitrary zero point on the film represents an arbitrary line in the level, and consequently the ϕ values refer to the angles which the reciprocal rows make with the line.

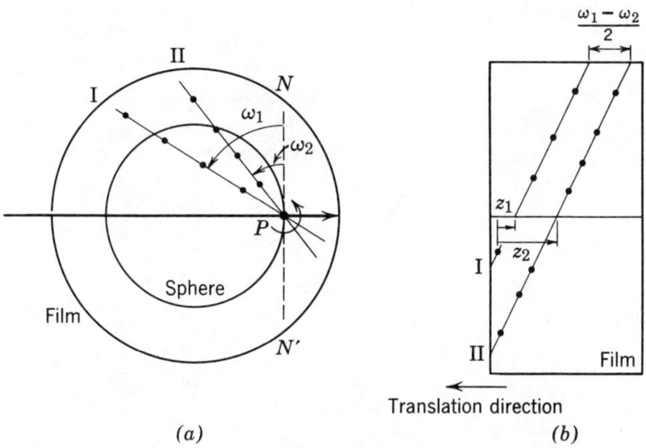

Fig. 11-5. Weissenberg projection of the angular relations between central rows in the zero level.

Weissenberg Projection of Upper-Level Central Rows

NORMAL-BEAM METHOD. Upper-level reflections may be recorded with the Weissenberg instrument by one of a number of methods. In the *normal-beam* arrangement, as the name implies, the x-ray beam continues to be directed at right angles to the rotation axis of the crystal, but the layer line screen is shifted to pass the Laue cone of the reciprocal level under investigation, while screening the film from the other cones. The method is illustrated in Fig. 11-6. Figure 11-6*a* is a section through the Weissenberg camera that includes the x-ray beam and the rotation axis PP_1 of the reciprocal lattice. The zero level, and one n level cutting the sphere in a circle of diameter AB, are shown in the diagram along with their Laue cones. Figure 11-6*b* is a section in the plane of the upper level. According to Fig. 11-6*a*, the screen must be shifted a distance

$$s = r_s \tan \nu$$

(in which r_s is the radius of the screen) to pass the n level. But $\nu =$

$\sin^{-1}(\zeta_{n\,\text{level}})$ and therefore

$$s = r_s \tan[\sin^{-1}(\zeta_{n\,\text{level}})]$$

The value of $\zeta_{n\,\text{level}}$ is obtained from a rotation photograph with the same crystal mount.

When a central lattice line (such as P_1L) in this level intersects the sphere, it does so in general at two points. Consequently, two potential reflections are continuously being developed from O, the center of the sphere, as a central line migrates through the circle. This means that the

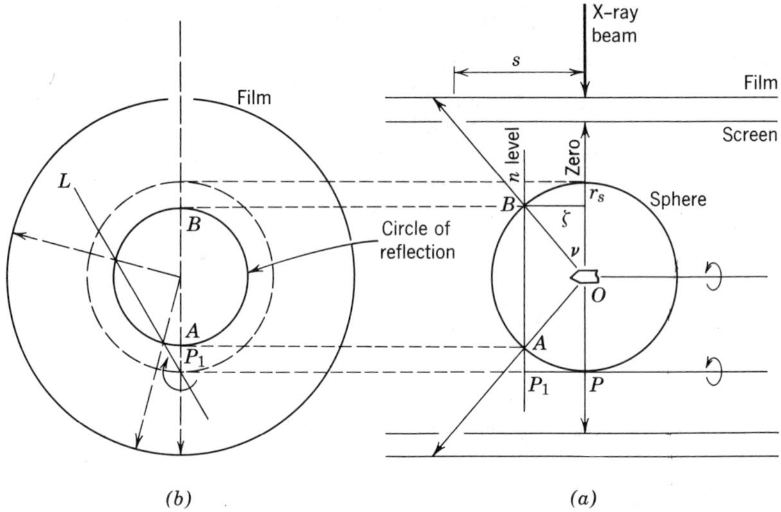

Fig. 11-6. Normal-beam method of recording upper levels.

potential reflections trace out a curve and not a straight line, as does a central line in the zero level. As a consequence of this, upper-level records cannot be directly compared with the zero-level record. Still another disadvantage is inherent in the method. Because the rotation point P_1 lies outside the circle, a circular portion of the level, of radius P_1A, remains outside the circle as the level rotates, and therefore planes represented by reciprocal points in this region of the level do not have the opportunity to reflect. The area of this region increases rapidly with the spacing between the zero level and the level under investigation.

FLAT-CONE METHOD. In the *flat-cone* method, the reciprocal planes are inclined at an angle μ to the beam such that the n level under investigation intersects the sphere in the equatorial circle (Fig. 11-7). Then the reflection directions for this level are all normal to the rotation axis, and the Laue cone is flat. The inclination angle is given by the relation

$$\frac{PP_1}{OP} = \zeta_{n\,\text{level}} = \sin \mu$$

The position of the layer line screen is the same for all levels, including the zero level.

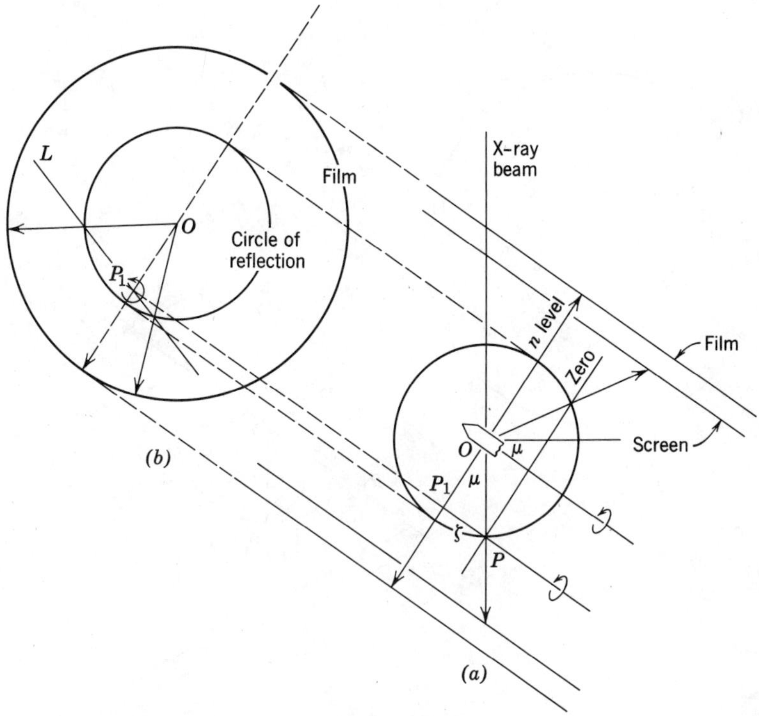

Fig. 11-7. Flat-cone method of recording upper levels.

Flat-cone photographs have two disadvantages in common with normal-beam photographs. In general, an upper-level central line intersects the sphere at two points, and the potential reflections trace out a curve. Upper- and zero-level films cannot, therefore, be directly compared. Second, in any upper level, a circular area which rapidly increases in size with increase in ζ cannot reflect because the rotation point P_1 lies within the sphere, hence the area remains within the sphere as the crystal rotates. One important advantage of the method arises from the circumstance that the reflected beams move along radial directions in the camera. The beams meet the film at right angles—the spots are uniformly well-shaped and eminently suitable for the quantitative measurement of intensities.

EQUI-INCLINATION METHOD. Figures 11-6 and 11-7 show that an upper-level central-lattice line develops two potential reflections if the point P_1, at which the rotation axis intersects the level, lies either outside or within the reflection circle. The point must lie on the circumference of the circle

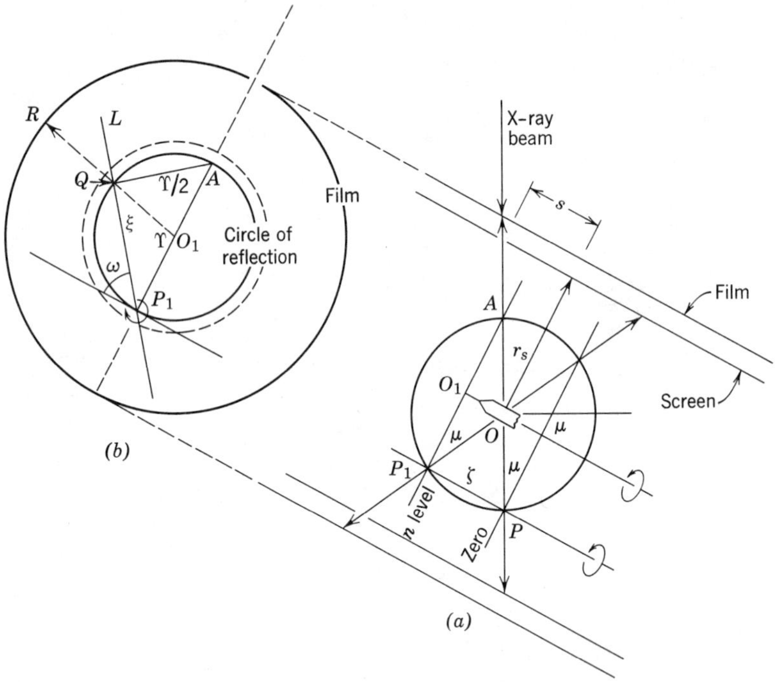

Fig. 11-8. Equi-inclination method of recording upper levels.

if the line is to continuously develop only one potential reflection and, like zero-level lattice lines, trace out a straight line on the film as it rotates through the circle. This situation can be attained by setting the inclination angle μ of the instrument such that the rotation axis of the reciprocal lattice intersects the level under investigation on the surface of the sphere (at P_1 in Fig. 11-8). Then any central-lattice line P_1L in the level develops only one potential reflection as it rotates through the reflection circle. The relation $\Upsilon = 2\omega$ holds, and therefore the situation is identical in principle to the method for recording the zero level.

Upper- and zero-level films will differ in scale, however, because the scale of a film depends on the diameter of the circle of reflection.

This method of recording an upper level is known as the *equi-inclination* method because the undeviated beam and the diffracted beams (such as OP_1 of Fig. 11-8) are inclined at the same angle to the reciprocal levels. The angle is equal to the angle of inclination μ, through which the rotation axis must be pivoted to bring the point P_1 onto the surface of the sphere.

The equi-inclination arrangement is the most useful of the Weissenberg film methods. In addition to the advantage that photographs of different levels are identical in geometry, if not scale, no blank region mars the record of any level, and the method allows twice as many levels to be recorded as the flat-cone method. Inclination angles up to 45° are used in some laboratories to record the maximum number of levels.

Two instrumental settings must be determined for each equi-inclination upper level photograph.

The inclination angle μ. From Fig. 11-8*a*

$$\mu = \sin^{-1}\left(\frac{PP_1}{AP}\right) = \sin^{-1}\left(\frac{\zeta}{2}\right) \tag{6}$$

It was shown in the Chapter 10 that the ζ coordinate of any level normal to the rotation axis is easily obtained from the measurement of the corresponding film coordinate y on a rotation photograph

$$\zeta = \cos\cot^{-1}\frac{y}{r_f}$$

Consequently

$$\mu = \sin^{-1}\left[\frac{\cos\cot^{-1}(y/r_f)}{2}\right] \tag{7}$$

Figure 11-9 offers a convenient graphical solution of relations (6) and (7).

The Layer Line Screen Setting, s. Figure 11-8 shows the layer line screen in position to permit the reflections from a particular upper level to reach the film. The screen has been moved a distance of s mm from its zero-level position. The distance bears a simple relation to the radius of the screen (r_s) and the inclination angle μ

$$s = r_s \tan\mu$$

A screen with an effective radius of 25.5 mm satisfactorily clears the interior of a camera with a film radius of $90/\pi = 28.6$ mm. This relation, too, is conveniently solved with the aid of Fig. 11-9.

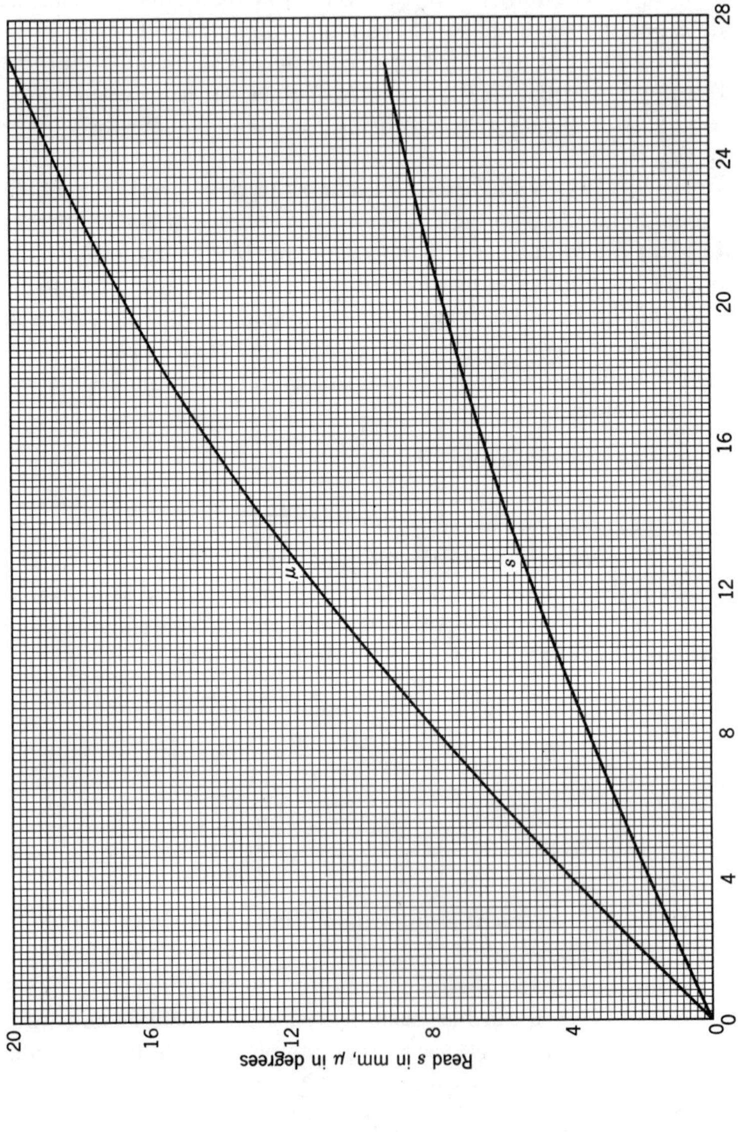

Fig. 11-9. Chart for deriving μ and s from y, the height of the layer line on a rotation photograph, when the apparatus has a film radius of $90/\pi = 28.6$ mm.

Reconstructing the Reciprocal Lattice from Equi-inclination Weissenberg Photographs

Three types of coordinates have been used to describe the position of a reciprocal point relative to other points in the level, and the location of its reflection relative to other reflections on the Weissenberg film:

Type:	Film	Reflection	Cylindrical
	x	Υ	ξ
	z	ω	ϕ

For convenience, the coordinates are collected in the three diagrams of Fig. 11-10. The x or vertical film coordinate of a reflection R gives the azimuth angle Υ of the corresponding reciprocal point Q

$$1 \text{ mm measured along } x \text{ represents } 2^\circ\ \Upsilon \text{ or } 1^\circ \frac{\Upsilon}{2} \tag{2}$$

for a film of radius $90/\pi$ mm. The relation between the ξ cylindrical coordinate and Υ is readily deduced from Fig. 11-8

$$\xi = AP_1 \sin \frac{\Upsilon}{2}$$

However $AP_1 = AP \cos \mu = 2 \cos \mu$, and therefore

$$\xi = 2 \sin \frac{\Upsilon}{2} \cos \mu \tag{8}$$

The maximum value of ξ that can be recorded in any level is equal to AP_1, the diameter of the circle of reflection. Consequently

$$\xi_{max} = 2 \cos \mu \tag{9}$$

The z or horizontal film coordinate of R is the distance from an arbitrarily chosen zero position on the center line of the film to the lattice line projection in which the reflection occurs. For a Weissenberg instrument with a translation/rotation constant of $\frac{1}{2}$

$$1 \text{ mm } z = 2^\circ\ \phi \tag{5}$$

Graphical Plot of a Reciprocal Level. The coordinates ξ and ϕ provide all of the data which are necessary to fix the location of a reciprocal point with reference to all other points in its level—in other words, to reconstruct the level.

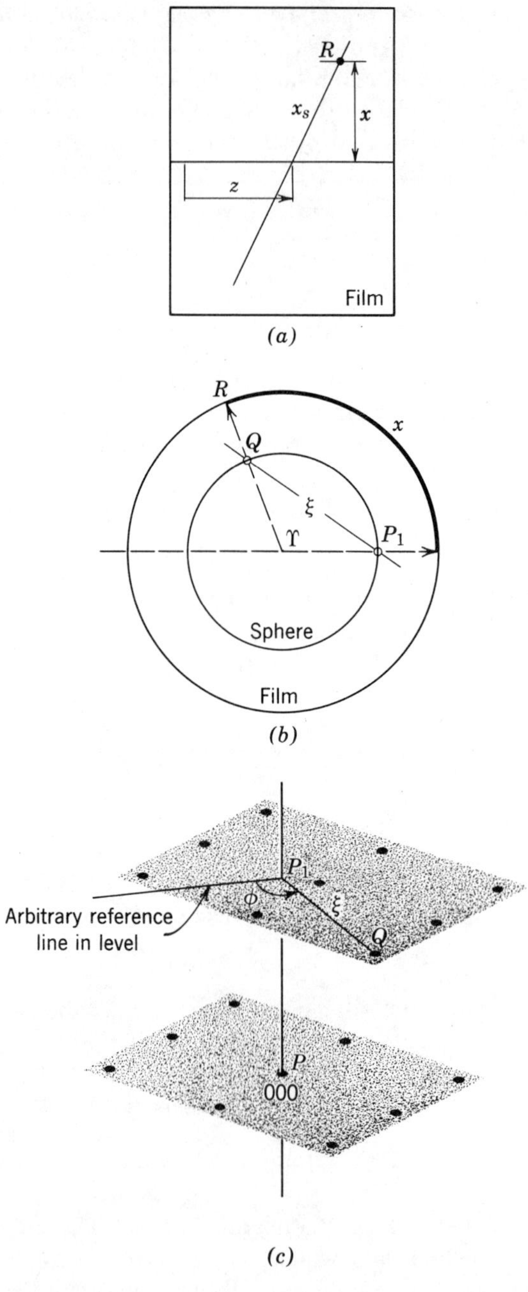

Fig. 11-10. Relation between film, reflection, and cylindrical coordinates.

It is sometimes difficult to picture the arrangement of points in a level from a visual examination of the film. Then the x and z coordinates of a representative number of reflections may be measured, the ξ and ϕ coordinates of the corresponding reciprocal points found, and the data plotted on a suitable scale. The labor of calculating the ξ and ϕ coordinates may be avoided by the use of two types of scales, depicted in Fig. 11-11, which were devised by Buerger[1] to give these coordinates directly in terms of film measurements.

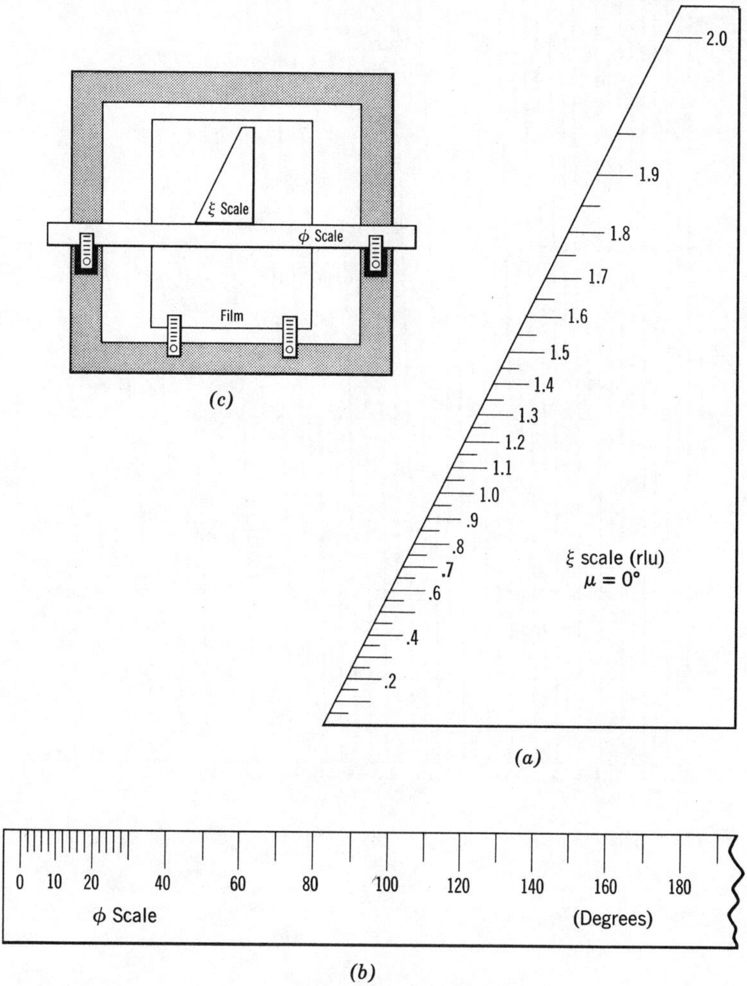

Fig. 11-11. Instruments for measuring cylindrical coordinates.

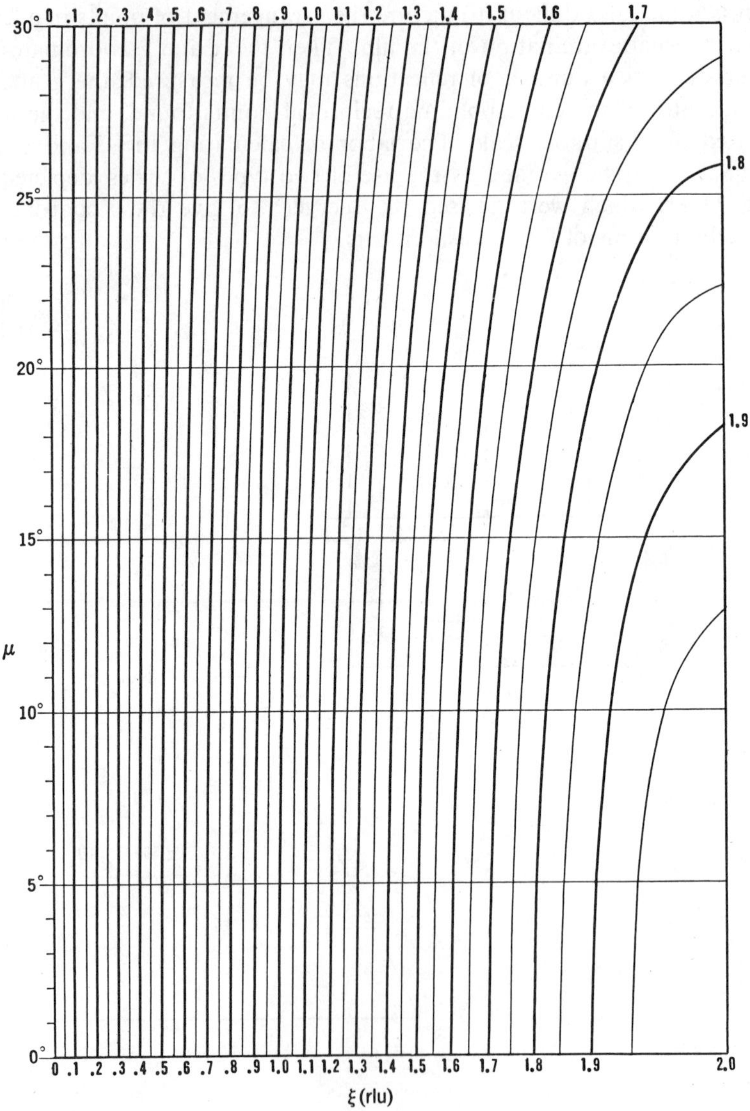

Fig. 11-12. Variation of ξ scale along x (Fig. 11-10c) with μ for film radius of $90/\pi$ mm (Buerger[1]).

ξ *Scales.* The relation

$$\xi = 2 \sin \frac{\Upsilon}{2} \cos \mu \tag{8}$$

contains the variable μ, which precludes the construction of a ξ scale suitable for all levels. Hence it becomes necessary to prepare a series of ξ scales for regular intervals of this angle. Because $\xi_{max} = 2\cos\mu$, each scale must give values of ξ which range from zero to a maximum of $2\cos\mu$. The absolute distances between ξ divisions on any scale naturally depend on the radius of the film, as well as on the value of μ. Thus the scale distance for a given value of ξ is found by solving (8) for $\Upsilon/2$ and by converting this to x. The relation

$$1\text{ mm } x = 1^\circ \frac{\Upsilon}{2} \tag{2}$$

applies when the film radius is $90/\pi$ mm. Figure 11-12 gives in graphical form, the variation in ξ scale with inclination angle μ for this radius.

It is found convenient, in practice, to determine the ξ coordinate of a reciprocal point by measuring the film coordinate along the slope of a central-lattice line projection (x_s of Fig. 11-10*a*) with the aid of a triangular scale whose measuring edge makes an angle of 63° 26′ with the horizontal. The absolute scale distances along the slope x_s are related to the vertical scale distances x as

$$x_s = \frac{x}{\sin 63^\circ\, 26'}$$

Figure 11-13 shows how the ξ scale varies along x_s, with μ for a film radius of $90/\pi$ mm. The ξ divisions of the actual-size zero-level scale reproduced in Fig. 11-11*a* were found by placing the long edge of the triangle on the 0° μ horizontal line of Fig. 11-13.

The ϕ Scale. The ϕ scale (Fig. 11-11*b*) is designed to express z directly in degrees ϕ. The absolute scale distances depend on the translation-rotation coupling of the instrument. A translation-rotation constant of $\frac{1}{2}$ requires that a crystal rotation of 2° be represented by an absolute scale distance of one millimeter.

MEASUREMENT OF THE FILM. One method of measuring a Weissenberg film makes use of a device of the type illustrated in Fig. 11-11*c*. This device includes a sheet of opal glass illuminated from the back, a bar ϕ scale, and clips to hold the scale and the film firmly in place. The ξ scale rides on top of the ϕ scale, and consequently the latter scale must coincide exactly with the position of zero ξ on the film. A consideration of Fig. 11-8 will show that the direct beam can be interpreted as the reflection from a reciprocal point with a ξ coordinate of zero. Therefore, a line joining direct-beam spots at the two edges of the film is a line where ξ is constant at zero. Direct beam fiducial spots can be produced by allowing a weak beam to fall for an instant on the unprotected film.

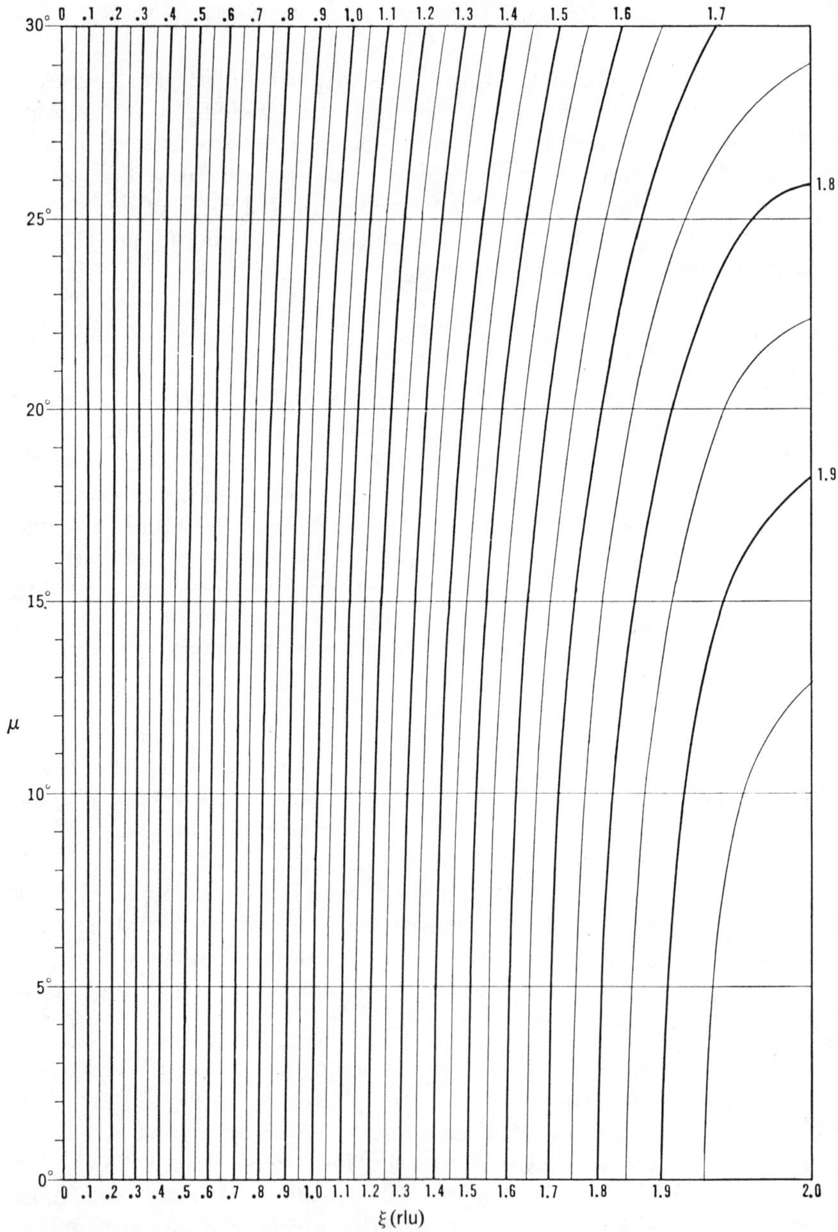

Fig. 11-13. Variation of ξ scale along x_s (Fig. 11-10c) with μ for film radius of $90/\pi$ mm (Buerger[1]).

The film is correctly adjusted on the measuring device when the centers of the spots are aligned with the top edge of the ϕ scale.

PLOTTING THE MEASUREMENTS. A method of plotting the measurements is illustrated in Fig. 11-14. This figure shows the upper half of an actual-size zero-level Weissenberg photograph obtained by rotating a calcite crystal about a cleavage edge. The ξ and ϕ measurements for the reflections are given in Table 11-1. The corresponding scales are affixed to the photograph to aid the reader in correlating the measurements with the reflection positions on the film. In this case the zero position on the scale was made coincident with the left vertical edge of the film. Figure 11-14*b* shows the reconstructed level which appears when the measurements are plotted. In practice it is convenient to plot the level on a scale of 1 rlu = 100 mm, which gives a useful size to the reconstructed level.

A picture of the entire lattice can be gained by superimposing the various reconstructed levels, one above the other, in correct orientation. The orientation of each reconstructed level relative to the others can be predetermined by introducing a correlation between the photographs of the various levels. This is accomplished by observing two precautions. First, each film is given a fiducial mark which relates the Weissenberg pattern to the film's orientation in the camera, and second, the coupling between the rotating crystal and the translating camera carriage is not

Table 11-1. ϕ and ξ Values

Measured on a Weissenberg film of calcite (Fig. 11-14)

Reflection Number	ϕ Measured	ξ Measured	Reflection Number	ϕ Measured	ξ Measured
1	2.2°	1.86 rlu	16	130.4°	0.51 rlu
2	10.6	1.61	17	130.4	0.65
3	25.4	1.04	18	130.4	0.76
4	36.0	1.49	19	130.4	1.02
5	54.6	0.52	20	130.4	1.53
6	54.6	0.66	21	149.8	1.48
7	54.6	0.78	22	159.2	1.02
8	54.6	1.04	23	167.8	1.60
9	54.6	1.54	24	183.2	0.61
10	71.8	1.74	25	183.2	1.24
11	78.0	1.26	26	183.2	1.85
12	92.4	0.82	27	197.4	1.60
13	92.4	1.62	28	206.2	1.01
14	107.0	1.23	29	235.4	0.51
15	114.0	1.73	30	235.4	0.63

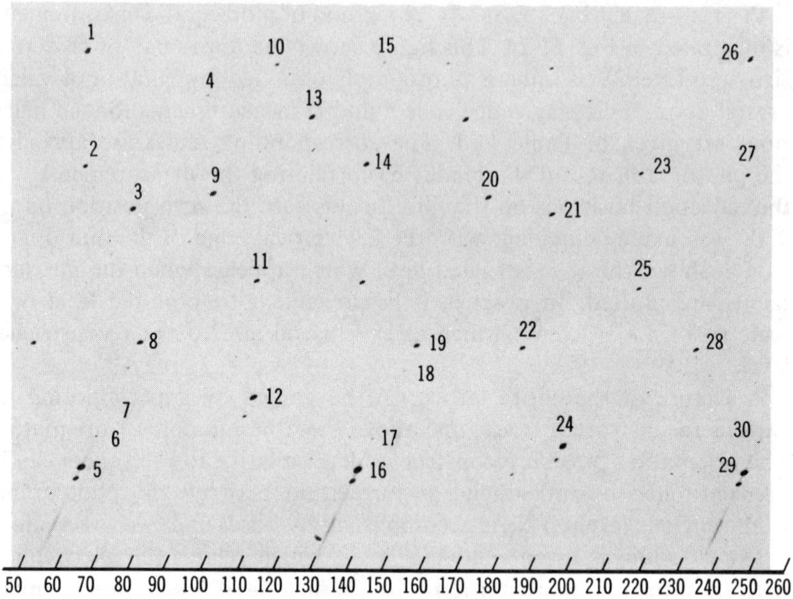

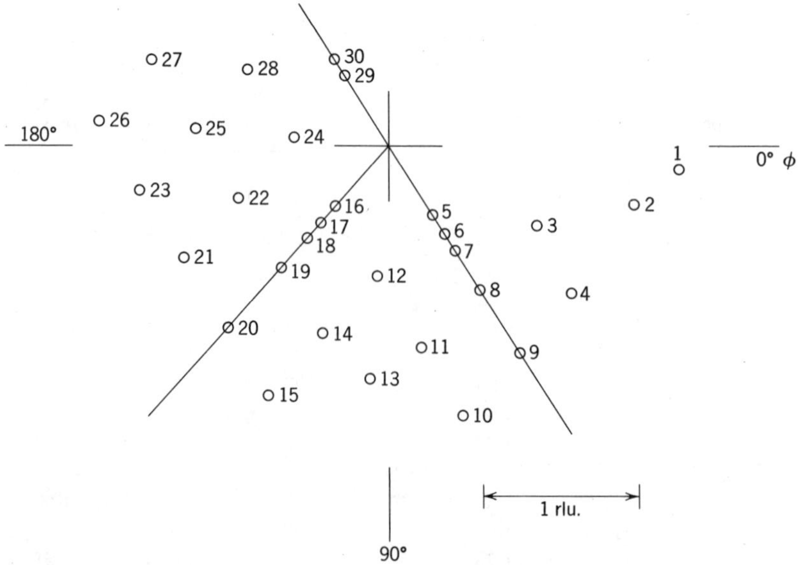

Fig. 11-14. (*a*) Actual-size reproduction of one half of zero level photograph of calcite $CaCO_3$ rotated about a cleavage edge (Cu$K\alpha$, film radius $90/\pi$ mm) and (*b*) its projection.

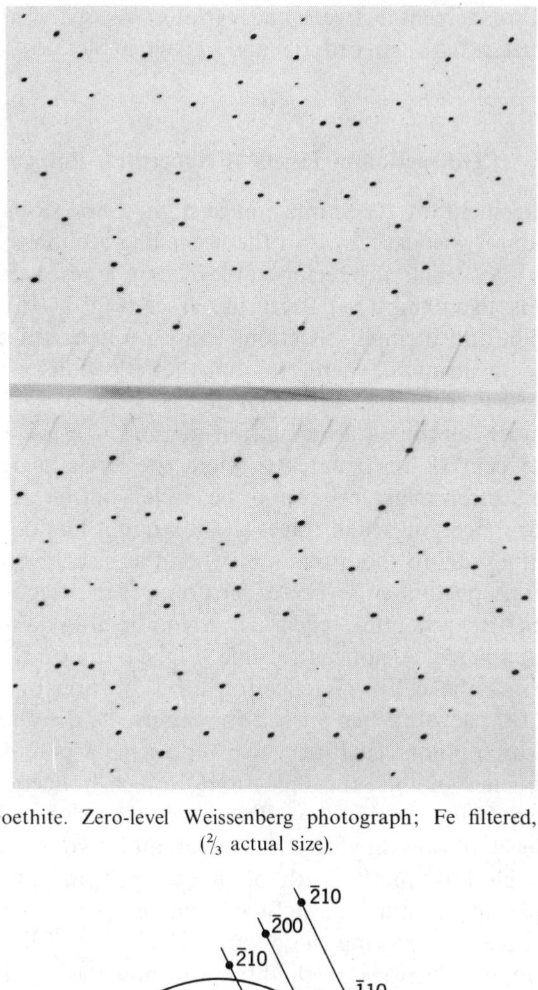

Fig. 11-15. Goethite. Zero-level Weissenberg photograph; Fe filtered, Co radiation ($\frac{2}{3}$ actual size).

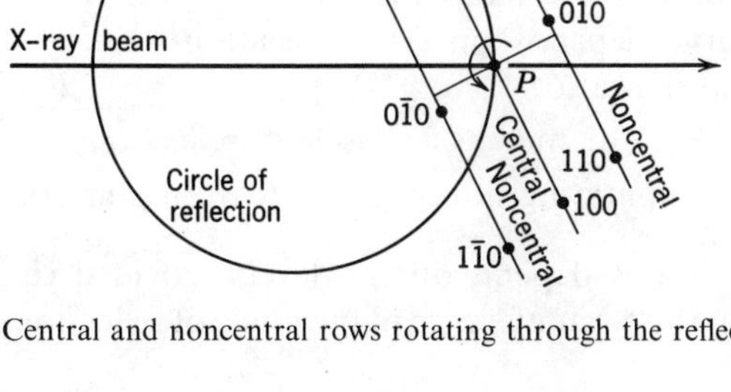

Fig. 11-16. Central and noncentral rows rotating through the reflection sphere.

altered between films. Then the films can be superimposed—there is no lateral displacement between the various Weissenberg records, and the reflections exhibit an orderly and recognizable relation from one film to the next.

Diffraction in Terms of Noncentral Rows

The arrangement of reflections on straight lines sloping at 63° 26′ to the horizontal was explained in the preceding section in terms of the central rows into which any reciprocal level can be resolved. Frequently, the reflections also exhibit a pattern of curves (Fig. 11-15) which sweep across the film and include reflections from a number of central-lattice rows. It may be surmised, therefore, that the reflections which comprise a curve represent reciprocal rows that are not central.

It was shown for the equi-inclination method in Figs. 11-4 and 11-8 that, when a central line rotates through the circle of reflection, one point of intersection migrates around the circle, continuously developing one potential reflection which traces out a straight line on the film. The line slopes at 63° 26′ to the horizontal when the instrument has a translation/rotation constant of $\frac{1}{2}$. The other point of intersection lies on the rotation axis—its potential reflection direction always coincides with the undeviated beam. A noncentral line (Fig. 11-16), on the other hand, remains outside the circle of reflection for a portion of one complete revolution of the level. When such a row intersects the circle, it does so in general at two points, each of which represents a potential reflection. As a consequence of this, the reflections from a noncentral row must lie on a curve. Figure 11-17 shows the development of such curves as a reciprocal level, represented by a central and two noncentral lines, rotates through 180° in the path of the x-ray beam; the central line traces out the now-familiar straight line on the two halves of the film. For convenience in following the diagrams, the central line of the level, and its potential reflections on the film, are shown as solid lines in contrast to the noncentral lines and their potential reflections which are represented by dotted and dashed lines.

A consideration of the diagrams shows that the shape of a reciprocal lattice line curve depends on the translation/rotation constant of the instrument and the ratio:

$$\frac{\text{diameter of a circle of reflection}}{\text{distance separating noncentral row from point of rotation}}$$

The shape does not depend on the level, provided that the point of rotation lies on the circumference of the circle of reflection. Consequently,

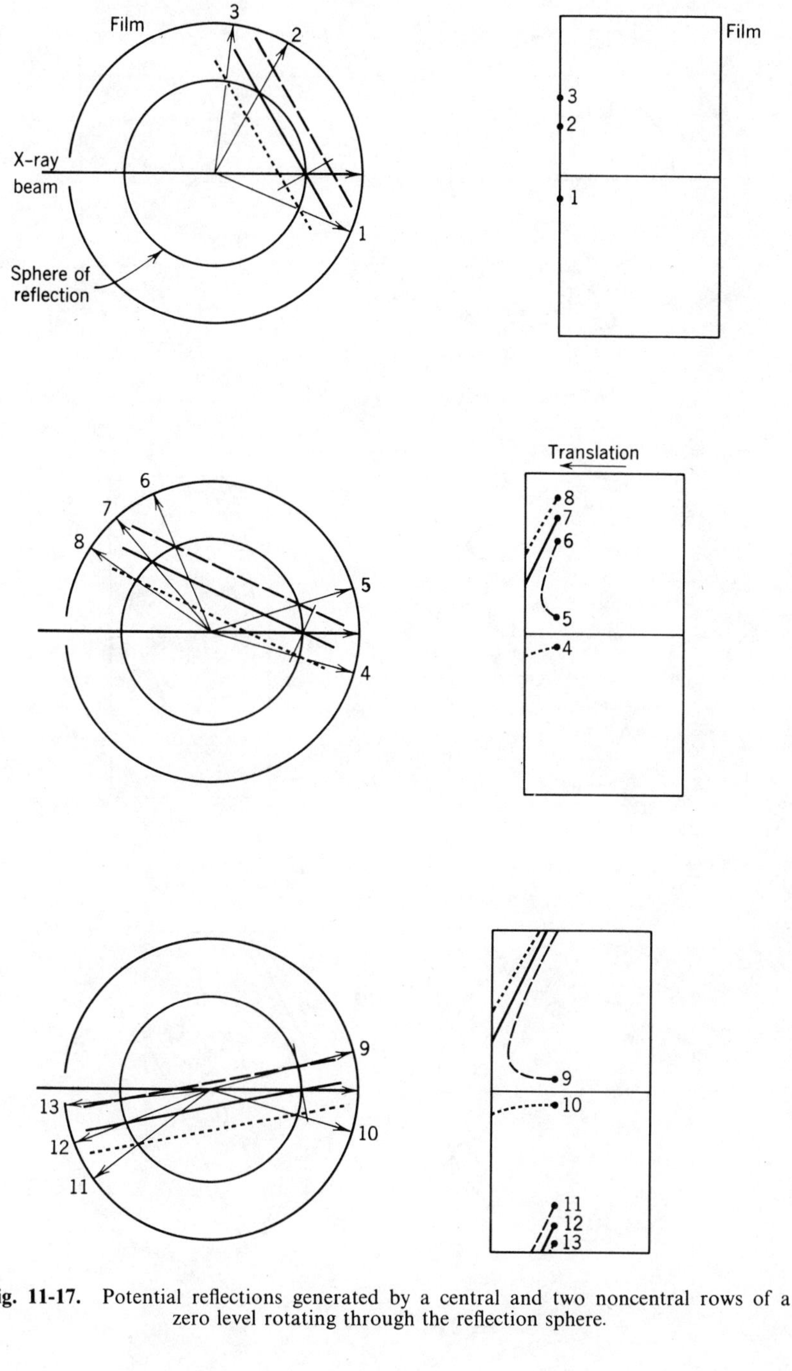

Fig. 11-17. Potential reflections generated by a central and two noncentral rows of a zero level rotating through the reflection sphere.

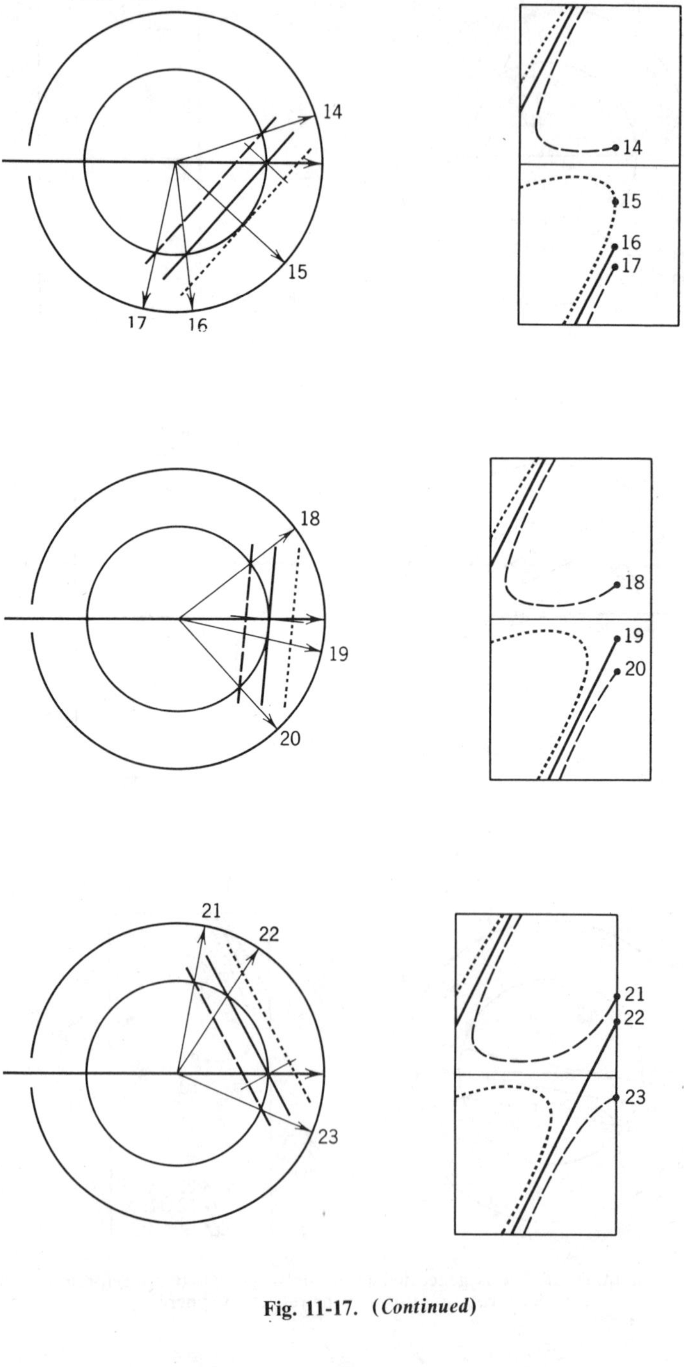

Fig. 11-17. (*Continued*)

with the equi-inclination method, the shapes of all curves representing identical values of this ratio are identical, regardless of the level. Thus a set of curves, constructed for a range of values of this ratio can serve as a guide for identifying the reciprocal lattice lines on the Weissenberg

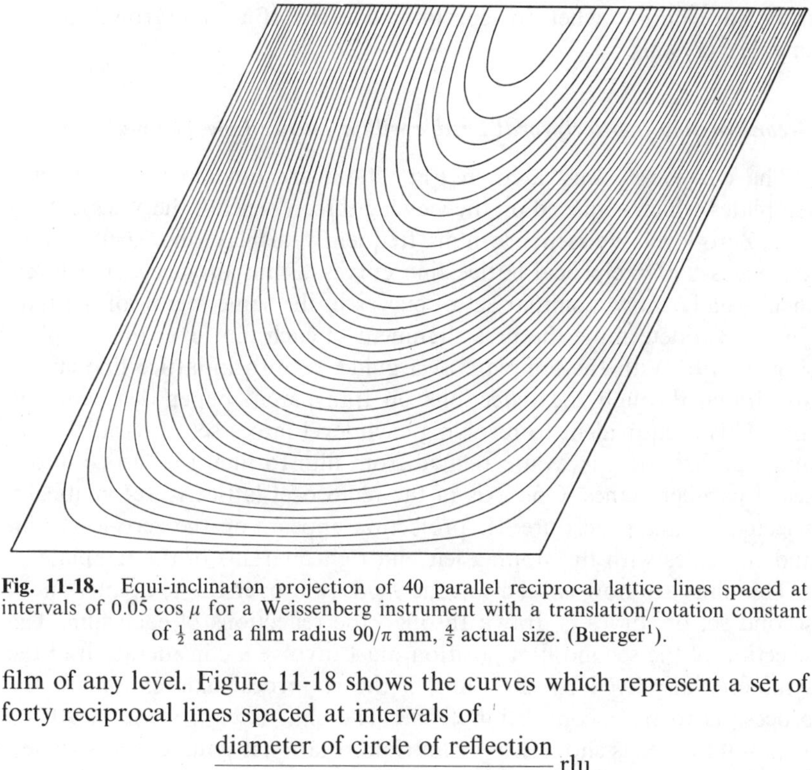

Fig. 11-18. Equi-inclination projection of 40 parallel reciprocal lattice lines spaced at intervals of $0.05 \cos \mu$ for a Weissenberg instrument with a translation/rotation constant of $\frac{1}{2}$ and a film radius $90/\pi$ mm, $\frac{4}{5}$ actual size. (Buerger[1]).

film of any level. Figure 11-18 shows the curves which represent a set of forty reciprocal lines spaced at intervals of

$$\frac{\text{diameter of circle of reflection}}{40} \text{ rlu}$$

for a Weissenberg instrument with a translation/rotation constant of $\frac{1}{2}$ and a film diameter of $180/\pi = 57.3$ mm. It was previously demonstrated in connection with Fig. 11-8 that the diameter of the circle of reflection is $2 \cos \mu$. The spacing interval of the reciprocal lines is therefore

$$\frac{2 \cos \mu}{40} = 0.05 \cos \mu$$

The term $\cos \mu$ is a variable and indicates that the magnitude of the interval between the curves varies with the level. This does not detract from the practical value of the curves, however. The important function of the curves is to aid in identifying reciprocal lattice lines on Weissenberg films. In most cases they eliminate the need to measure reflections with the ξ and ϕ scales previously described and enable the experienced crystallographer to visualize the arrangement of reciprocal points in each level, to select appropriate axes to represent the reciprocal lattice, and to index the reflections. The method of measuring reflections with ξ and ϕ scales and plotting the reciprocal levels is useful in doubtful cases, for example when the specimen consists of an intergrowth of two or more crystals.

Reconstructing the Reciprocal Lattice with a Lattice Line Template

The curves, as illustrated in Fig. 11-18, are used in the form of a template[2] which is placed on the viewing screen beneath the Weissenberg film. Zero-level Weissenberg films frequently display one easily distinguished set of reciprocal-lattice line curves. Films of succeeding levels then usually show a similar set of curves in the same region of the film. The curves determine the first alignment of each film over the template (Fig. 11-19). With the template as a guide, reciprocal-lattice line curves are drawn through the reflections on transparent paper overlying the film. This is equivalent to resolving each level into a set of parallel rows and, because the curves are related from film to film, the lattice into a set of parallel planes. One axis of the reciprocal lattice is automatically selected by this procedure. Its projection appears on the zero-level film and coincides with the sloping left and right margins of the template.

The films are now shifted to a new position over the template, and a second set of curves is drawn through the reflections of each film. The selection of the second film position must involve a consideration of the symmetry displayed by the films, if the reciprocal axis chosen in the process is to be a conventional one. The shift is equivalent to 90° or 45 mm if the levels show two- or fourfold rotational symmetry associated with the rotation axis, 60° or 30 mm if the symmetry is sixfold, and so on. The amount of shift may not be obvious if the films show less than orthogonal symmetry, especially on upper-level films of triclinic crystals, because the axes are noncentral. Under these circumstances it may be necessary to measure and plot the reflections with the aid of ϕ and ξ scales.

The act of drawing a second set of curves through the reflections is equivalent to resolving the level into a second set of parallel rows. When this is done for each film of a set, it is equivalent to resolving the lattice into a second set of parallel planes. Each reciprocal point can now be referred to three axes, two of which are in the zero level and were selected with the lattice line template. The third axis passes from the

[2] Commercially available from a number of sources.

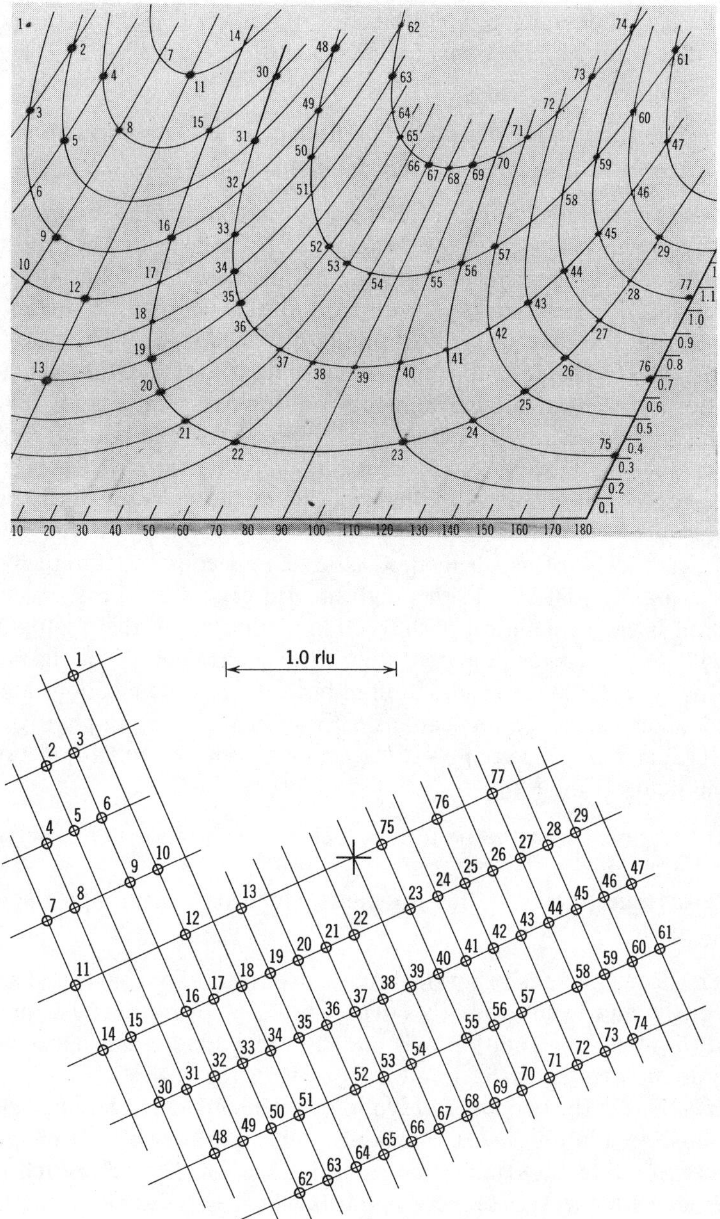

(b)

Fig. 11-19. Reconstruction of reciprocal level with a lattice line template.

origin in the zero level through symmetry-equivalent points in each of the levels. After appropriately labeling the axes, the reciprocal points (and therefore the reflections) can be indexed.

The Recognition of Diffraction Symmetry on Equi-inclination Weissenberg Photographs

It has been shown in the preceding sections that the three-dimensional array of reciprocal points can be derived by measuring the reflections with ξ and ϕ scales and plotting the points, or by drawing lattice line curves through the reflections with the aid of a template. A preliminary step to indexing the network of points obtained by the first procedure consists of choosing an appropriate set of axes for the array. The act of positioning the lattice line template on the Weissenberg photograph in the second method includes the selection of reference axes, after which only the labeling of the axes remains. The rules that govern the choice of reference axes, as outlined in Chapter 1, are based partly on the geometry of the lattice, but largely on the symmetry of the crystal. Thus, regardless of which method is used to reconstruct the reciprocal lattice, an essential step in the analysis of a crystal by the Weissenberg method is the recognition of diffraction symmetry on the photographs. It will soon become apparent that the determination of diffraction symmetry includes a consideration of both the relative positions and the relative intensities of all reflections on a set of Weissenberg photographs.

The symmetry of a reciprocal level may be referred to two operations in the plane of the level:

(1) rotation about the point at which the rotation axis intersects the level and
(2) reflection across a line (or lines) which passes through the same point.

When the symmetries of several consecutive levels are considered simultaneously and combined, the two symmetry operations become (1) rotation about the rotation axis and (2) reflection across planes which include the axis, and refer to all the points in the lattice.

Examples of the two types of symmetry elements that operate in a level are illustrated in Fig. 11-20. The figure shows portions of four reciprocal levels, with different symmetries, rotating through the circle of reflection together with the Weissenberg patterns which they develop. The levels are assumed to have rotated through 360°.

The point at which the rotation axis intersects the level, depicted in Fig. 11-20*a*, is evidently a onefold rotor. Consequently, the Weissenberg

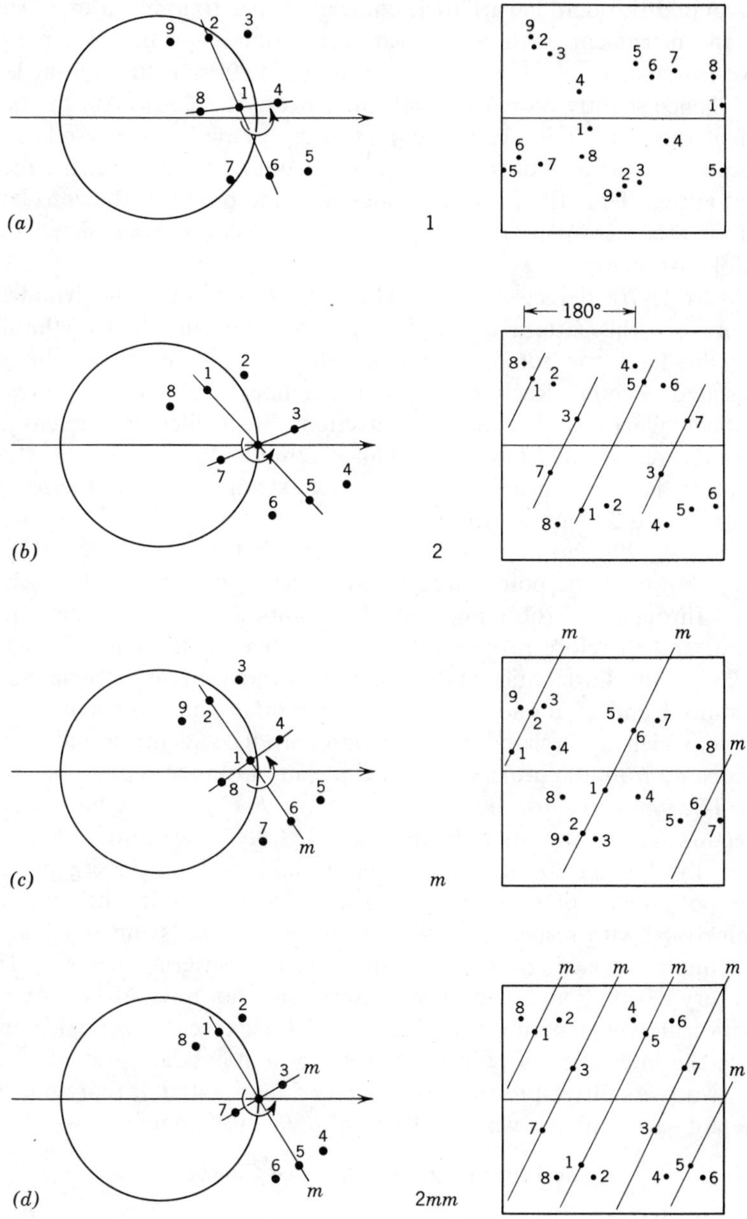

Fig. 11-20. Diffraction symmetry on equi-inclination Weissenberg photographs.

pattern begins to repeat on the same half of the film only after the level has rotated through 360°. This is equivalent to a translation of 180 mm for an instrument with a translation/rotation constant of 1 mm/2°. Most instruments are constructed to rotate the crystal through at least 200°, hence slightly more than half the pattern is recorded on the same half of the film. It is an inherent property of the sphere-of-reflection concept that the axis of rotation passes through the origin of the reciprocal lattice. Thus the rotor coincides with the origin in the zero level, and for this reason the rotor in this level always possesses higher than onefold symmetry.

Figure 11-20*b* depicts a level in which the rotor has twofold symmetry. In this case the pattern begins to repeat on the same half of the film after the level has rotated through 180°—that is, after the film has translated 90 mm. Reciprocal points numbered 1 and 5 represent identical planes and produce reflections with identical intensities. Similarly, reflections from points 3 and 7 are identical.

Rotors that have three-, four- and sixfold symmetry repeat the pattern in 60, 45, and 30 mm, respectively.

Figure 11-20*c* shows a level with a rotor that has only onefold symmetry; however, the points are related by reflection across a row which passes through the rotor and includes points 2, 1, and 6. The row is central and therefore projects as a straight line on the film, sloping at 63° 26′ to the horizontal, and is marked by reflections corresponding to points 1 and 2 on one end and 6 on the other end. Points, such as 3 and 9, which are related by reflection, produce identical reflections, equidistant from the projection of the line and from the center line of the film. The symmetry line recurs every 180° (or 90 mm) on the film, but the reflection pattern repeats only at intervals of 360° (180 mm).

Finally, Fig. 11-20*d* illustrates a level, and its Weissenberg pattern, with both types of symmetry elements. The points in the level are symmetrical with respect to a twofold rotor and two symmetry lines at right angles to each other. Consequently, the pattern repeats in 180° or every 90 mm, and shows symmetry lines at intervals of 45 mm. Reciprocal points numbered 2, 4, 6, and 8 represent identical planes, hence the intensities of reflections from these points are also identical. The two pairs of reciprocal points, 1 and 5, and 3 and 7, represent two different sets of planes whose reflection intensities differ, in general.

The Determination of Lattice Constants

Linear Constants

The cylindrical coordinate ξ of a reciprocal point specifies its right-angle distance from the rotation axis of the lattice. Because the axis passes through the origin, the ξ coordinates of points on axial rows which lie in the zero level are identical with, or simple multiples of, the corresponding axial periods. This relation may be expressed as

$$a^* = \frac{\xi_{h00}}{h}, \quad b^* = \frac{\xi_{0k0}}{k}, \quad c^* = \frac{\xi_{00l}}{l} \tag{10}$$

where ξ refers to coordinates in the zero level. The ξ scales described previously, are not meant to give particularly accurate values of ξ, hence the coordinates must be determined in another way to obtain satisfactory values for the edges of the unit reciprocal cell.

The relation $\xi = 2\sin(\Upsilon/2)\cos\mu$ (8), reduces to

$$\xi = 2\sin\frac{\Upsilon}{2}$$

for the zero level. However, according to (3), $\Upsilon/2 = \theta$ in this level, and thus

$$\xi_{\text{zero level}} = 2\sin\theta = \frac{\lambda}{d} \tag{11}$$

Now from (4), 1 mm $x = 1°\ \theta$, and therefore ξ coordinates (hence the reciprocal cell edges a^*, b^*, and c^*) and direct-lattice spacings d, are readily deduced from measurements of x on zero-level Weissenberg photographs.

The accuracy which can be attained from the careful measurement of good Weissenberg photographs justifies the use of a special measuring device for the determination of linear constants. The device, described for powder films in connection with Fig. 5-7, has a scale and vernier reading to ± 0.05 mm, and is suitable for precise measurements. The film is fastened into position with the hairline on this device parallel to the film's center line (Fig. 11-21).

The β reflections often yield more reliable measurements near the center of the film, where the α reflections are unresolved and frequently are broad. The best measurements on any film are obtained from the high-angle, resolved α_1, α_2 reflections.

The systematic and random errors which affect the accuracy of measured spacings from powdered specimens also apply to single-crystal measurements. The largest systematic errors in a Weissenberg camera that has been carefully constructed to eliminate specimen eccentricity, arise from two factors:

1. *Film Shrinkage.* The direct beam enters the Weissenberg camera between the top and bottom edges of the film. This kind of film mount

was described as the *regular* or Bradley–Jay arrangement in connection with Fig. 5-16 in Chapter 5 (polycrystal methods). It was pointed out that this arrangement, unlike the Straumanis method of mounting the film in the camera, is not self-calibrating against film shrinkage. Unfortunately, the Straumanis scheme cannot be applied to the ordinary

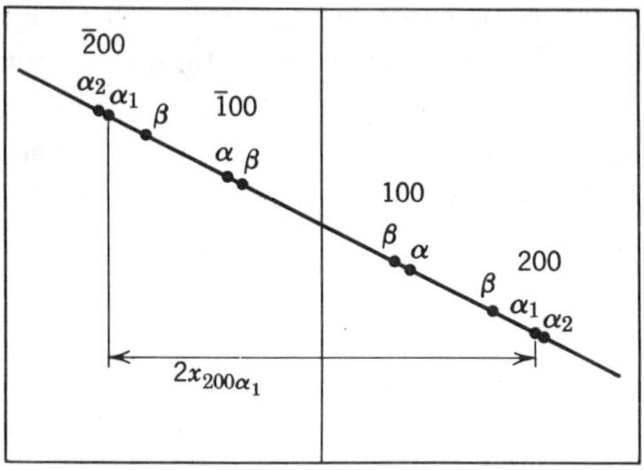

Fig. 11-21. Measurement of the x film coordinate for the $200\alpha_1$ reflection.

Weissenberg method. The hole punched in the film to admit the collimator must be an elongated slot to accommodate the translating action of the method. This would seriously weaken the film, and make it difficult to maintain a perfectly cylindrical shape while it is in the camera, and almost impossible to avoid distorting the exposed film while processing it in the darkroom.

2. *Absorption of X-rays by the Crystal.* It has been shown in Chapter 5 that absorption by a powder specimen increases the apparent angle of reflection, but the effect diminishes with increasing θ and vanishes at $\theta = 90°$. Consequently, lattice spacings or cell edges are plotted against some function of θ and extrapolated to $\theta = 90°$ to obtain true values. Exactly the same effect influences the cell constants obtained from single crystals.

Errors due to film shrinkage and absorption can be minimized on an ordinary zero-level Weissenberg film, by a method of comparison—that is, by recording on the same film the diffraction pattern of a substance whose constants are accurately known. The method seeks to eliminate the error by comparing the position of a reflection with the positions of reflections of a substance with known constants, in the same region of the film. The calibrating substance is usually a polycrystalline specimen in the form of a powder or wire. Its pattern is subsequently recorded as powder arcs on the sides of the Weissenberg film (Christ[3]), or simultaneously as continuous powder lines superimposed on the Weissenberg pattern over the whole width of the film by combining crystal and calibrating substance into one mount (Fridrichsons[4]). The calibrating substance may also be an oriented crystal whose zero

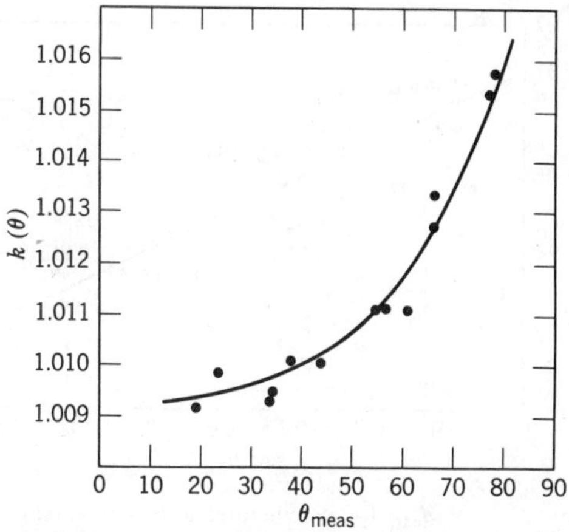

Fig. 11-22. Calibration curve: plot of $k(\theta)$ versus θ for powdered silver (after Christ[3]).

layer line rotation pattern is subsequently recorded on the sides of the Weissenberg film (Pabst[5]). Christ found that powdered silver, annealed in air at 350°C for 18 hours, acted as a satisfactory comparison standard—its pattern is easily interpreted and has strong, well-resolved back reflections. He plotted for this pattern the ratio $(\theta_{\text{theoretical}}/\theta_{\text{measured}})$, which he called $k(\theta)$, against θ_{measured} to obtain a calibration curve for the film (Fig. 11-22). The shape of the curve proved to vary considerably from film to film, even for the same camera. Each measured value for the material under study was then multiplied by the appropriate value of the constant $k(\theta)$ to obtain θ values corrected for film shrinkage. Apparent values of lattice constants were calculated from the corrected θ values with the aid of relation (11), plotted against $\sin^2\theta$, and linearly extrapolated to $\sin^2\theta = 1$ to obtain the true value corrected for absorption (Fig. 11-23). Christ estimated that his procedure gave an accuracy of two to four parts in 10,000.

[3] Christ, C. L., *Am. Mineralogist.*, **41**, 569 (1956).
[4] Fridrichsons, J., *Am. Mineralogist.*, **44**, 200 (1959).
[5] Pabst, A., *Am. Mineralogist.*, **42**, 664 (1957).

Angular Constants

BY MEASUREMENT OF THE DISTANCES BETWEEN LATTICE ROW PROJECTIONS. It was previously shown that the angle ϕ between two central lattice rows is related to the horizontal film distance z between the projections of the rows by the relation 1 mm $z = 2° \phi$ (5), when the translation/rotation constant of the instrument is 1 mm/2°. This suggests

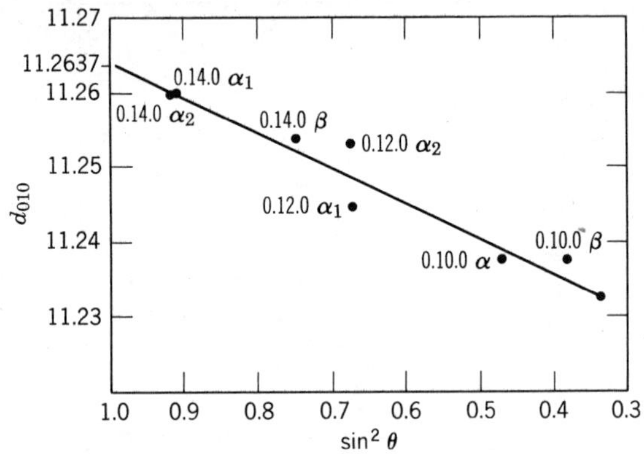

Fig. 11-23. Determination of d_{010} for colemanite; apparent values of d_{0k0}, calibrated against silver (Fig. 11-22) extrapolated to $\sin^2\theta = 1$ (after Christ[3]).

a simple way of determining ϕ, namely, by measuring the distance between the two points at which the projections cross the center line of the film. The angle can be more accurately determined with the film aligned to bring its lattice row projections parallel to the hairline of the measuring device, as illustrated in Fig. 11-24. The figure shows the projections of two axial rows, a^* and c^*, which are marked, respectively, by $h00$ and $00l$ reflections. The interaxial angle β^* is represented by the distance s between a^* and c^*. The distance between a^* and $\bar{a}^*$, namely, t, represents 180°. Consequently, $s/t = \beta^*/180$ and

$$\beta^* = \frac{s}{t}180$$

Angles can be determined to within at least $\frac{1}{2}$° by this method, if the reflections spots are sharp and small and if care is exercised in aligning the crystal on the instrument.

BY THE METHOD OF TRIANGULATION. Figure 11-25 shows the zero level of a reciprocal lattice with the axes a^* and c^* enclosing the interaxial angle β^*. A triangle, formed by the reciprocal points 000*, $\bar{3}$00*,

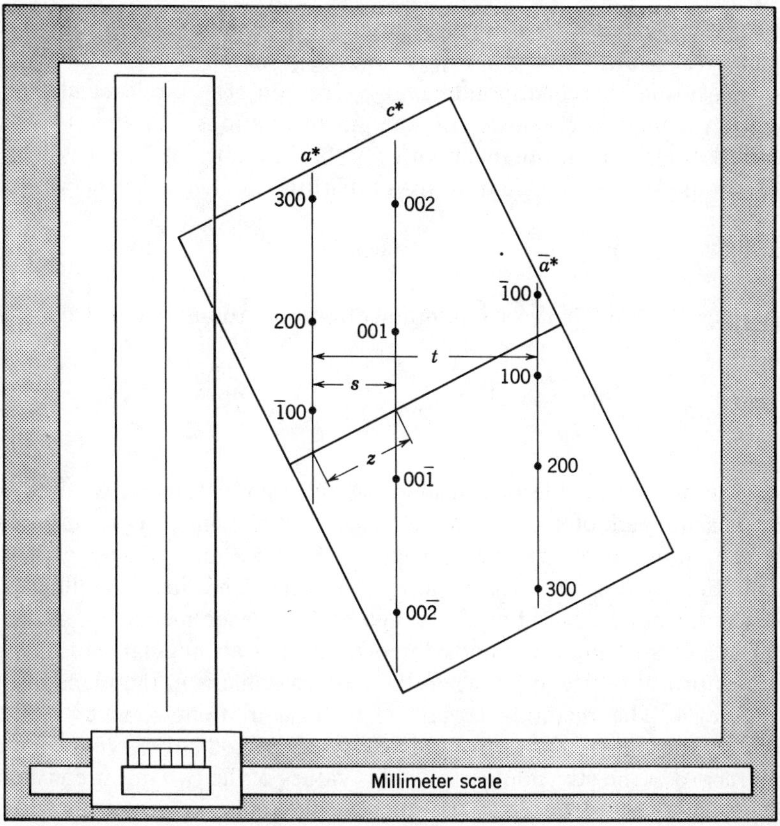

Fig. 11-24. Method of determining an angle by measurement of the distance between lattice-row projections.

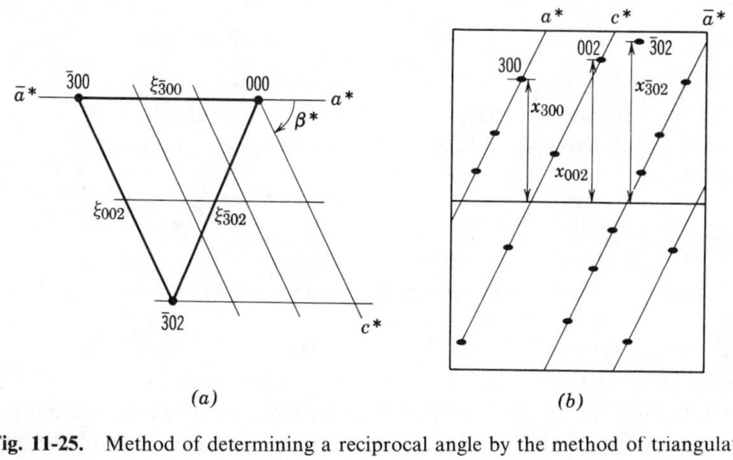

Fig. 11-25. Method of determining a reciprocal angle by the method of triangulation.

and $\bar{3}02^*$, includes this angle. The triangle is bounded by the three cylindrical coordinates, $\xi_{\bar{3}00}$, ξ_{002}, and $\xi_{\bar{3}02}$, which can be determined by measuring the corresponding $2x$ values on the zero-level film and by calculating the ξ values with the aid of relations (10) and (11). The desired angle is then obtained with one of a number of formulas which give an angle of a triangle in terms of its three sides. The cosine formula

$$\cos\frac{A}{2} = \frac{s(s-a)}{bc} \qquad \left(\text{where } s = \frac{a+b+c}{2}\right)$$

provides the angle with a minimum of computation. Hence, from Fig. 11-25

$$\cos\frac{\beta^*}{2} = \frac{s(s-\xi_{\bar{3}02})}{\xi_{\bar{3}00}\cdot\xi_{002}} \qquad \left(\text{where } s = \frac{\xi_{\bar{3}02}+\xi_{\bar{3}00}+\xi_{002}}{2}\right).$$

The method is capable of considerable accuracy because it is based on the measurement of linear constants which can be refined, as, for example, by the comparison method of Christ.[3] To take full advantage of the method, it is necessary to select a triangle which lends itself to refinement or whose sides are based on three high-angle measurements.

The two preceding methods for determining an angular constant of the reciprocal lattice apply when the desired angle is in the plane of the zero level. The methods require that (say) a monoclinic crystal be rotated about its b[001] axis if the desired angle is β^* (the zero-level film is a record of the $h0l^*$ points and yields values of the two linear constants a^* and c^*, as well as β^*). Triclinic crystals must be rotated, in turn, about the three direct-lattice axes to provide the three reciprocal angles, although any two zero-level films give the three linear constants (Table 11-2).

Table 11-2

Rotation Axis	Zero-Level Record	Linear Constants	Angular Constant
a[100]	$0kl^*$	b^*, c^*	α^*
b[010]	$h0l^*$	a^*, c^*	β^*
c[001]	$hk0^*$	a^*, b^*	γ^*

BY MEASUREMENT OF THE "SHEAR" BETWEEN LEVELS. If it is impossible or undesirable to rotate a triclinic crystal about more than one axis, a knowledge of two of the direct-lattice angles can be gained from a measurement of the displacement between successive levels and can be used to calculate those constants that cannot be measured directly. The angles α, β, and γ separate the direct-lattice axes and, because the unit reciprocal planes are normal to these axes, the angles are also the angles between the unit reciprocal planes. Some of the relations are shown in Fig. 11-26 for a crystal rotating about the $c[001]$ axis and

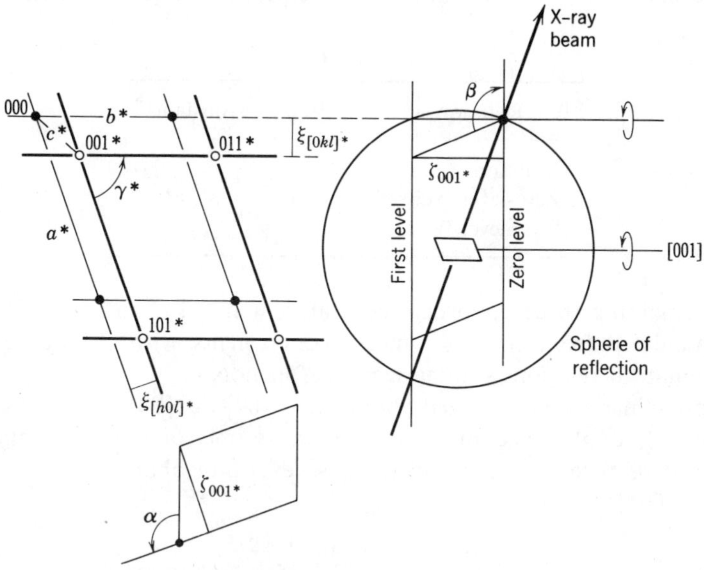

Fig. 11-26. Method of determining an angle by measurement of the "shear" between levels.

inclined to the beam at an angle designed to record the first level by the equi-inclination method. From this figure

$$\alpha = 180 - \tan^{-1}\frac{\zeta_{001^*}}{\xi_{[h0l]^*}}$$

$$\beta = 180 - \tan^{-1}\frac{\zeta_{001^*}}{\xi_{[0kl]^*}}$$

in which ζ_{001^*}, the cylindrical coordinate of points in the first level, can be determined by measurement on a rotation photograph, and $\xi_{[h0l]^*}$ and $\xi_{[0kl]^*}$ are the offsets in the first level, of the $[h0l]^*$ and $[0kl]^*$ rows from the rotation axis., Whereas the $[h00]^*$ and $[0k0]^*$ rows are central rows and project as straight lines with a slope of two on the zero-level film, the $[h0l]^*$ and $[0kl]^*$ rows are noncentral and project as lattice line curves on the first-level film. The height (value of x) of the lowest part of each curve represents the amount of each offset. With the aid of relations (2) and (8), the measured values of x for the two curves are converted to the coordinates $\xi_{[h0l]^*}$ and $\xi_{[0kl]^*}$, which in turn can be used to calculate α and β from the two equations derived above. The rotation and zero-level Weissenberg photographs provide four additional lattice constants, of which three are reciprocal constants (Table 11-4).

Table 11-4

Film (Rotation Axis $c[001]$)	Constants
Rotation	c
Zero-level Weissenberg	a^*, b^*, γ^*
First-level Weissenberg	α, β

The remaining three reciprocal constants, namely, c^*, α^*, and β^*, can be calculated from these six measured elements with the aid of the mathematical relations summarized in Appendix 6.

Monoclinic crystals rotated about the $a[100]$ or $c[001]$ axes present a similar, although less complex case, because only one angle, namely, β needs to be determined. From the first-level film obtained by rotation about $c[001]$

$$\beta = 180 - \tan^{-1}\frac{\zeta_{001^*}}{\xi_{[0k1]^*}}$$

and for rotation about $a[100]$

$$\beta = 180 - \tan^{-1}\frac{\zeta_{100^*}}{\xi_{[1k0]^*}}$$

For a number of reasons, the method is capable of determining the angles with only limited accuracy. The precision of an upper level recording is sensitive to instrumental settings which are not needed for recording the zero level. Weissenberg procedures are not designed to refine the measurements which lead to these settings. Furthermore, the "offset curves" are not continuous on the film and must be inferred, in the critical region, with the aid of a lattice curve template. Finally, if the angles in question are within a few degrees of 90, the curves come very close to the center line of the film (region of $\xi = 0$), and the accuracy depends on measuring very small distances which are, consequently, subject to large percentage errors.

Monoclinic crystals are therefore best studied on the Weissenberg instrument with the $b[010]$ axis as rotation axis. The zero-level film records $h0l$ reflections from which the reciprocal elements a^*, c^*, and β^* can be obtained. Needle-shaped crystals, with no other morphological characteristic, are most easily and accurately mounted with the needle axis parallel to the axis of the goniometer head. When such crystals are elongated parallel to c[001] (or a[100]), the β angle cannot be accurately determined by Weissenberg analysis, as indicated above, and the crystal is better studied on the precession instrument. If the c[001] axis has first been accurately aligned with the axis of the crystal holder (for example, by means of oscillation–orientation photographs), it is only necessary to locate the a^*c^* plane and arrange it normal to the beam with the aid of precession–orientation photographs to put the crystal into the most favorable orientation for the precession instrument.

The three angular constants—α^*, β^*, and γ^*—which are needed to define triclinic cells, require that the crystal be reoriented twice on the Weissenberg instrument, to make each of the crystallographic axes coincide, in turn, with the rotation axis. Special mounting techniques (Fig. 6-4) may permit two of the orientations to be attained from one mount, although the labor of locating the orientation is hardly lessened. The precession method is more versatile in this respect. A triclinic crystal, mounted with a unit plane of the direct lattice normal to the dial axis (which is equivalent to arranging a reciprocal axis parallel to the dial axis) is oriented for precession about two axial directions, merely, by adjusting the setting of the dial axis. This means that only two mounts are needed to obtain the three reciprocal angles on zero-level precession films. The functions of the two instruments can in general be successfully combined to obtain all three angles without remounting the crystal. If the crystal is adjusted to set one of the unit planes, say, the (001) or ab plane, normal to the axis of the goniometer head, the crystal is easily oriented on the precession instrument for precession with the a[100] and b[010] axes to provide the two zero-level films which will yield α^* and β^*. In general, in triclinic crystals, a direct-lattice axis is within 30° of being normal to a unit plane of the crystal. Thus it should be possible to change the orientation of the crystal sufficiently, with the arc adjustments of the goniometer head, to bring the third axis, namely, c[001], parallel to the axis of the head for a zero-level Weissenberg photograph and a measurement of γ^*.

CHAPTER TWELVE

SINGLE-CRYSTAL X-RAY DIFFRACTOMETRY

The Laue photographic methods used by W. L. Bragg to solve the first crystal structures in 1912 were superseded almost immediately by W. H. Bragg's ionization spectrometer for the quantitative measurement of reflection intensities. This instrument had certain limitations, however—it was difficult to operate, and crystals measuring an inch or more in diameter were needed to obtain sufficient intensity. The latter requirement dictated that small-size crystals could be satisfactorily examined with the spectrometer only if the structures were simple enough to be determined from relatively few, strong reflections. As investigators turned to more complex structures, the need to measure weak reflections grew, which naturally stimulated the development of methods suitable for small crystals (the rotation, oscillation, and Weissenberg film methods described in preceding chapters). The methods were rapidly refined, and they almost completely replaced the use of other methods for the determination of crystal structures. They continue to have a number of advantages: they are capable of registering the weakest reflections in reasonable exposure times, no highly specialized equipment is required, and the photographic record is obtained without the need for close supervision.

It had been demonstrated by 1940 that Geiger counters could be used to measure the intensities of reflections from small crystals and that they were capable of yielding more accurate intensity data than the photographic methods. Another advantage in the use of counters lay in the possibility of employing a guidance system to automatically set crystal and counter to record the hundreds of reflections which must be individually measured in a photographic method. Despite these apparent advantages quantum counters did not come into general use for diffraction studies until the North American Philips company marketed the first powder diffractometer, fitted with a Geiger counter, in 1945. The linearity of response of a Geiger counter is limited to a few hundred counts per second, and consequently the apparatus has to be continually adjusted to reduce the counting rate for strong reflections to the acceptable range. This makes the Geiger counter unsatisfactory for automatic recording. Proportional and scintillation counters have a linear response to very high counting rates and the additional advantage that they can be used in conjunction with a pulse height analyzer to discriminate against unwanted radiation. These counters have replaced the Geiger counter for single crystal diffractometry.

The use of quantum counter methods has developed rapidly in recent years, and several types of single crystal goniometers are now commercially available. Three kinds of instruments, based on three fundamentally different principles are described here:

(1) the Buerger equi-inclination diffractometer,
(2) the Eulerian cradle, and
(3) the linear diffractometer.

The methods of single-crystal x-ray diffractometry assume that an unknown crystal will first be studied with a film method to establish the cell constants and the space group. The main purpose of the diffractometer is to set the crystal and counter at the predetermined angular positions for a selection of reflections and to measure the intensities.

The Buerger Equi-inclination Diffractometer

This is an adaptation of the Weissenberg equi-inclination film method. Reflections are generated according to the equi-inclination principle, but the intensities of the reflections are measured with a quantum counter which can be inclined at any angle ν to the x-ray beam (Fig. 12-1) and can be rotated about the crystal axis to trace the outline of any Laue cone coaxial with this axis.

Four angular values must be known to generate and detect an *hkl* reflection corresponding to a reciprocal point in the *n*th level (Fig. 12-2):

1. μ, the inclination angle of the rotation axis at which the *n*-level reflections and the primary beam are equally inclined to the level. According to relation 6 in Chapter 11

$$\mu = \sin^{-1}\left(\frac{\zeta}{2}\right)$$

where ζ refers to the *n*th level.

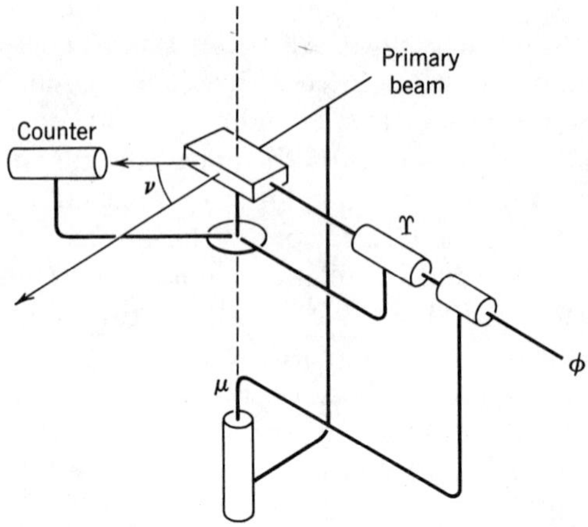

Fig. 12-1. Principle of the Buerger single-crystal diffractometer.

2. ν, the angle that reflected beams in the nth Laue cone make with the level. This angle is equal to μ in the equi-inclination method (Fig. 12-2).

3. Υ, the azimuthal angle of the reflection. From (8) in Chapter 11

$$\Upsilon = 2 \sin^{-1} \frac{\xi}{2 \cos \mu}$$

4. ϕ, the angle through which the crystal must be turned from some

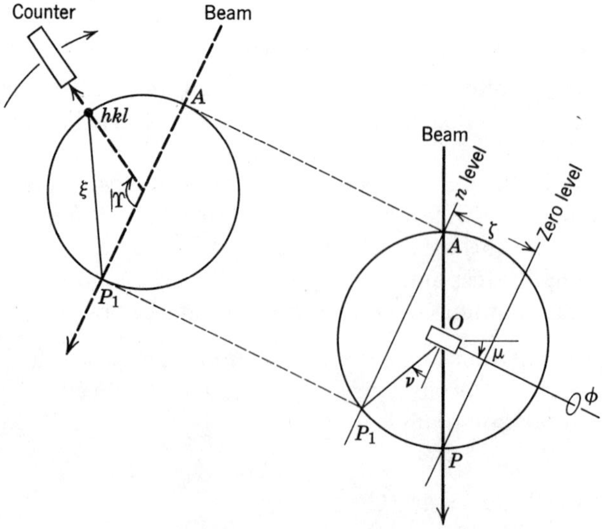

Fig. 12-2. The four setting angles (μ, ν, Υ, ϕ) for a Weissenberg diffractometer. In the equi-inclination method $\mu = \nu$.

reference position to bring the reciprocal point *hkl* onto the surface of the reflection sphere.

The μ and ν settings of the crystal and counter, respectively, are the same for all the reflections in a Laue cone. Only two adjustments, namely, Υ and ϕ, are needed for a particular reflection after the initial cone (or level) settings have been made. The Υ adjustment to the counter sets it into position to receive the reflection; the ϕ adjustment to the crystal rotates the reciprocal point onto the surface of the sphere. One method of measuring the intensity is to count the number of pulses generated while the crystal is oscillated through the ϕ setting a fixed number of times.

The Buerger instrument can be set and operated by hand, or guided by a commercial electronic control system, to set the Υ and ϕ angles and to record the intensities for a level of reciprocal points in accordance with computer-prepared instructions.

The Eulerian Cradle

Furnas and Harker[1] devised an accessory for orienting a crystal on a powder diffractometer. The combination makes use of the inherently simple and stable design of the powder diffractometer in which the counter rotates in only one plane. The penalty for this simplicity of counter movement is that each reciprocal point must be rotated into the counter's plane (the equatorial plane) for measurement. This requires, in general, that the crystal be reoriented for each reflection. The Furnas–Harker device meets this complex condition by providing rotation of the crystal about each of the Eulerian axes. These motions and the shape of the instrument (Fig. 12-3) are responsible for describing it as an Eulerian cradle.

The combination of diffractometer and cradle brings the axes of four circles into play, as indicated in Fig. 12-3. The counter rotates about the axis of the horizontal 2θ circle of the diffractometer. The ω circle of the cradle is also horizontal, and coaxial with the 2θ circle. The χ circle is vertical. The ϕ circle, whose axis is a radius of the χ circle and a selected axis of the crystal, is generally inclined. A device that combines these motions into a single instrument is properly called a four-circle single-crystal diffractometer. The axes of the four circles intersect at a point in the x-ray beam, and this common intersection must always be accurately occupied by the crystal (or, in terms of the reciprocal-lattice concept, by the origin of the lattice, as shown in Fig. 12-4). This requires

[1] T. C. Furnas, Jr., and D. Harker, *Rev. Sci. Instr.*, **26**, No. 5 (1955).

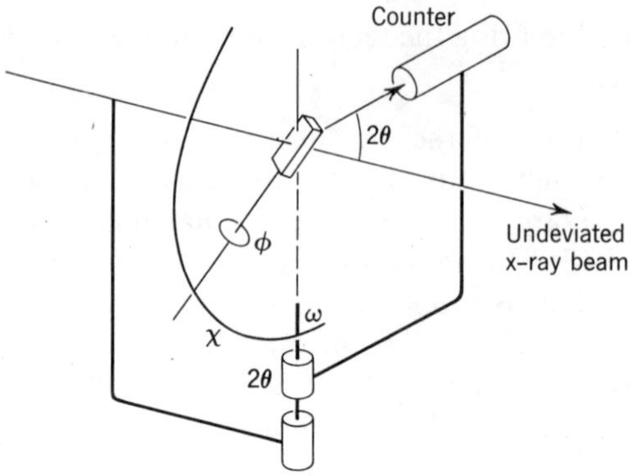

Fig. 12-3. Principle of the Eulerian cradle. The 2θ motion of the counter is provided by the diffractometer.

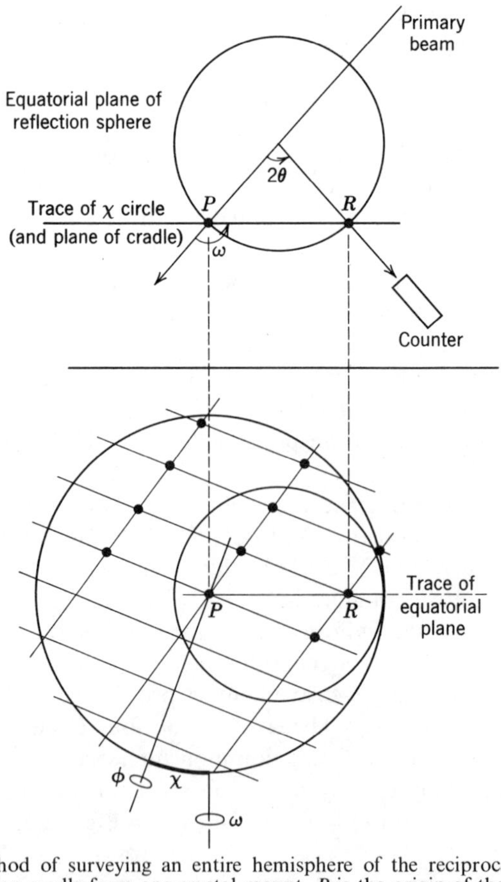

Fig. 12-4. Method of surveying an entire hemisphere of the reciprocal lattice with an Eulerian cradle from one crystal mount. P is the origin of the lattice.

high-grade workmanship in the construction of the cradle and its alignment on the diffractometer, and great care in setting the crystal.

As in all methods of collecting single-crystal intensity data, the lattice constants must be known. In general, the four angles 2θ, ω, χ, and ϕ differ for each reflection. If the orientation of the crystal relative to the four circles is known, values of these angles can be calculated to set the circles, either manually or by an automated system, and to measure the intensities of the reflections. Unless the χ setting is less than 53° in the Furnas–Harker instrument, the x-ray beam interferes with the χ arc and prevents a complete rotation of the cradle about its ω axis. This need not be a disadvantage if special procedures are adopted. If one of the most useful of these procedures, a selected axis of the crystal is made coincident with the ϕ axis (hence it is always in the plane of the χ circle), and ω is adjusted for each reflection so that the χ circle is always normal to the bisector of the calculated 2θ angle (Fig. 12-4). Then there is no interference with the beam to about $2\theta = 163°$. Having set the χ circle for a particular reflection by rotating the cradle about its ω axis, the corresponding reciprocal point (say, R in Fig. 12-4) is brought into the plane of the χ circle by rotating the crystal about the ϕ axis, and then onto the circumference of the equatorial circle with an adjustment of the χ setting. By systematically varying ϕ from 0 to 360° and χ from 0 to 90°, an entire hemisphere of the reciprocal lattice can be surveyed from a single mounting of the crystal. The radius of this hemisphere is limited by the condition that 2θ cannot exceed a value of about 163°. The maximum radius is evidently the value of PR (in Fig. 12-4) at $2\theta = 163°$ and is equal to $2 \sin 81\frac{1}{2}° = 1.98$ rlu. An alternative scheme is to set the χ circle parallel to the 2θ bisector for each reflection. Then the beam interferes with the χ circle (that is, the cradle) only at low 2θ values. The two methods can be usefully combined to survey an even greater range of 2θ values.

The Linear Diffractometer

Arndt and Phillips[2] have recently described a diffractometer based on a completely novel design. To use the methods described in the preceding sections, crystal and counter settings must be calculated for each reflection. The Arndt and Phillips instrument requires only a knowledge of the reciprocal lattice dimensions of the crystal. Once the crystal is correctly oriented on this instrument, the crystal and counter take up their positions automatically for any reflection *hkl* when the coordinates ha^*, kb^*, and lc^* of the corresponding reciprocal point are set on three sliding adjustments. One systematic way of surveying a reciprocal-lattice

[2] U. W. Arndt and D. C. Phillips, *Acta Cryst.* **14**, 807 (1961).

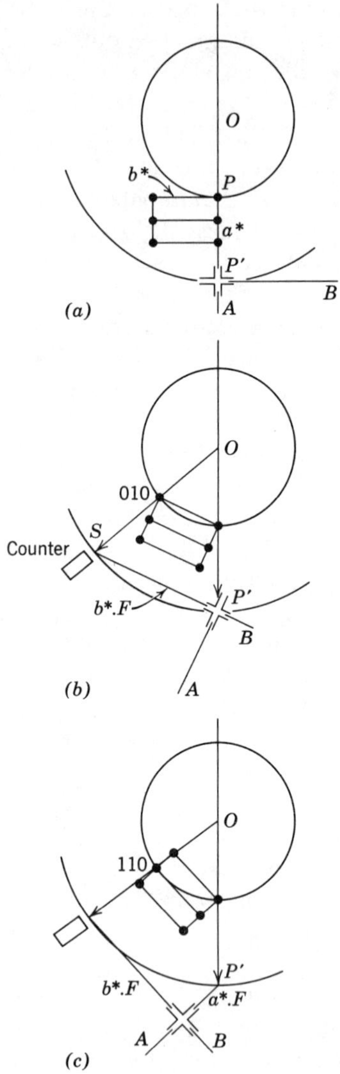

Fig. 12-5. The principle of zero-level measurements with the linear diffractometers.

level with the instrument consists of traversing along rows in reciprocal space, hence the name *linear diffractometer*.

According to the reciprocal-lattice concept a reflection, for example, 110 in Fig. 12-5, is generated when the corresponding reciprocal point is on the surface of the sphere of reflection. To receive the reflected beam, a counter must be aimed along the line joining the reciprocal point to the center of the sphere. The design of the linear diffractometer is based directly on this concept. A reciprocal-lattice level is represented by two slides—A, which pivots at a point P' corresponding to the origin P, and B, which can be moved along A. In a commercial version of the instrument, the angle between the slides can be varied from 90° to match the interaxial angle of nonorthogonal levels. The slides operate against a saddle S, which pivots about O at the end of a link of fixed length OS.

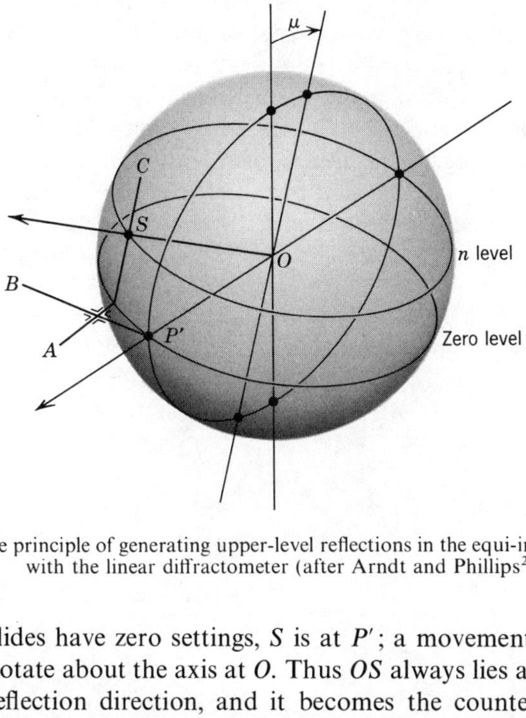

Fig. 12-6. The principle of generating upper-level reflections in the equi-inclination setting with the linear diffractometer (after Arndt and Phillips[2].)

When the slides have zero settings, S is at P'; a movement of the slides forces S to rotate about the axis at O. Thus OS always lies along a potential x-ray reflection direction, and it becomes the counter arm of the diffractometer. The counter will be in position to receive the 110 reflection of Fig. 12-5, for example, when the slides have settings which correspond to $1a^*$ and $1b^*$, respectively. The absolute values of the settings depend on a factor which is related to the length of the link OS. The values may be designated as $1a^* \cdot F$, if F is this factor.

The crystal is mounted on an axis at O, and can rotate independently of the link OS. However, a linkage system allows any rotation of the slide system at P' to be transmitted to the crystal at O. Thus, if the crystal is oriented with its a^* and b^* axes in the equatorial plane and a^* parallel to the beam, as in (a), when the slides are set at zero, movements of

$1a^* \cdot F$ along A and B will rotate the reciprocal point 110 onto the circumference of the reflection circle and the counter into position to receive the reflection. A systematic method of scanning consists of zigzagging through a level along reciprocal rows. To do this, the A slide (say) is set at $1a^*$ and the B slide is successively set at $1b^*$, $2b^*$, $3b^*, \ldots,$ nb^* until the slide is fully extended. The A slide is then set at $2a^*$, and the procedure with the B slide is repeated in reverse. The entire procedure can be arranged for automatic operation.

In the Arndt–Phillips diffractometer access to upper levels of the reciprocal lattice is gained by a third slide C, perpendicular to the other two. The slide system and crystal axis, held parallel by linkages, are free to tilt towards the x-ray source in a plane that includes the beam and is normal to the equatorial plane. Because the link between O and S is fixed in length, the saddle is constrained to remain on the surface of a sphere corresponding to the sphere of reflection, and the counter continues to be aimed at O, along a potential reflection direction. With the C slide set at a value corresponding to the ζ coordinate of an n level, the angle of tilt is equal to μ, and the instrument has the equi-inclination setting for this level (Fig. 12-6). Reflections in this level can then be generated and detected by operating slides A and B.

CHAPTER THIRTEEN

A GENERAL METHOD FOR ORIENTING A CRYSTAL

The Laue, oscillation, and precession methods all include systematic techniques for detecting and correcting errors in the orientation of a crystal. However, the techniques are limited to the evaluation of small misalignments and are therefore not satisfactory for orienting unknown crystals, or even known crystals with no diagnostic morphological traits.

Most practical x-ray methods for determining the orientation of a randomly set crystal rely on Laue photographs. The methods are generally based on a preliminary trial-and-error approach involving a search through a number of photographs of the crystal in different arbitrary orientations for some recognizable feature which will indicate the attitude of a symmetry plane or a prominent zone. They are therefore best suited to crystals of high symmetry and have not found much use in the general field of crystallography. The success of these methods depends on the aptitude of the worker and comes more readily to those who are fully experienced. Some of the methods have been described in Chapter 7.

The method[1] described here is a systematic method for ascertaining the orientation of a randomly set known crystal and adjusting it to within about 1° of a preferred orientation. The final adjustment is made with one of the techniques especially suited to correcting small misalignments. The data are plotted in such a manner that the principal zones in the crystal become apparent, and thus the method can be applied in the general case to locate a principal zone axis in an unknown, randomly oriented crystal and to set it for rotation or precession in one of the standard single-crystal x-ray methods of analysis. In contrast to

[1] E. J. Brooker and E. W. Nuffield, *Acta Cryst.*, **20**, 496 (1966).

procedures that employ Laue photographs, application of this method is simplified by a decrease in crystal symmetry.

An Expression for the Orientation of a Reflecting plane

The locus of a reciprocal point Q, representing a randomly oriented direct-lattice plane (hkl), is confined to the surface of a sphere, which is

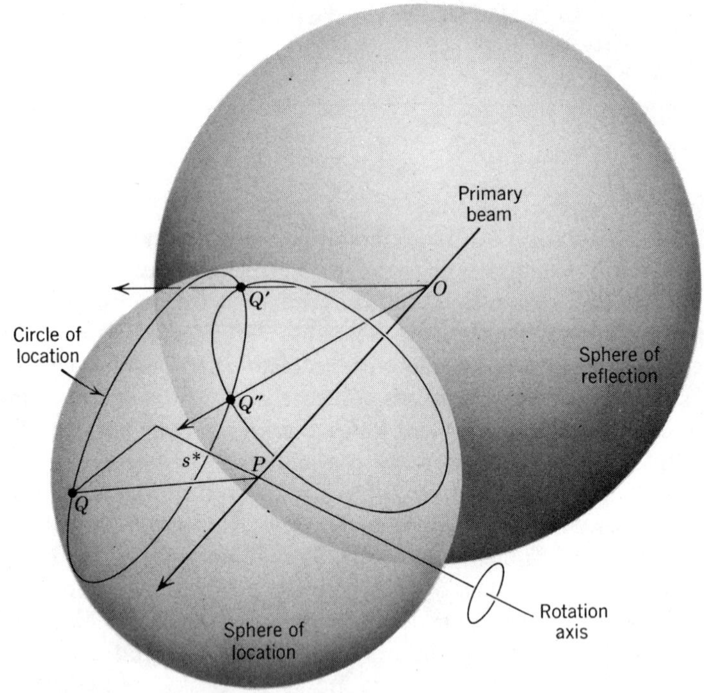

Fig. 13-1. Reflections generated by a random reciprocal point Q rotating about an axis normal to x-ray beam.

centered at the origin P and has a radius of $PQ = \lambda/d_{hkl}$. This may be called the sphere of location. If the point is rotated about an axis, it will trace out a circle (the circle of location) on the surface of the sphere of location. Figure 13-1 shows the case for a rotation axis situated normal to the incident beam. In general, the point will pass through the surface of the reflection sphere twice (at Q' and Q'') in one complete rotation, and therefore will satisfy the condition for reflection twice. If the reflections (R' and R'') are intercepted on a flat film placed normal to the beam, they

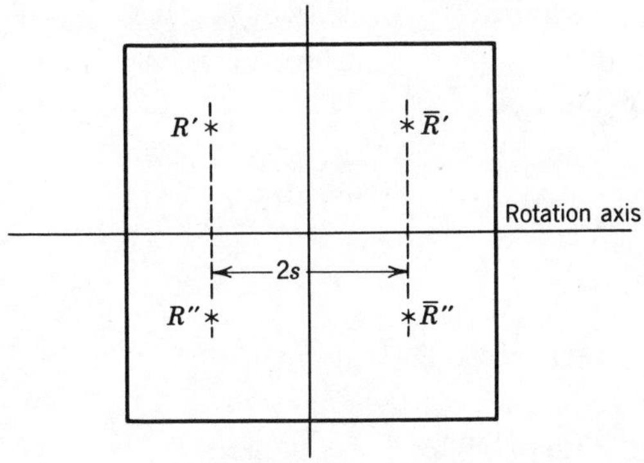

Fig. 13-2. Flat-film rotation photograph from a randomly oriented plane. X-ray beam normal to page.

will lie on a line normal to the rotation axis (Fig. 13-2). For every reciprocal point Q there is an opposite point $\bar{Q}$, and consequently the film will show a second pair of reflections $\bar{R}'$ and $\bar{R}''$. The half-distance s between the two sets of reflections is equal to $s^* \cdot F$, where s^* is the projection of the vector PQ onto the rotation axis, and F is the crystal-to-film distance. Therefore, $s^* = s/F$.

Suppose that one of the arcs (say, the small or S arc) of the goniometer head on which the crystal is mounted is moved through an angle α_S, whereas the setting of the other arc is maintained at zero to confine motion of the crystal to the plane of the arc that is moved. Then the vector's projection in this plane will also move through an angle α_S from PQ_1' to PQ_2', and its projection on the rotation axis will change from

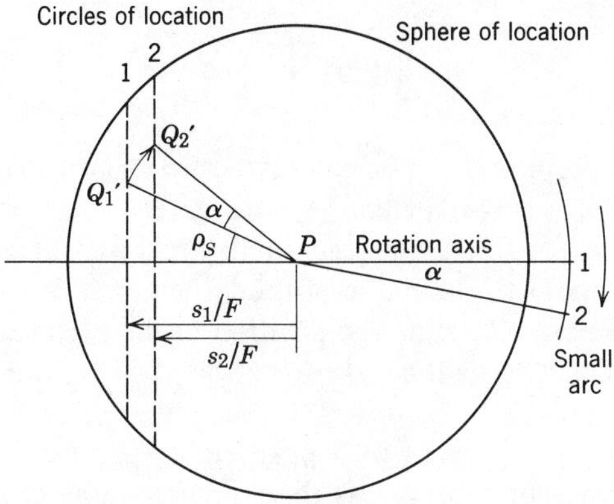

Fig. 13-3. Displacement of a reciprocal point resulting from a change in an arc setting in the plane of the page.

s_1/F to s_2/F (Fig. 13-3). The angle ρ_S between PQ_1' and the rotation axis defines the original position of the vector as measured in the plane of the small arc. From Fig. 13-3

$$\cos \rho_S = \frac{s_1/F}{PQ_1'}$$

and

$$\cos(\rho_S + \alpha_S) = \frac{s_2/F}{PQ_2'} = \frac{s_2/F}{PQ_1'}$$

because $PQ_1' = PQ_2'$. Hence

$$PQ_1' = \frac{s_1/F}{\cos \rho_S} = \frac{s_2/F}{\cos(\rho_S + \alpha_S)}$$

and

$$\begin{aligned}\frac{s_2}{s_1} &= \frac{\cos(\rho_S + \alpha_S)}{\cos \rho_S} \\ &= \frac{\cos \rho_S \cos \alpha_S - \sin \rho_S \sin \alpha_S}{\cos \rho_S} \\ &= \frac{\cos \rho_S \cos \alpha_S}{\cos \rho_S} - \frac{\sin \rho_S \sin \alpha_S}{\cos \rho_S} \\ &= \cos \alpha_S - \tan \rho_S \sin \alpha_S\end{aligned}$$

This can be rearranged into a more useful form

$$\rho_S = \tan^{-1}\left(\frac{\cos \alpha_S - (s_2/s_1)}{\sin \alpha_S}\right)$$

The expression permits ρ_S to be calculated from an instrumental setting (α_S) and two film measurements ($2s_1$ and $2s_2$). A similar relation holds for the large arc provided that the small arc is set at zero to confine motion of the crystal to the plane of the large arc. The values ρ_S and ρ_L for a reciprocal point Q define the position of the point on the sphere of location, as measured against the rotation axis in the planes of the two arcs.

The expression for ρ leads to a positive, a negative, or a zero value depending on whether s_2/s_1 is less than, greater than, or equal to $\cos \alpha$. The signs of ρ_L and ρ_s signify the quadrant of the reflection sphere in which Q occurs. Figure 13-4 illustrates the three possibilities (for one

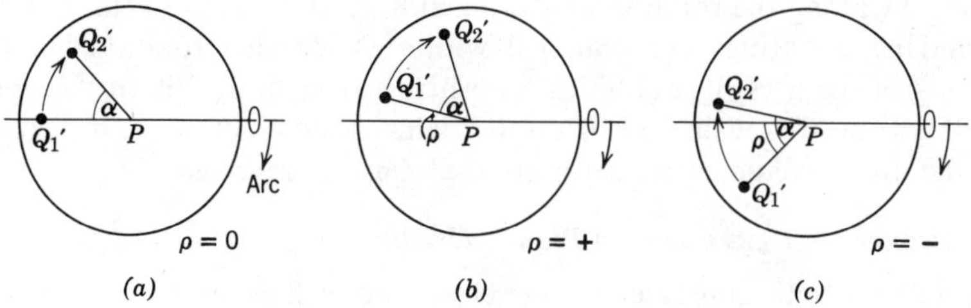

Fig. 13-4. The sign of ρ; arc movement is right-handed and in plane of page.

arc) which result from a right rotation of the arc in the plane of the page.

The sphere of location may be represented by a stereographic sphere in which the rotation axis coincides with the vertical or polar axis of the stereogram and the planes of the two arcs are normal to the projection plane and occupy the N–S, E–W directions. The reciprocal point Q (when the two arcs have zero settings) is located at the intersection of two great circles which make angles of ρ_L and ρ_S with the planes of the small and large arcs respectively (Fig. 13-5). For purposes of standardization,

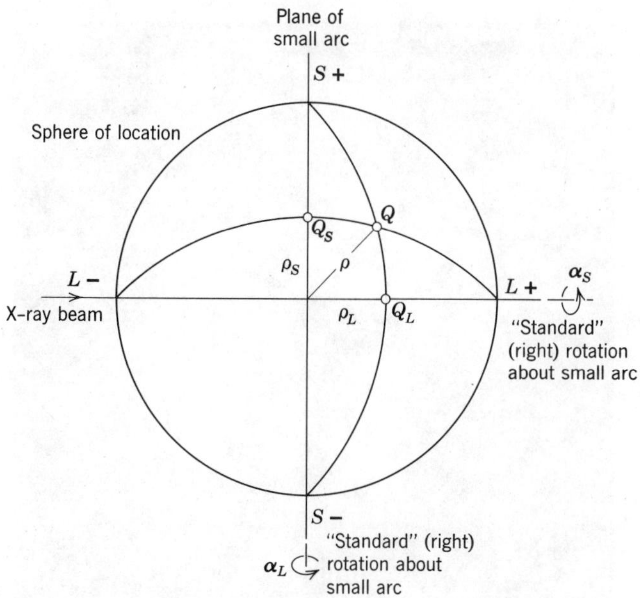

Fig. 13-5. Relation between "standard" (right) arc movement and sign of ρ, ρ_S, and ρ_L are the projections of ρ for a random point Q in the planes of the small and large arcs, respectively. View is along rotation axis from crystal to arcs.

the view along the rotation axis is taken from the crystal to the arcs and standard α settings are obtained with right-handed rotational movements of the arcs. It follows that positive values of ρ_S fall on the upper half and positive values of ρ_L on the right-handed half of the projection. If the arc movements are reversed, the signs are reversed.

Procedure for Determining a Plane's Orientation

1. With both goniometer arcs set at zero, a flat-film rotation photograph (Fig. 13-6) is prepared and the s_1 distance for the reflection measured.
2. With one of the arcs (say, the large arc) set at an angle α (15° is a practical setting), a second rotation photograph is prepared and the $s_2(L)$ distance measured.
3. The value of ρ_L is calculated from the relation

$$\rho_L = \tan^{-1}\left(\frac{\cos\alpha - s_2/s_1}{\sin\alpha}\right)$$

4. With the large arc set at zero and the small arc at an angle α, a third rotation photograph is prepared and the distance $s_2(S)$ measured.
5. The value of ρ_S is calculated.

Fig. 13-6. Flat-film rotation photograph of $CuSO_4 \cdot 5H_2O \cdot \alpha$-zero for both arcs. $F = 50$ cm. Filtered Co radiation. Exposure two hours.

6. The intersection of great circles plotted (as in Fig. 13-5) at angles of ρ_L and ρ_s is the stereographic projection of the plane represented by Q when both arcs have zero settings.

Not all of a crystal's planes will necessarily give values for ρ. Reciprocal-lattice vectors within a few degrees of the rotation axis may not enter the sphere of reflection as the crystal is revolved, or they may be moved permanently outside the sphere by the α setting of one of the arcs.

All equivalent reciprocal points reflect to positions equidistant from the film center—that is, to the circumference of a circle. In general then, the higher the symmetry of the crystal, the more densely populated is the circle and the more difficulty may be encountered in sorting out the movement of reflections caused by the α settings. Thus the method is best suited to crystals of low symmetry.

Application of the Method

The Buerger precession goniometer is a convenient instrument for applying the new method. The instrument is readily converted to the required flat-film rotation technique by setting the precession angle at zero and by rotating the drum axis independently by means of a belt drive from a small 1-rpm motor. Furthermore, crystal settings derived from application of the method can be readily tested and precisely adjusted with the aid of precession-orientation photographs on the same instrument.

Important direct-lattice rows are axes of zones that include a high proportion of simple-index planes. Reflections from these planes are generally well-represented in the low-2θ region of a film obtained with monochromatic radiation. Therefore, a systematic procedure for locating such an axis is to plot the stereographic poles of the low-2θ reflecting planes according to the scheme described in the preceding section, and to analyze the projection for important zones.

A randomly set crystal of the triclinic compound $CuSO_4 \cdot 5H_2O$ provides a test of the method in the general case for locating a principal zone in the crystal and defining its orientation in a useful way. Table 13-1 gives the s measurements and calculated ρ values obtained for the first fifteen reflections from such a crystal, beginning with the reflection nearest the film center and working outward to larger 2θ values.[2]

[2] The simple relation $r/F = \tan 2\theta$ (in which r is the film center-to-reflection distance and F is the crystal-to-film distance) holds for all reflections in a monochromatic normal-beam, flat-film rotation method. It enables the d spacing and, if the cell dimensions are known, the index of each reflecting plane to be determined with a single-film measurement on any of the three preliminary rotation photographs.

Table 13-1. Values of ρ Calculated from Film[a] Measurements of A $CuSO_4 \cdot 5H_2O$ Crystal

Reflection	s_1	$s_2(L)$	ρ_L	$s_2(S)$	ρ_S
1	3.6 mm	2.8 mm	$36\frac{1}{2}°$	1.9 mm	$59\frac{1}{2}°$
2	10.2	10.1	−1	12.3	$-42\frac{1}{2}$
3	0.3	3.1	$-88\frac{1}{2}$	1.7	87
4	13.8	12.6	$12\frac{1}{2}$	14.3	−14
5	7.4	5.7	38	4.0	59
6	3.4	5.6	−69	0.3	$73\frac{1}{2}$
7	4.1	0.3	74	3.7	14
8	13.8	15.9	−35	12.9	$7\frac{1}{2}$
9	10.1	13.1	−52	10.8	−21
10	6.7	7.3	−25	10.5	$-66\frac{1}{2}$
11	18.1	15.7	$21\frac{1}{2}$	16.8	9
12	17.9	19.7	$-26\frac{1}{2}$	15.0	$26\frac{1}{2}$
13	7.3	8.9	−44	2.3	68
14	6.5	10.5	−68	9.0	−58
15	8.0	2.7	67	6.0	$40\frac{1}{2}$

[a] Cu radiation, $F = 50$ cm.

Figure 13-7 shows the ρ values plotted on a stereogram according to the scheme illustrated in Fig. 13-5. The most notable feature of the projection is the appearance of three prominent zones. The zones are numbered from one to three in decreasing order of importance, the estimation of importance being based on the abundance of low-2θ (larger spacing, simple index) reflecting planes.

Table 13-2 shows a comparison between the d values determined from measurements of r for the fifteen reflections and d values calculated for the $CuSO_4 \cdot 5H_2O$ cell. The table establishes that the three zones are composed of planes with general indices ($hk0$), ($0kl$), and (hkh), respectively. The zone axes are therefore [001], [100], and [101]. It is noteworthy that the two most important zones are axial zones. The orientation of the crystal when both arcs are set at zero is now established, and the pole of any additional plane or axis may be plotted with the aid of standard crystallographic data for the compound.

Although fifteen reflections were used in this illustration, the three zones could have been discovered by measurement of the first nine reflections. The two axial zones are defined by the first four and seven poles, respectively. This suggests that a principal zone in an unknown crystal should be located by measurement of the first five or six reflections. If the crystal is known, one or two *selected* reflections indicate the

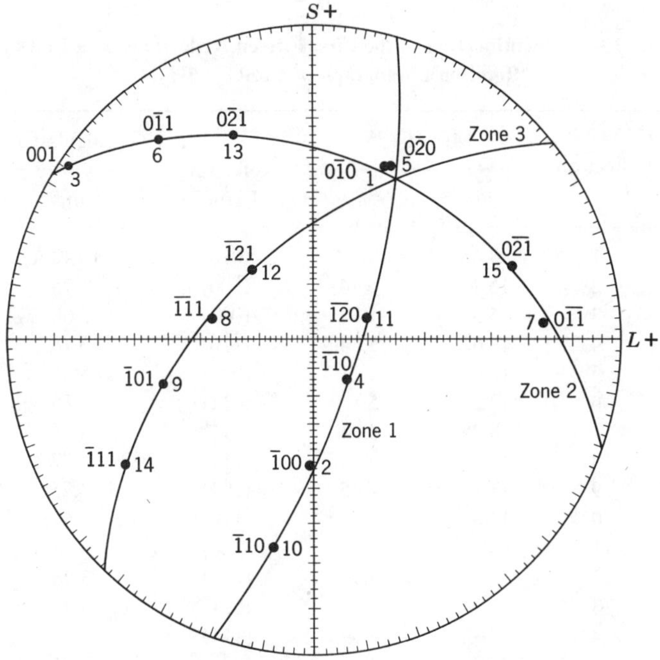

Fig. 13-7. Stereogram of randomly oriented $CuSO_4 \cdot 5H_2O$ crystal. Arcs and beam are of the same orientation as in Fig. 13-5.

orientation. In practice, it is essential to ensure that the very low angle reflections are recorded. This may involve the use of long-wavelength radiation and extended exposure times for some crystals.

Setting the Crystal

The next step, whether the crystal is known or not, is to orient one or more chosen axes for further investigation with the precession, Weissenberg, or other method. To photograph the reciprocal lattice, it is necessary to orient an important direct-lattice row (preferably an edge of the conventional unit cell) parallel to the beam (which is left-to-right in the standard orientation adopted in Fig. 13-5) for precession studies, and normal to it for Weissenberg work. The precession condition is met with arc adjustments to bring the row into the plane of projection and with a drum movement to rotate the axis into the beam. An advantage of plotting orientation data in the manner illustrated in Fig. 13-7 is that the arc and drum adjustments can be read directly from the stereogram. An arc adjustment moves a point along a small circle, parallel to the plane of the arc; a drum adjustment rotates the point about the center

Table 13-2. Identification of the First Fifteen Reflections on a Flat-Film Rotation Photograph of $CuSO_4 \cdot 5H_2O$

Reflection	Measured[a] 2θ	Measured[a] d	Reflecting Plane[b]	Calculated d_{hkl}
1	8.45	10.4 Å	$(0\bar{1}0)$	10.42 Å
2	15.45	5.74	$(\bar{1}00)$	5.75
3	15.5	5.72	(001)	5.68
4	16.6	5.34	$(\bar{1}\bar{1}0)$	5.48
5	16.95	5.27	$(0\bar{2}0)$	5.21
6	17.2	5.16	$(0\bar{1}1)$	5.15
7	18.35	4.84	$(0\bar{1}\bar{1})$	4.84
8	18.7	4.75	$(\bar{1}\bar{1}1)$	4.72
9	18.7	4.75	$(\bar{1}01)$	4.75
10	19.0	4.67	$(\bar{1}10)$	4.68
11	20.7	4.29	$(\bar{1}\bar{2}0)$	4.28
12	22.3	3.99	$(\bar{1}\bar{2}1)$	3.96
13	22.3	3.99	$(0\bar{2}\bar{1})$	3.99
14	22.3	3.99	$(\bar{1}11)$	4.01
15	23.9	3.72	$(0\bar{2}\bar{1})$	3.71

[a] On a film obtained with unfiltered Cu radiation and an F setting of 50 cm.

[b] For the cell with $a = 6.122$, $b = 10.695$, $c = 5.962$ Å; $\alpha = 97.58°$, $\beta = 107.17°$, $\gamma = 77.55°$.

of the projection. A consideration of Fig. 13-8*a*, which shows only the ($hk0$) zone and the pole representing its axis, indicates that the large arc must be given a 15° left rotation to bring the axis into the plane of projection. An anticlockwise rotation of 18° of the drum aligns the axis with the beam. The orientation photograph for this setting (Fig. 13-9) revealed an error of 15′ in the drum setting and 40′ in the large-arc setting.

The axis of zone 2 is clearly beyond reach of the arc adjustments for precession studies. However, the Weissenberg method requires the chosen axis to be coincident with the axis of the goniometer head (that is, normal to the plane of projection in Fig. 13-8b). This condition is easily within reach. It requires a right rotation of 14° about the large arc and a right rotation of 24° about the small arc.

Zone 3 presents another situation. It requires a right rotation of 41° about the large arc to bring the axis into the plane of projection. This is

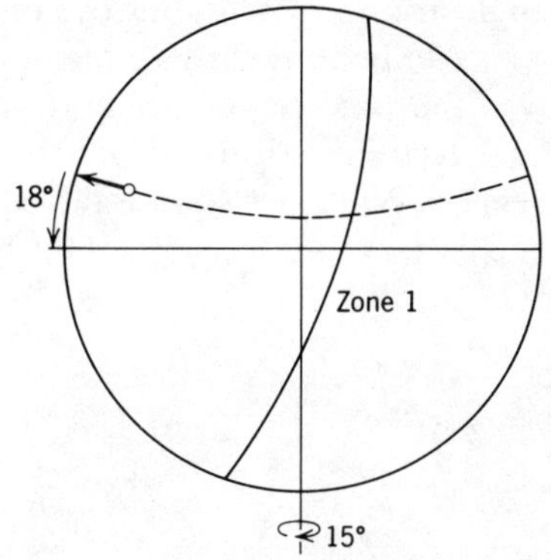

(a)

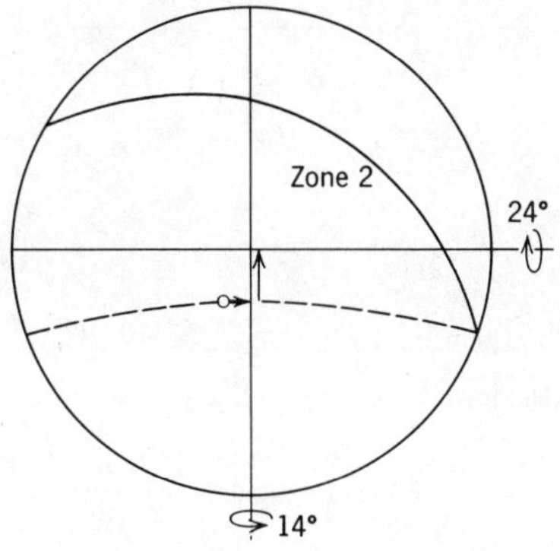

(b)

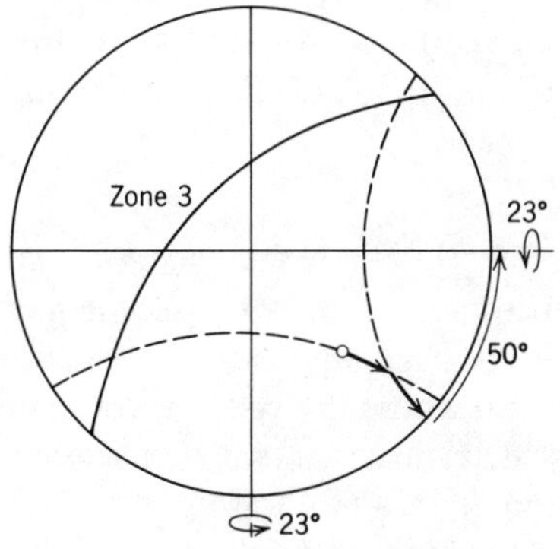

(c)

Fig. 13-8. Instrumental adjustments to orient the axes of zones in Fig. 13-7 for single-crystal study.

beyond the limit of a single arc movement, but by combining the movements of both arcs the axis can be brought into the desired position. The rotation of an arc moves the pole representing an axis along a small circle parallel to the arc. A trial investigation with the pole marked on a transparent sheet overlying a Wulff net soon establishes the magnitude of equal angular movements to bring the pole onto the primitive circle.

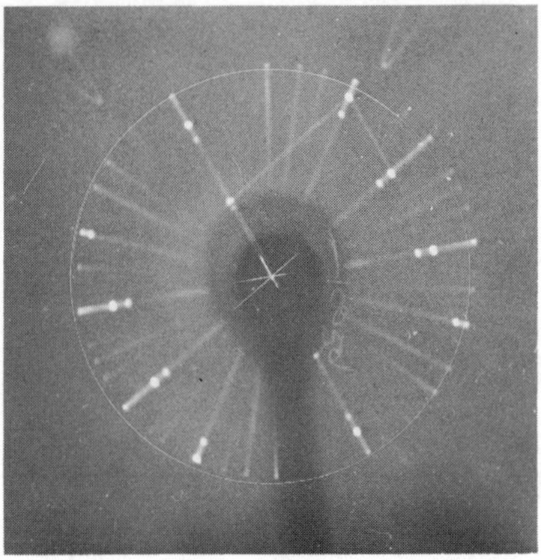

Fig. 13-9. Precession-orientation photograph of $CuSO_4 \cdot 5H_2O$ crystal after a 15° left rotation of large arc and an 18° anticlockwise rotation of drum from the random orientation of Fig. 13-7 to make the axis of zone 1 coincident with beam. The settings are within 40′ of achieving a perfect orientation.

It was found in the present case that this procedure called for a right rotation of 23° about the large arc and a left rotation of 23° about the small arc. A left rotation of 50° of the drum brought the axis into near-coincidence with the beam.

The Recognition of Crystal Symmetry

The principal advantage of the orientation method described above derives from instrumental arrangements (normal-beam monochromatic radiation, flat film) that allow the spacing corresponding to each reflection to be rapidly determined. As a consequence, the stereographic projection of reflecting planes has an important advantage over a similar plot obtained from a Laue photograph. It is possible, as shown in the preceding section, to estimate the relative importance of zones and thereby locate simple-index direct-lattice rows. The angles between these rows are conveniently measured with the Wulff net. They provide information about the geometry of the cell and, indirectly, its symmetry. This is naturally of great value in the preliminary investigation of an unknown crystal.

The use of these data, in the case of $CuSO_4 \cdot 5H_2O$, is illustrated by the analysis of the stereogram obtained from measurement of the first 30 reflections of this compound (Fig. 13-10). The stereogram shows four

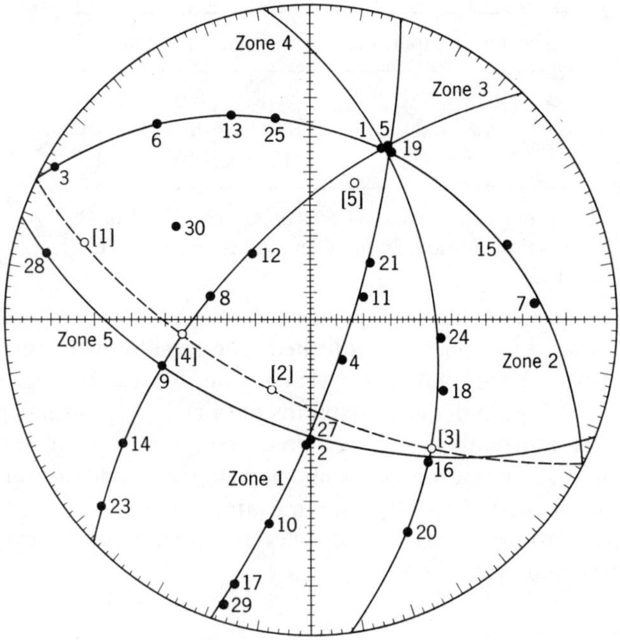

Fig. 13-10. Stereogram of randomly oriented $CuSO_4 \cdot 5H_2O$ crystal.

prominent zones. The zone axes lie in a common plane, identified in the projection by the dashed great circle. The pole of this plane coincides with the intersection of the four zone circles. It represents an important crystal plane as shown by the multiple-order reflections 1, 5, and 19, but its normal is not a direct-lattice row because no stereographic poles fall on the dashed great circle. If the axes of zones 1 and 2 are crystal axes, it follows that the third crystal axis is not at right angles and the crystal is triclinic (the angle between the first two axes is clearly not 90°). Because the interedge angles in conventionally selected triclinic cells are generally within 15° of being right angles, the pole of the third axis should occur near the intersection of the zones. The zone of the axis should lie in the vicinity of the dashed great circle. A fifth zone, comprising poles 2, 9, 16, 27, and 28, does indeed occur in this region of the stereogram.

The angles between the axes of zones 1, 2, and 5, as measured graphically on the projection, confirm the deduction that they are the principal axes in the crystal (Table 13-3).

Table 13-3

Measured (Graphically)	Standard Angles[a]
$1 \wedge 2 = 72\frac{1}{2}°$	$[001] \wedge [\bar{1}00] = 72° 50'$
$1 \wedge 5 = 82\frac{1}{2}$	$[001] \wedge [0\bar{1}0] = 82\ 25$
$2 \wedge 5 = 78$	$[\bar{1}00] \wedge [0\bar{1}0] = 77\ 33$

[a] From E. S. Dana, *System of Mineralogy*, Vol. 2, Seventh Edition, John Wiley and Sons, New York (1951).

The example of copper sulfate cited above calls attention to the similarity between the stereogram that results from application of the orientation method and the stereogram obtained by plotting the face-normals of a morphologically well-developed crystal. The important zones are immediately evident—from the relative abundance of low-2θ reflections in the first case and from an examination of the crystal in the second. The relations between these zones are equally useful for deducing cell geometry and symmetry.

APPENDIX ONE

THE 230 SPACE GROUPS

Table A1-1. Triclinic

Point group	1	$\bar{1}$
Schoenflies symbols	C_1^1	C_i^1
	1 $P1$	2 $P\bar{1}$

Table A1-2. Monoclinic

Point group	Setting with Twofold Axis as the b Axis and Centered Lattice as C		
	2	m	$2/m$
Schoenflies symbols	C_2^{1-3}	C_s^{1-4}	C_{2h}^{1-6}
	3 $P2$	6 Pm	10 $P2/m$
	4 $P2_1$	7 Pc	11 $P2_1/m$
	5 $C2$	8 Cm	12 $C2/m$
		9 Cc	13 $P2/c$
			14 $P2_1/c$
			15 $C2/c$
	Setting with Twofold Axis as the c Axis and Centered Lattice as B		
	3 $P2$	6 Pm	10 $P2/m$
	4 $P2_1$	7 Pb	11 $P2_1/m$
	5 $B2$	8 Bm	12 $B2/m$
		9 Bb	13 $P2/b$
			14 $P2_1/b$
			15 $B2/b$

Table A1-3. Orthorhombic

Point group	222	mm2	mmm
Schoenflies symbols	D_2^{1-9}	C_{2v}^{1-22}	D_{2h}^{1-28}
	16 $P222$	25 $Pmm2$	47 $Pmmm$
	17 $P222_1$	26 $Pmc2_1$	48 $Pnnn$
	18 $P2_12_12$	27 $Pcc2$	49 $Pccm$
	19 $P2_12_12_1$	28 $Pma2$	50 $Pban$
	20 $C222_1$	29 $Pca2_1$	51 $Pmma$
	21 $C222$	30 $Pnc2$	52 $Pnna$
	22 $F222$	31 $Pmn2_1$	53 $Pmna$
	23 $I222$	32 $Pba2$	54 $Pcca$
	24 $I2_12_12_1$	33 $Pna2_1$	55 $Pbam$
		34 $Pnn2$	56 $Pccn$
		35 $Cmm2$	57 $Pbcm$
		36 $Cmc2_1$	58 $Pnnm$
		37 $Ccc2$	59 $Pmmn$
		38 $Amm2$	60 $Pbcn$
		39 $Abm2$	61 $Pbca$
		40 $Ama2$	62 $Pnma$
		41 $Aba2$	63 $Cmcm$
		42 $Fmm2$	64 $Cmca$
		43 $Fdd2$	65 $Cmmm$
		44 $Imm2$	66 $Cccm$
		45 $Iba2$	67 $Cmma$
		46 $Ima2$	68 $Ccca$
			69 $Fmmm$
			70 $Fddd$
			71 $Immm$
			72 $Ibam$
			73 $Ibca$
			74 $Imma$

Table A1-4. Tetragonal

Point group	4	$\bar{4}$	$4/m$	422	$4mm$	$\bar{4}2m$	$4/mmm$
Schoenflies symbols	C_4^{1-6}	$S_4^{1,2}$	C_{4h}^{1-6}	D_4^{1-10}	C_{4v}^{1-12}	D_{2d}^{1-12}	D_{4h}^{1-20}
	75 $P4$	81 $P\bar{4}$	83 $P4/m$	89 $P422$	99 $P4mm$	111 $P\bar{4}2m$	123 $P4/mmm$
	76 $P4_1$	82 $I\bar{4}$	84 $P4_2/m$	90 $P42_12$	100 $P4bm$	112 $P\bar{4}2c$	124 $P4/mcc$
	77 $P4_2$		85 $P4/n$	91 $P4_122$	101 $P4_2cm$	113 $P\bar{4}2_1m$	125 $P4/nbm$
	78 $P4_3$		86 $P4_2/n$	92 $P4_12_12$	102 $P4_2nm$	114 $P\bar{4}2_1c$	126 $P4/nnc$
	79 $I4$		87 $I4/m$	93 $P4_222$	103 $P4cc$	115 $P\bar{4}m2$	127 $P4/mbm$
	80 $I4_1$		88 $I4_1/a$	94 $P4_22_12$	104 $P4nc$	116 $P\bar{4}c2$	128 $P4/mnc$
				95 $P4_322$	105 $P4_2mc$	117 $P\bar{4}b2$	129 $P4/nmm$
				96 $P4_32_12$	106 $P4_2bc$	118 $P\bar{4}n2$	130 $P4/ncc$
				97 $I422$	107 $I4mm$	119 $I\bar{4}m2$	131 $P4_2/mmc$
				98 $I4_122$	108 $I4cm$	120 $I\bar{4}c2$	132 $P4_2/mcm$
					109 $I4_1md$	121 $I\bar{4}2m$	133 $P4_2/nbc$
					110 $I4_1cd$	122 $I\bar{4}2d$	134 $P4_2/nnm$
							135 $P4_2/mbc$
							136 $P4_2/mnm$
							137 $P4_2/nmc$
							138 $P4_2/ncm$
							139 $I4/mmm$
							140 $I4/mcm$
							141 $I4_1/amd$
							142 $I4_1/acd$

Table A1-5. Trigonal

Point group	3	$\bar{3}$	32	$3m$	$\bar{3}m$
Schoenflies symbols	C_3^{1-4}	$C_{3i}^{1,1}$	D_3^{1-7}	C_{3v}^{1-6}	D_{3d}^{1-6}
	143 $P3$	147 $P\bar{3}$	149 $P312$	156 $P3m1$	162 $P\bar{3}1m$
	144 $P3_1$	148 $R\bar{2}$	150 $P321$	157 $P31m$	163 $P\bar{3}1c$
	145 $P3_2$		151 $P3_112$	158 $P3c1$	164 $P\bar{3}m1$
	146 $R3$		152 $P3_121$	159 $P31c$	165 $P\bar{3}c1$
			153 $P3_212$	160 $R3m$	166 $R\bar{3}m$
			154 $P3_221$	161 $R3c$	167 $R\bar{3}c$
			155 $R32$		

Table A1-6. Hexagonal

Point group	6	$\bar{6}$	$6/m$	622	$6mm$	$\bar{6}m2$	$6/mmm$
Schoenflies symbols	C_6^{1-6}	C_{3h}^1	$C_{6h}^{1,2}$	D_6^{1-6}	C_6^{1-6}	D_{3h}^{1-4}	D_{6h}^{1-4}
	168 $P6$	174 $P\bar{6}$	175 $P6/m$	177 $P622$	183 $P6mm$	187 $P\bar{6}m2$	191 $P6/mmm$
	169 $P6_1$		176 $P6_3/m$	178 $P6_122$	184 $P6cc$	188 $P\bar{6}c2$	192 $P6/mcc$
	170 $P6_5$			179 $P6_522$	185 $P6_3cm$	189 $P\bar{6}2m$	193 $P6_3/mcm$
	171 $P6_2$			180 $P6_222$	186 $P6_3mc$	190 $P\bar{6}2c$	194 $P6_3/mmc$
	172 $P6_4$			181 $P6_422$			
	173 $P6_3$			182 $P6_322$			

Table A1-7. Cubic

Point group	23		$m3$		432		$\bar{4}3m$		$m3m$	
Schoenflies symbols	T^{1-5}		T_h^{1-7}		O^{1-8}		T_d^{1-6}		O_h^{1-10}	
	195	$P23$	200	$Pm3$	207	$P432$	215	$P\bar{4}3m$	221	$Pm3m$
	196	$F23$	201	$Pn3$	208	$P4_232$	216	$F\bar{4}3m$	222	$Pn3n$
	197	$I23$	202	$Fm3$	209	$F432$	217	$I\bar{4}3m$	223	$Pm3n$
	198	$P2_13$	203	$Fd3$	210	$F4_132$	218	$P\bar{4}3n$	224	$Pn3m$
	199	$I2_13$	204	$Im3$	211	$I432$	219	$F\bar{4}3c$	225	$Fm3m$
			205	$Pa3$	212	$P4_332$	220	$I\bar{4}3d$	226	$Fm3c$
			206	$Ia3$	213	$P4_132$			227	$Fd3m$
					214	$I4_132$			228	$Fd3c$
									229	$Im3m$
									230	$Ia3d$

APPENDIX TWO

THE SIX PERMUTATIONS OF THE ORTHORHOMBIC SPACE GROUPS

	abc	*cab*	*bca*	$a\bar{c}b$	$ba\bar{c}$	$\bar{c}ba$
16	$P222$	$P222$	$P222$	$P222$	$P222$	$P222$
17	$P222_1$	$P2_122$	$P22_12$	$P22_12$	$P222_1$	$P2_122$
18	$P2_12_12$	$P22_12_1$	$P2_122_1$	$P2_122_1$	$P2_12_12$	$P22_12_1$
19	$P2_12_12_1$	$P2_12_12_1$	$P2_12_12_1$	$P2_12_12_1$	$P2_12_12_1$	$P2_12_12_1$
20	$C222_1$	$A2_122$	$B22_12$	$B22_12$	$C222_1$	$A2_122$
21	$C222$	$A222$	$B222$	$B222$	$C222$	$A222$
22	$F222$	$F222$	$F222$	$F222$	$F222$	$F222$
23	$I222$	$I222$	$I222$	$I222$	$I222$	$I222$
24	$I2_12_12_1$	$I2_12_12_1$	$I2_12_12_1$	$I2_12_12_1$	$I2_12_12_1$	$I2_12_12_1$
25	$Pmm2$	$P2mm$	$Pm2m$	$Pm2m$	$Pmm2$	$P2mm$
26	$Pmc2_1$	$P2_1ma$	$Pb2_1m$	$Pm2_1b$	$Pcm2_1$	$P2_1am$
27	$Pcc2$	$P2aa$	$Pb2b$	$Pb2b$	$Pcc2$	$P2aa$
28	$Pma2$	$P2mb$	$Pc2m$	$Pm2a$	$Pbm2$	$P2cm$
29	$Pca2_1$	$P2_1ab$	$Pc2_1b$	$Pb2_1a$	$Pbc2_1$	$P2_1ca$
30	$Pnc2$	$P2na$	$Pb2n$	$Pn2b$	$Pcn2$	$P2an$
31	$Pmn2_1$	$P2_1mn$	$Pn2_1m$	$Pm2_1n$	$Pnm2_1$	$P2_1nm$
32	$Pba2$	$P2cb$	$Pc2a$	$Pc2a$	$Pba2$	$P2cb$
33	$Pna2_1$	$P2_1nb$	$Pc2_1n$	$Pn2_1a$	$Pbn2_1$	$P2_1cn$
34	$Pnn2$	$P2nn$	$Pn2n$	$Pn2n$	$Pnn2$	$P2nn$
35	$Cmm2$	$A2mm$	$Bm2m$	$Bm2m$	$Cmm2$	$A2mm$
36	$Cmc2_1$	$A2_1ma$	$Bb2_1m$	$Bm2_1b$	$Ccm2_1$	$A2_1am$
37	$Ccc2$	$A2aa$	$Bb2b$	$Bb2b$	$Ccc2$	$A2aa$
38	$Amm2$	$B2mm$	$Cm2m$	$Am2m$	$Bmm2$	$C2mm$
39	$Abm2$	$B2cm$	$Cm2a$	$Ac2m$	$Bma2$	$C2mb$
40	$Ama2$	$B2mb$	$Cc2m$	$Am2a$	$Bbm2$	$C2cm$
41	$Aba2$	$B2cb$	$Cc2a$	$Ac2a$	$Bba2$	$C2cb$
42	$Fmm2$	$F2mm$	$Fm2m$	$Fm2m$	$Fmm2$	$F2mm$
43	$Fdd2$	$F2dd$	$Fd2d$	$Fd2d$	$Fdd2$	$F2dd$
44	$Imm2$	$I2mm$	$Im2m$	$Im2m$	$Imm2$	$I2mm$

	abc	*cab*	*bca*	*a$\bar{c}$b*	*ba$\bar{c}$*	*$\bar{c}$ba*
45	*Iba2*	*I2cb*	*Ic2a*	*Ic2a*	*Iba2*	*I2cb*
46	*Ima2*	*I2mb*	*Ic2m*	*Im2a*	*Ibm2*	*I2cm*
47	*Pmmm*	*Pmmm*	*Pmmm*	*Pmmm*	*Pmmm*	*Pmmm*
48	*Pnnn*	*Pnnn*	*Pnnn*	*Pnnn*	*Pnnn*	*Pnnn*
49	*Pccm*	*Pmaa*	*Pbmb*	*Pbmb*	*Pccm*	*Pmaa*
50	*Pban*	*Pncb*	*Pcna*	*Pcna*	*Pban*	*Pncb*
51	*Pmma*	*Pbmm*	*Pmcm*	*Pmam*	*Pmmb*	*Pcmm*
52	*Pnna*	*Pbnn*	*Pncn*	*Pnan*	*Pnnb*	*Pcnn*
53	*Pmna*	*Pbmn*	*Pncm*	*Pman*	*Pnmb*	*Pcnm*
54	*Pcca*	*Pbaa*	*Pbcb*	*Pbab*	*Pccb*	*Pcaa*
55	*Pbam*	*Pmcb*	*Pcma*	*Pcma*	*Pbam*	*Pmcb*
56	*Pccn*	*Pnaa*	*Pbnb*	*Pbnb*	*Pccn*	*Pnaa*
57	*Pbcm*	*Pmca*	*Pbma*	*Pcmb*	*Pcam*	*Pmab*
58	*Pnnm*	*Pmnn*	*Pnmn*	*Pnmn*	*Pnnm*	*Pmnn*
59	*Pmmn*	*Pnmm*	*Pmnm*	*Pmnm*	*Pmmn*	*Pnmm*
60	*Pbcn*	*Pnca*	*Pbna*	*Pcnb*	*Pcan*	*Pnab*
61	*Pbca*	*Pbca*	*Pbca*	*Pcab*	*Pcab*	*Pcab*
62	*Pnma*	*Pbnm*	*Pmcn*	*Pnam*	*Pmnb*	*Pcmn*
63	*Cmcm*	*Amma*	*Bbmm*	*Bmmb*	*Ccmm*	*Amam*
64	*Cmca*	*Abma*	*Bbcm*	*Bmab*	*Ccmb*	*Acam*
65	*Cmmm*	*Ammm*	*Bmmm*	*Bmmm*	*Cmmm*	*Ammm*
66	*Cccm*	*Amaa*	*Bbmb*	*Bbmb*	*Cccm*	*Amaa*
67	*Cmma*	*Abmm*	*Bmcm*	*Bmam*	*Cmmb*	*Acmm*
68	*Ccca*	*Abaa*	*Bbcb*	*Bbab*	*Cccb*	*Acaa*
69	*Fmmm*	*Fmmm*	*Fmmm*	*Fmmm*	*Fmmm*	*Fmmm*
70	*Fddd*	*Fddd*	*Fddd*	*Fddd*	*Fddd*	*Fddd*
71	*Immm*	*Immm*	*Immm*	*Immm*	*Immm*	*Immm*
72	*Ibam*	*Imcb*	*Icma*	*Icma*	*Ibam*	*Imcb*
73	*Ibca*	*Ibca*	*Ibca*	*Icab*	*Icab*	*Icab*
74	*Imma*	*Ibmm*	*Imcm*	*Imam*	*Immb*	*Icmm*

APPENDIX THREE

SPACE-GROUP DETERMINATIVE TABLES

Table A3-1. Triclinic

Possible Reflections \ Diffraction Symmetry:	$\bar{1}$	
Point Group:	$\bar{1}$	1
hkl all orders	$P\bar{1}$	$P1$

Table A3-2. Monoclinic

Setting with Twofold Axis as the *b* Axis and Centered Lattice as *C*

Possible Reflections \ Diffraction Symmetry:	$2/m$		
Point Group:	$2/m$	2	m
A. *hkl* all orders			
I. *h0l* all orders			
1. *0k0* all orders	$P2/m$	$P2$	Pm
2. *0k0* with $k = 2n$	$P2_1/m$	$P2_1$	
II. *h0l* with $l = 2n$			
1. *0k0* all orders	$P2/c$		Pc
2. *0k0* with $k = 2n$	$P2_1/c$		
B. *hkl* with $h + k = 2n$			
I. *h0l* with $h = 2n$	$C2/m$	$C2$	Cm
II. *h0l* with $h = 2n$ and $l = 2n$	$C2/c$		Cc

Table A3-2 (Continued)

Setting with Twofold Axis as the c Axis and Centered Lattice as B

Possible Reflections \ Diffraction Symmetry:	2/m		
Point Group:	2/m	2	m
A. *hkl* all orders			
I. *hk*0 all orders			
1. 00*l* all orders	$P2/m$	$P2$	Pm
2. 00*l* with $l = 2n$	$P2_1/m$	$P2_1$	
II. *hk*0 with $k = 2n$			
1. 00*l* all orders	$P2/b$		Pb
2. 00*l* with $l = 2n$	$P2_1/b$		
B. *hkl* with $h + l = 2n$			
I. *hk*0 with $h = 2n$	$B2/m$	$B2$	Bm
II. *hk*0 with $h = 2n$ and $k = 2n$	$B2/b$		Bb

Table A3-3. Orthorhombic

Possible Reflections \ Diffraction Symmetry:	mmm		
Point Group:	mmm	(mm2, m2m, 2mm)	222
A. *hkl* all orders			
I. 0*kl* all orders			
1. *h*0*l* all orders			
(*a*) *hk*0 all orders			
*h*00 all orders			
0*k*0 all orders			
00*l* all orders	$Pmmm$	$Pmm2$ $Pm2m$ $P2mm$	$P222$
00*l* with $l = 2n$			$P222_1$
0*k*0 with $k = 2n$			
00*l* all orders			$P22_12$
00*l* with $l = 2n$			$P22_12_1$
*h*00 with $h = 2n$			
0*k*0 all orders			
00*l* all orders			$P2_122$

Table A3-3 (Continued)

Possible Reflections / Diffraction Symmetry:	mmm		
Point Group:	mmm	$\begin{pmatrix} mm2 \\ m2m \\ 2mm \end{pmatrix}$	222
$00l$ with $l = 2n$			$P2_122_1$
$0k0$ with $k = 2n$			
$00l$ all orders			$P2_12_12$
$00l$ with $l = 2n$			$P2_12_12_1$
(*b*) $hk0$ with $h + k = 2n$	$Pmmn$	$P2_1mn$ $Pm2_1n$	
(*c*) $hk0$ with $h = 2n$	$Pmma$	$P2_1ma$ $Pm2a$	
(*d*) $hk0$ with $k = 2n$	$Pmmb$	$Pm2_1b$ $P2mb$	
2. $h0l$ with $h + l = 2n$			
(*a*) $hk0$ all orders	$Pmnm$	$Pmn2_1$ $P2_1nm$	
(*b*) $hk0$ with $h + k = 2n$	$Pmnn$	$P2nn$	
(*c*) $hk0$ with $h = 2n$	$Pmna$	$P2na$	
(*d*) $hk0$ with $k = 2n$	$Pmnb$	$P2_1nb$	
3. $h0l$ with $h = 2n$			
(*a*) $hk0$ all orders	$Pmam$	$P2_1am$ $Pma2$	
(*b*) $hk0$ with $h + k = 2n$	$Pman$	$P2an$	
(*c*) $hk0$ with $h = 2n$	$Pmaa$	$P2aa$	
(*d*) $hk0$ with $k = 2n$	$Pmab$	$P2_1ab$	
4. $h0l$ with $l = 2n$			
(*a*) $hk0$ all orders	$Pmcm$	$Pmc2_1$ $P2cm$	
(*b*) $hk0$ with $h + k = 2n$	$Pmcn$	$P2_1cn$	
(*c*) $hk0$ with $h = 2n$	$Pmca$	$P2_1ca$	
(*d*) $hk0$ with $k = 2n$	$Pmcb$	$P2cb$	
II. $0kl$ with $k + l = 2n$			
1. $h0l$ all orders			
(*a*) $hk0$ all orders	$Pnmm$	$Pnm2_1$ $Pn2_1m$	
(*b*) $hk0$ with $h + k = 2n$	$Pnmn$	$Pn2n$	
(*c*) $hk0$ with $h = 2n$	$Pnma$	$Pn2_1a$	
(*d*) $hk0$ with $k = 2n$	$Pnmb$	$Pn2b$	

Table A3-3 (Continued)

Possible Reflections — Diffraction Symmetry: / Point Group:	*mmm*	*mmm* (*mm*2, *m*2*m*, 2*mm*)	222
2. *h*0*l* with $h + l = 2n$			
(*a*) *hk*0 all orders	*Pnnm*	*Pnn*2	
(*b*) *hk*0 with $h + k = 2n$	*Pnnn*		
(*c*) *hk*0 with $h = 2n$	*Pnna*		
(*d*) *hk*0 with $k = 2n$	*Pnnb*		
3. *h*0*l* with $h = 2n$			
(*a*) *hk*0 all orders	*Pnam*	$Pna2_1$	
(*b*) *hk*0 with $h + k = 2n$	*Pnan*		
(*c*) *hk*0 with $h = 2n$	*Pnaa*		
(*d*) *hk*0 with $k = 2n$	*Pnab*		
4. *h*0*l* with $l = 2n$			
(*a*) *hk*0 all orders	*Pncm*	*Pnc*2	
(*b*) *hk*0 with $h + k = 2n$	*Pncn*		
(*c*) *hk*0 with $h = 2n$	*Pnca*		
(*d*) *hk*0 with $k = 2n$	*Pncb*		
III. 0*kl* with $k = 2n$			
1. *h*0*l* all orders			
(*a*) *hk*0 all orders	*Pbmm*	$Pb2_1m$ *Pbm*2	
(*b*) *hk*0 with $h + k = 2n$	*Pbmn*	*Pb*2*n*	
(*c*) *hk*0 with $h = 2n$	*Pbma*	$Pb2_1a$	
(*d*) *hk*0 with $k = 2n$	*Pbmb*	*Pb*2*b*	
2. *h*0*l* with $h + 1 = 2n$			
(*a*) *hk*0 all orders	*Pbmn*	$Pbn2_1$	
(*b*) *hk*0 with $h + k = 2n$	*Pbnn*		
(*c*) *hk*0 with $h = 2n$	*Pbna*		
(*d*) *hk*0 with $k = 2n$	*Pbnb*		
3. *h*0*l* with $h = 2n$			
(*a*) *hk*0 all orders	*Pbam*	*Pba*2	
(*b*) *hk*0 with $h + k = 2n$	*Pban*		
(*c*) *hk*0 with $h = 2n$	*Pbaa*		
(*d*) *hk*0 with $k = 2n$	*Pbab*		
4. *h*0*l* with $l = 2n$			
(*a*) *hk*0 all orders	*Pbcm*	$Pbc2_1$	

Table A3-3 (Continued)

Possible Reflections \ Diffraction Symmetry: / Point Group:	mmm: *mmm*	mmm: (*mm*2, *m*2*m*, 2*mm*)	mmm: 222
(*b*) *hk*0 with $h + k = 2n$	*Pbcn*		
(*c*) *hk*0 with $h = 2n$	*Pbca*		
(*d*) *hk*0 with $k = 2n$	*Pbcb*		
IV. 0*kl* with $1 = 2n$			
1. *h*0*l* all orders			
(*a*) *hk*0 all orders	*Pcmm*	$Pcm2_1$ *Pc*2*m*	
(*b*) *hk*0 with $h + k = 2n$	*Pcmn*	$Pc2_1n$	
(*c*) *hk*0 with $h = 2n$	*Pcma*	*Pc*2*a*	
(*d*) *hk*0 with $k = 2n$	*Pcmb*	$Pc2_1b$	
2. *h*0*l* with $h + 1 = 2n$			
(*a*) *hk*0 all orders	*Pcnm*	*Pcn*2	
(*b*) *hk*0 with $h + k = 2n$	*Pcnn*		
(*c*) *hk*0 with $h = 2n$	*Pcna*		
(*d*) *hk*0 with $k = 2n$	*Pcnb*		
3. *h*0*l* with $h = 2n$			
(*a*) *hk*0 all orders	*Pcam*	$Pca2_1$	
(*b*) *hk*0 with $h + k = 2n$	*Pcan*		
(*c*) *hk*0 with $h = 2n$	*Pcaa*		
(*d*) *hk*0 with $k = 2n$	*Pcab*		
4. *h*0*l* with $1 = 2n$			
(*a*) *hk*0 all orders	*Pccm*	*Pcc*2	
(*b*) *hk*0 with $h + k = 2n$	*Pccn*		
(*c*) *hk*0 with $h = 2n$	*Pcca*		
(*d*) *hk*0 with $k = 2n$	*Pccb*		
B. *hkl* with $h + k = 2n$			
I. 0*kl* with $k = 2n$			
1. *h*0*l* with $h = 2n$			
(*a*) *hk*0 with $h + k = 2n$			
00*l* all orders	*Cmmm*	*Cmm*2 *Cm*2*m* *C*2*mm*	*C*222
00*l* with $l = 2n$			$C222_1$

Table A3-3 (Continued)

Possible Reflections \ Diffraction Symmetry: / Point Group:	*mmm*	*mmm*: (*mm*2, *m*2*m*, 2*mm*)	222
(*b*) *hk*0 with $h = 2n$ and $k = 2n$	*Cmma* *Cmmb*	*Cm*2*a* *C*2*mb*	
2. *h*0*l* with $h = 2n$ and $l = 2n$			
(*a*) *hk*0 with $h + k = 2n$	*Cmcm*	$Cmc2_1$ *C*2*cm*	
(*b*) *hk*0 with $h = 2n$ and $k = 2n$	*Cmca*	*C*2*cb*	
II. 0*kl* with $k = 2n$ and $l = 2n$			
1. *h*0*l* with $h = 2n$			
(*a*) *hk*0 with $h + k = 2n$	*Ccmm*	$Ccm2_1$ *Cc*2*m*	
(*b*) *hk*0 with $h = 2n$ and $k = 2n$	*Ccmb*	*Cc*2*a*	
2. *h*0*l* with $h = 2n$ and $l = 2n$			
(*a*) *hk*0 with $h + k = 2n$	*Cccm*	*Ccc*2	
(*b*) *hk*0 with $h = 2n$ and $k = 2n$	*Ccca* *Cccb*		
C. *hkl* with $k + l = 2n$			
I. 0*kl* with $k + 1 = 2n$			
1. *h*0*l* with $l = 2n$			
(*a*) *hk*0 with $k = 2n$			
*h*00 all orders	*Ammm*	*Amm*2 *Am*2*m* *A*2*mm*	*A*222
*h*00 with $h = 2n$			$A2_122$
(*b*) *hk*0 with $h = 2n$ and $k = 2n$	*Amma*	$A2_1ma$ *Am*2*a*	
2. *h*0*l* with $h = 2n$ and $l = 2n$			
(*a*) *hk*0 with $k = 2n$	*Amam*	$A2_1am$ *Ama*2	
(*b*) *hk*0 with $h = 2n$ and $k = 2n$	*Amaa*	*A*2*aa*	
II. 0*kl* with $k = 2n$ and $l = 2n$			
1. *h*0*l* with $l = 2n$			
(*a*) *hk*0 with $k = 2n$	*Abmm* *Acmm*	*Abm*2 *Ac*2*m*	
(*b*) *hk*0 with $h = 2n$ and $k = 2n$	*Abma*	*Ac*2*a*	

Table A3-3 (Continued)

Possible Reflections — Diffraction Symmetry:	mmm		
Point Group:	*mmm*	*mm*2 / *m*2*m* / 2*mm*	222
2. $h0l$ with $h = 2n$ and $l = 2n$			
(*a*) $hk0$ with $k = 2n$	*Acam*	*Aba*2	
(*b*) $hk0$ with $h = 2n$ and $k = 2n$	*Abaa* *Acaa*		
D. hkl with $h + l = 2n$			
I. $0kl$ with $l = 2n$			
1. $h0l$ with $h + l = 2n$			
(*a*) $hk0$ with $h = 2n$			
$0k0$ all orders	*Bmmm*	*Bmm*2 *Bm*2*m* *B*2*mm*	*B*222
$0k0$ with $k = 2n$			$B22_12$
(*b*) $hk0$ with $h = 2n$ and $k = 2n$	*Bmmb*	$Bm2_1b$ *B*2*mb*	
2. $h0l$ with $h = 2n$ and $l = 2n$			
(*a*) $hk0$ with $h = 2n$	*Bmcm* *Bmam*	*B*2*cm* *Bma*2	
(*b*) $hk0$ with $h = 2n$ and $k = 2n$	*Bmab*	*B*2*cb*	
II. $0kl$ with $k = 2n$ and $l = 2n$			
1. $h0l$ with $h + l = 2n$			
(*a*) $hk0$ with $h = 2n$	*Bbmm*	$Bb2_1m$ *Bbm*2	
(*b*) $hk0$ with $h = 2n$ and $k = 2n$	*Bbmb*	*Bb*2*b*	
2. $h0l$ with $h = 2n$ and $l = 2n$			
(*a*) $hk0$ with $h = 2n$	*Bbcm*	*Bba*2	
(*b*) $hk0$ with $h = 2n$ and $k = 2n$	*Bbcb* *Bbab*		
E. hkl with $h + k + l = 2n$			
I. $0kl$ with $k + l = 2n$			
1. $h0l$ with $h + l = 2n$			
(*a*) $hk0$ with $h + k = 2n$	*Immm*	*Imm*2 *Im*2*m* *I*2*mm*	*I*222 $I2_12_12_1$

Table A3-3 (Continued)

Possible Reflections \ Diffraction Symmetry:	mmm		
Point Group:	*mmm*	(*mm*2, *m*2*m*, 2*mm*)	222
(*b*) *hk*0 with $h = 2n$ and $k = 2n$	*Imma* *Immb*	*Im*2*a* *I*2*mb*	
2. *h*0*l* with $h = 2n$ and $l = 2n$			
(*a*) *hk*0 with $h + k = 2n$	*Imam* *Imcm*	*Ima*2 *I*2*cm*	
(*b*) *hk*0 with $h = 2n$ and $k = 2n$	*Imcb*	*I*2*cb*	
II. 0*kl* with $k = 2n$ and $l = 2n$			
1. *h*0*l* with $h + l = 2n$			
(*a*) *hk*0 with $h + k = 2n$	*Ibmm* *Icmm*	*Ibm*2 *Ic*2*m*	
(*b*) *hk*0 with $h = 2n$ and $k = 2n$	*Icma*	*Ic*2*a*	
2. *h*0*l* with $h = 2n$ and $l = 2n$			
(*a*) *hk*0 with $h + k = 2n$	*Ibam*	*Iba*2	
(*b*) *hk*0 with $h = 2n$ and $k = 2n$	*Ibca* *Icab*		
F. *hkl* with $h + k = 2n$ and $k + l = 2n$			
I. 0*kl* with $k = 2n$ and $l = 2n$			
1. *h*0*l* with $h = 2n$ and $l = 2n$			
*hk*0 with $h = 2n$ and $k = 2n$	*Fmmm*	*Fmm*2 *Fm*2*m* *F*2*mm*	*F*222
2. *h*0*l* with $h = 2n$, $l = 2n$ and $h + l = 4n$			
*hk*0 with $h = 2n$, $k = 2n$ and $h + k = 4n$		*F*2*dd*	
II. 0*kl* with $k = 2n$, $l = 2n$ and $k + l = 4n$			
1. *h*0*l* with $h = 2n$ and $l = 2n$			
*hk*0 with $h = 2n$, $k = 2n$ and $h + k = 4n$		*Fd*2*d*	
2. *h*0*l* with $h = 2n$, $l = 2n$ and $h + l = 4n$			
(*a*) *hk*0 with $h = 2n$, and $k = 2n$		*Fdd*2	
(*b*) *hk*0 with $h = 2n$, $k = 2n$ and $h + k = 4n$	*Fddd*		

Table A3-4. Tetragonal

Diffraction Symmetry:	4/mmm				4/m		
Possible Reflections / Point Group:	4/mmm	422	4mm	$\bar{4}2m$ $\bar{4}m2$	4/m	4	$\bar{4}$
A. *hkl* all orders							
I. *hk*0 all orders							
1. *hhl* all orders							
(*a*) 0*kl* all orders							
0*k*0 all orders							
00*l* all orders	$P4/mmm$	$P422$	$P4mm$	$P\bar{4}2m$ $P\bar{4}m2$	$P4/m$	$P4$	$P\bar{4}$
00*l* with $l = 2n$		$P4_222$			$P4_2/m$	$P4_2$	
00*l* with $l = 4n$		$P4_{1,3}22$				$P4_{1,3}$	
0*k*0 with $k = 2n$							
00*l* all orders		$P42_12$		$P\bar{4}2_1m$			
00*l* with $l = 2n$		$P4_22_12$					
00*l* with $l = 4n$		$P4_{1,3}2_12$					
(*b*) 0*kl* with $k = 2n$	$P4/mbm$		$P4bm$	$P\bar{4}b2$			
(*c*) 0*kl* with $l = 2n$	$P4_2/mcm$		$P4_2cm$	$P\bar{4}c2$			
(*d*) 0*kl* with $k + l = 2n$	$P4_2/mnm$		$P4_2nm$	$P\bar{4}n2$			
2. *hhl* with $l = 2n$							
(*a*) 0*kl* all orders							
0*k*0 all orders	$P4_2/mmc$		$P4_2mc$	$P\bar{4}2c$			
0*k*0 with $k = 2n$				$P\bar{4}2_1c$			
(*b*) 0*kl* with $k = 2n$	$P4_2/mbc$		$P4_2bc$				
(*c*) 0*kl* with $l = 2n$	$P4/mcc$		$P4cc$				
(*d*) 0*kl* with $k + l = 2n$	$P4/mnc$		$P4nc$				

Possible Reflections \ Diffraction Symmetry:	4/mmm				4/m		
Point Group:	4/mmm	422	4mm	$\bar{4}2m$ $\bar{4}m2$	4/m	4	$\bar{4}$
II. *hk*0 with $h + k = 2n$							
1. *hhl* all orders							
(*a*) 0*kl* all orders							
00*l* all orders	*P*4/*nmm*				*P*4/*n*		
00*l* with $l = 2n$					$P4_2/n$		
(*b*) 0*kl* with $k = 2n$	*P*4/*nbm*						
(*c*) 0*kl* with $l = 2n$	$P4_2/ncm$						
(*d*) 0*kl* with $k + l = 2n$	$P4_2/nnm$						
2. *hhl* with $l = 2n$							
(*a*) 0*kl* all orders	$P4_2/nmc$						
(*b*) 0*kl* with $k = 2n$	$P4_2/nbc$						
(*c*) 0*kl* with $l = 2n$	*P*4/*ncc*						
(*d*) 0*kl* with $k + l = 2n$	*P*4/*nnc*						
B. *hkl* with $h + k + l = 2n$							
I. *hk*0 with $h + k = 2n$							
1. *hhl* with $l = 2n$							
(*a*) 0*kl* with $k + l = 2n$							
00*l* with $l = 2n$	*I*4/*mmm*	*I*422	*I*4*mm*	$I\bar{4}2m$ $I\bar{4}m2$	*I*4/*m*	*I*4	$I\bar{4}$
00*l* with $l = 4n$		$I4_122$				$I4_1$	
(*b*) 0*kl* with $k = 2n$ and $l = 2n$	*I*4/*mcm*		*I*4*cm*	$I\bar{4}c2$			

Table A3-7 (Continued)

Possible Reflections \ Diffraction Symmetry:	4/mmm				4/m		
Point Group:	$4/mmm$	422	$4mm$	$\bar{4}2m$ $\bar{4}m2$	$4/m$	4	$\bar{4}$
2. hhl with $2h + l = 4n$							
(*a*) $0kl$ with $k + l = 2n$			$I4_1md$	$I\bar{4}2d$			
(*b*) $0kl$ with $k = 2n$ and $l = 2n$			$I4_1cd$				
II. $hk0$ with $h = 2n$ and $k = 2n$							
1. hhl with $l = 2n$							
$00l$ with $l = 4n$					$I4_1/a$		
2. hhl with $2h + l = 4n$							
(*a*) $0kl$ with $k + l = 2n$	$I4/amd$						
(*b*) $0kl$ with $k = 2n$ and $l = 2n$	$I4_1/acd$						

Table A3-5. Hexagonal

Diffraction Symmetry:	6/*mmm*				6/*m*			$\bar{3}m$			$\bar{3}$	
Possible Reflections / Point Group:	6/*mmm*	$\bar{6}m2$ $\bar{6}2m$	6*mm*	622	6/*m*	$\bar{6}$	6	$\bar{3}m$	3*m*	32	$\bar{3}$	3
HEXAGONAL AXES												
A. *hkīl* all orders												
I. $hh\bar{2h}l$ all orders												
1. $h\bar{h}0l$ all orders												
(*a*) 000*l* all orders	*P*6/*mmm*	$P\bar{6}m2$ $P\bar{6}2m$	*P*6*mm*	*P*622	*P*6/*m*	$P\bar{6}$	*P*6	$P\bar{3}m1$ $P\bar{3}1m$	*P*3*m*1 *P*31*m*	*P*321 *P*312	$P\bar{3}$	*P*3
(*b*) 000*l* with $l = 2n$				$P6_322$	$P6_3/m$		$P6_3$					
(*c*) 000*l* with $l = 3n$				$P6_{2,4}22$			$P6_{2,4}$			$P3_{1,2}21$ $P3_{1,2}12$		$P3_{1,2}$
(*d*) 000*l* with $l = 6n$				$P6_{1,5}22$			$P6_{1,5}$					
2. $h\bar{h}0l$ with $l = 2n$	$P6_3/mcm$	$P\bar{6}c2$	$P6_3cm$					$P\bar{3}c1$	*P*3*c*1			
II. $hh\bar{2h}l$ with $l = 2n$												
1. $h\bar{h}0l$ all orders	$P6_3/mmc$	$P\bar{6}2c$	$P6_3mc$					$P\bar{3}1c$	*P*31*c*			
2. $h\bar{h}0l$ with $l = 2n$	*P*6/*mcc*		*P*6*cc*									
B. *hkīl* with $-h + k + l = 3n$												
I. $hh\bar{2h}l$ with $l = 3n$												
1. $h\bar{h}0l$ with $2h + l = 3n$								$R\bar{3}m$	*R*3*m*	*R*32	$R\bar{3}$	*R*3
2. $h\bar{h}0l$ with $2h + l = 3n$ and $l = 2n$								$R\bar{3}c$	*R*3*c*			
RHOMBOHEDRAL AXES												
hkl all orders												
hhl all orders								$R\bar{3}m$	*R*3*m*	*R*32	$R\bar{3}$	*R*3
hhl with $l = 2n$								$R\bar{3}c$	*R*3*c*			

Table A3-6. Cubic

Possible Reflections \ Diffraction Symmetry:	$m3m$			$m3$	
Possible Reflections \ Point Group:	$m3m$	432	$\bar{4}3m$	$m3$	23
A. *hkl* all orders					
I. *hhl* all orders					
1. 0*kl* all orders					
(*a*) 00*l* all orders	$Pm3m$	$P432$	$P\bar{4}3m$	$Pm3$	$P23$
(*b*) 00*l* with $l = 2n$		$P4_232$			$P2_13$
(*c*) 00*l* with $l = 4n$		$P4_{1,3}32$			
2. 0*kl* with $k + l = 2n$	$Pn3m$			$Pn3$	
3. 0*kl* with $k = 2n$ (*h*0*l* with $l = 2n$, *hk*0 with $h = 2n$)				$Pa3$	
II. *hhl* with $l = 2n$					
1. 0*kl* all orders	$Pm3n$		$P\bar{4}3n$		
2. 0*kl* with $k + l = 2n$	$Pn3n$				
B. *hkl* with $h + k + l = 2n$					
I. *hhl* with $l = 2n$					
1. 0*kl* with $k + l = 2n$					
(*a*) 00*l* with $l = 2n$	$Im3m$	$I432$	$I\bar{4}3m$	$Im3$	$I23$ $I2_13$
(*b*) 00*l* with $l = 4n$		$I4_132$			
2. 0*kl* with $k = 2n$ and $l = 2n$				$Ia3$	
II. *hhl* with $l = 2n$ and $2h + l = 4n$					
1. 0*kl* with $k + l = 2n$			$I\bar{4}3d$		
2. 0*kl* with $k = 2n$ and $l = 2n$	$Ia3d$				
C. *hkl* with $h + k = 2n$, $k + l = 2n$ and $h + l = 2n$					
I. *hhl* with $h + l = 2n$					
1. 0*kl* with $k = 2n$ and $l = 2n$					
(*a*) 00*l* with $l = 2n$	$Fm3m$	$F432$	$F\bar{4}3m$	$Fm3$	$F23$
(*b*) 00*l* with $l = 4n$		$F4_132$			
2. 0*kl* with $k = 2n$, $l = 2n$ and $k + l = 4n$	$Fd3m$			$Fd3$	
II. *hhl* with $h = 2n$ and $l = 2n$					
1. 0*kl* with $k = 2n$ and $l = 2n$	$Fm3c$		$F\bar{4}3c$		
2. 0*kl* with $k = 2n$, $l = 2n$ and $k + l = 4n$	$Fd3c$				

APPENDIX FOUR

TABLE OF $\frac{1}{2}\left(\frac{\cos^2\theta}{\sin\theta}+\frac{\cos^2\theta}{\theta}\right)$[1]

θ	0.0	0.1	0.2	0.3	0.4	0.5	0.6	0.7	0.8	0.9
°										
10	5.572	5.513	5.456	5.400	5.345	5.291	5.237	5.185	5.134	5.084
11	5.034	4.986	4.939	4.892	4.846	4.800	4.756	4.712	4.669	4.627
12	4.585	4.544	4.504	4.464	4.425	4.386	4.348	4.311	4.274	4.238
13	4.202	4.167	4.133	4.098	4.065	4.032	3.999	3.967	3.935	3.903
14	3.872	3.842	3.812	3.782	3.753	3.724	3.695	3.667	3.639	3.612
15	3.584	3.558	3.531	3.505	3.479	3.454	3.429	3.404	3.379	3.355
16	3.331	3.307	3.284	3.260	3.237	3.215	3.192	3.170	3.148	3.127
17	3.105	3.084	3.063	3.042	3.022	3.001	2.981	2.962	2.942	2.922
18	2.903	2.884	2.865	2.847	2.828	2.810	2.792	2.774	2.756	2.738
19	2.921	2.704	2.687	2.670	2.653	2.636	2.620	2.604	2.588	2.572
20	2.556	2.540	2.525	2.509	2.494	2.479	2.464	2.449	2.434	2.420
21	2.405	2.391	2.376	2.362	2.348	2.335	2.321	2.307	2.294	2.280
22	2.267	2.254	2.241	2.228	2.215	2.202	2.189	2.177	2.164	2.152
23	2.140	2.128	2.116	2.104	2.092	2.080	2.068	2.056	2.045	2.034
24	2.022	2.011	2.000	1.989	1.978	1.967	1.956	1.945	1.934	1.924
25	1.913	1.903	1.892	1.882	1.872	1.861	1.851	1.841	1.831	1.821
26	1.812	1.802	1.792	1.782	1.773	1.763	1.754	1.745	1.735	1.726
27	1.717	1.708	1.699	1.690	1.681	1.672	1.663	1.654	1.645	1.637
28	1.628	1.619	1.611	1.602	1.594	1.586	1.577	1.569	1.561	1.553
29	1.545	1.537	1.529	1.521	1.513	1.505	1.497	1.489	1.482	1.474
30	1.466	1.459	1.451	1.444	1.436	1.429	1.421	1.414	1.407	1.400
31	1.392	1.385	1.378	1.371	1.364	1.357	1.350	1.343	1.336	1.329
32	1.323	1.316	1.309	1.302	1.296	1.289	1.282	1.276	1.269	1.263
33	1.256	1.250	1.244	1.237	1.231	1.225	1.218	1.212	1.206	1.200
34	1.194	1.188	1.182	1.176	1.170	1.164	1.158	1.152	1.146	1.140

[1] From J. B. Nelson and D. P. Riley, *Proc. Phys. Soc.* (*London*), **57**, 126 (1945).

θ	0.0	0.1	0.2	0.3	0.4	0.5	0.6	0.7	0.8	0.9
35	1.134	1.128	1.123	1.117	1.111	1.106	1.100	1.094	1.098	1.083
36	1.078	1.072	1.067	1.061	1.056	1.050	1.045	1.040	1.034	1.029
37	1.024	1.019	1.013	1.008	1.003	0.998	0.993	0.988	0.982	0.977
38	0.972	0.967	0.962	0.958	0.953	0.948	0.943	0.938	0.933	0.928
39	0.924	0.919	0.914	0.909	0.905	0.900	0.895	0.891	0.886	0.881
40	0.877	0.872	0.868	0.863	0.859	0.854	0.850	0.845	0.841	0.837
41	0.832	0.828	0.823	0.819	0.815	0.810	0.806	0.802	0.798	0.794
42	0.789	0.785	0.781	0.777	0.773	0.769	0.765	0.761	0.757	0.753
43	0.749	0.745	0.741	0.737	0.733	0.729	0.725	0.721	0.717	0.713
44	0.709	0.706	0.702	0.698	0.694	0.690	0.687	0.683	0.679	0.676
45	0.672	0.668	0.665	0.661	0.657	0.654	0.650	0.647	0.643	0.640
46	0.636	0.632	0.629	0.625	0.622	0.619	0.615	0.612	0.608	0.605
47	0.602	0.598	0.595	0.591	0.588	0.585	0.582	0.578	0.575	0.572
48	0.569	0.565	0.562	0.559	0.556	0.553	0.549	0.546	0.543	0.540
49	0.537	0.534	0.531	0.528	0.525	0.522	0.518	0.515	0.512	0.509
50	0.506	0.504	0.501	0.498	0.495	0.492	0.489	0.486	0.483	0.480
51	0.477	0.474	0.472	0.469	0.466	0.463	0.460	0.458	0.455	0.452
52	0.449	0.447	0.444	0.441	0.439	0.436	0.433	0.430	0.428	0.425
53	0.423	0.420	0.417	0.415	0.412	0.410	0.407	0.404	0.402	0.399
54	0.397	0.394	0.392	0.389	0.387	0.384	0.382	0.379	0.377	0.375
55	0.372	0.370	0.367	0.365	0.363	0.360	0.358	0.356	0.353	0.351
56	0.349	0.346	0.344	0.342	0.339	0.337	0.335	0.333	0.330	0.328
57	0.326	0.324	0.322	0.319	0.317	0.315	0.313	0.311	0.309	0.306
58	0.304	0.302	0.300	0.298	0.296	0.294	0.292	0.290	0.288	0.286
59	0.284	0.282	0.280	0.278	0.276	0.274	0.272	0.270	0.268	0.266
60	0.264	0.262	0.260	0.258	0.256	0.254	0.252	0.250	0.249	0.247
61	0.245	0.243	0.241	0.239	0.237	0.236	0.234	0.232	0.230	0.229
62	0.227	0.225	0.223	0.221	0.220	0.218	0.216	0.215	0.213	0.211
63	0.209	0.208	0.206	0.204	0.203	0.201	0.199	0.198	0.196	0.195
64	0.193	0.191	0.190	0.188	0.187	0.185	0.184	0.182	0.180	0.179
65	0.177	0.176	0.174	0.173	0.171	0.170	0.168	0.167	0.165	0.164
66	0.162	0.161	0.160	0.158	0.157	0.155	0.154	0.152	0.151	0.150
67	0.148	0.147	0.146	0.144	0.143	0.141	0.140	0.139	0.138	0.136
68	0.135	0.134	0.132	0.131	0.130	0.128	0.127	0.126	0.125	0.123
69	0.122	0.121	0.120	0.119	0.117	0.116	0.115	0.114	0.112	0.111
70	0.110	0.109	0.108	0.107	0.106	0.104	0.103	0.102	0.101	0.100
71	0.099	0.098	0.097	0.096	0.095	0.094	0.092	0.091	0.090	0.089
72	0.088	0.087	0.086	0.085	0.084	0.083	0.082	0.081	0.080	0.079
73	0.078	0.077	0.076	0.075	0.075	0.074	0.073	0.072	0.071	0.070
74	0.069	0.068	0.067	0.066	0.065	0.065	0.064	0.063	0.062	0.061

θ	0.0	0.1	0.2	0.3	0.4	0.5	0.6	0.7	0.8	0.9
75	0.060	0.059	0.059	0.058	0.057	0.056	0.055	0.055	0.054	0.053
76	0.052	0.052	0.051	0.050	0.049	0.048	0.048	0.047	0.046	0.045
77	0.045	0.044	0.043	0.043	0.042	0.041	0.041	0.040	0.039	0.039
78	0.038	0.037	0.037	0.036	0.035	0.035	0.034	0.034	0.033	0.032
79	0.032	0.031	0.031	0.030	0.029	0.029	0.028	0.028	0.027	0.027
80	0.026	0.026	0.025	0.025	0.024	0.023	0.023	0.023	0.022	0.022
81	0.021	0.021	0.020	0.020	0.019	0.019	0.018	0.018	0.017	0.017
82	0.017	0.016	0.016	0.015	0.015	0.015	0.014	0.014	0.013	0.013
83	0.013	0.012	0.012	0.012	0.011	0.011	0.010	0.010	0.010	0.010
84	0.009	0.009	0.009	0.008	0.008	0.008	0.007	0.007	0.007	0.007
85	0.006	0.006	0.006	0.006	0.005	0.005	0.005	0.005	0.005	0.004
86	0.004	0.004	0.004	0.003	0.003	0.003	0.003	0.003	0.003	0.002
87	0.002	0.002	0.002	0.002	0.002	0.002	0.001	0.001	0.001	0.001
88	0.001	0.001	0.001	0.001	0.001	0.001	0.001	0.000	0.000	0.000

APPENDIX FIVE

DATA FOR CONSTRUCTING STEREOGRAPHIC SCALES

Table A5-1. Data for Constructing Stereographic Scales for Transmission Laue Photographs

$F = 1$ cm; $r = 1$ cm

2θ	$F \tan 2\theta$	$r \tan \frac{1}{2}(90 - \theta)$	2θ	$F \tan 2\theta$	$r \tan \frac{1}{2}(90 - \theta)$
0°	0.000 cm	1.000 cm	22°	0.404 cm	0.824 cm
1	0.018	0.991	23	0.425	0.817
2	0.035	0.983	24	0.445	0.810
3	0.052	0.974	25	0.466	0.803
4	0.070	0.966	26	0.488	0.795
5	0.088	0.957	27	0.510	0.788
6	0.105	0.949	28	0.532	0.781
7	0.123	0.941	29	0.554	0.774
8	0.141	0.932	30	0.577	0.767
9	0.158	0.924	31	0.601	0.760
10	0.176	0.916	32	0.625	0.754
11	0.194	0.908	33	0.649	0.747
12	0.213	0.900	34	0.675	0.740
13	0.231	0.893	35	0.700	0.733
14	0.249	0.885	36	0.727	0.727
15	0.268	0.877	37	0.754	0.720
16	0.287	0.869	38	0.781	0.713
17	0.306	0.862	39	0.810	0.707
18	0.325	0.854	40	0.839	0.700
19	0.344	0.847	41	0.869	0.694
20	0.364	0.839	42	0.900	0.687
21	0.384	0.832	43	0.933	0.681

2θ	$F \tan 2\theta$	$r \tan \frac{1}{2}(90 - \theta)$	2θ	$F \tan 2\theta$	$r \tan \frac{1}{2}(90 - \theta)$
44°	0.966 cm	0.675 cm	53°	1.327 cm	0.619 cm
45	1.000	0.668	54	1.376	0.613
46	1.036	0.662	55	1.428	0.607
47	1.072	0.656	56	1.483	0.601
48	1.111	0.649	57	1.540	0.595
49	1.150	0.643	58	1.600	0.589
50	1.192	0.637	59	1.664	0.583
51	1.235	0.631	60	1.732	0.577
52	1.280	0.625			

Table A5-2. Data for Constructing Stereographic Scales for Back-reflection Laue Photographs

$F = 1$ cm; $r = 1$ cm

2θ	$F \tan(180 - 2\theta)$	$r \tan(90 - \theta)$	2θ	$F \tan(180 - 2\theta)$	$r \tan(90 - \theta)$
120°	1.732 cm	0.268 cm	150°	0.577 cm	0.132 cm
121	1.664	0.263	151	0.554	0.127
122	1.600	0.259	152	0.532	0.123
123	1.540	0.254	153	0.510	0.118
124	1.483	0.249	154	0.488	0.114
125	1.428	0.245	155	0.466	0.110
126	1.376	0.240	156	0.445	0.105
127	1.327	0.236	157	0.425	0.101
128	1.280	0.231	158	0.404	0.096
129	1.235	0.226	159	0.384	0.092
130	1.192	0.222	160	0.364	0.088
131	1.150	0.217	161	0.344	0.083
132	1.111	0.213	162	0.325	0.079
133	1.072	0.208	163	0.306	0.074
134	1.036	0.204	164	0.287	0.070
135	1.000	0.199	165	0.268	0.066
136	0.966	0.194	166	0.249	0.061
137	0.933	0.190	167	0.231	0.057
138	0.900	0.185	168	0.213	0.052
139	0.869	0.181	169	0.194	0.048
140	0.839	0.176	170	0.176	0.044
141	0.810	0.172	171	0.158	0.040
142	0.781	0.167	172	0.141	0.035
143	0.754	0.163	173	0.123	0.031
144	0.727	0.158	174	0.105	0.026
145	0.700	0.154	175	0.088	0.022
146	0.675	0.150	176	0.070	0.018
147	0.649	0.145	177	0.052	0.013
148	0.625	0.141	178	0.035	0.009
149	0.601	0.136	179	0.018	0.004

APPENDIX SIX

MATHEMATICAL RELATIONS BETWEEN RECIPROCAL- AND DIRECT-LATTICE CONSTANTS

Direct and reciprocal lattices are related by two fundamental equations

$$d^*_{hkl} = \frac{1}{p_{hkl}} \qquad d_{hkl} = \frac{1}{p^*_{hkl}}$$

where d represents an interplanar spacing and p a translation period along a lattice row. Additional mathematical relations are necessary to convert the experimentally determined reciprocal- and direct-lattice constants into the conventional form of presentation. A number of useful equations have been derived for triclinic cells on the following pages and are summarized at the end of the appendix. The equations are readily reduced to the simpler expressions for the more symmetrical systems with the aid of the relations in Table A6-1.

The Volume of a Triclinic Cell

Figure A6-1*a* shows a triclinic cell oriented with the a and b axes in the plane of the page. Figure A6-1*b* is a section normal to the $ab = (001)$ planes. The volume of this cell is the product of the area of the ab plane and the interval or spacing between this and the next identical plane

$$V = ab \sin \gamma \, d_{001} = \frac{ab \sin \gamma}{c^*} \tag{1}$$

It can be similarly shown that

$$V = \frac{ac \sin \beta}{b^*} = \frac{bc \sin \alpha}{a^*} \tag{2, 3}$$

Table A6-1

	Direct	Reciprocal
Monoclinic	$\alpha = \beta = 90°$	$\alpha^* = \beta^* = 90°$
Hexagonal	$a = b$ $\alpha = \beta = 90°$ $\gamma = 120°$	$a^* = b^*$ $\alpha^* = \beta^* = 90°$ $\gamma^* = 60°$
Rhombohedral	$a = b = c$ $\alpha = \beta = \gamma$	$a^* = b^* = c^*$ $\alpha^* = \beta^* = \gamma^*$
Orthorhombic	$\alpha = \beta = \gamma = 90°$	$\alpha^* = \beta^* = \gamma^* = 90°$
Tetragonal	$a = b$ $\alpha = \beta = \gamma = 90°$	$a^* = b^*$ $\alpha^* = \beta^* = \gamma^* = 90°$
Cubic	$a = b = c$ $\alpha = \beta = \gamma = 90°$	$a^* = b^* = c^*$ $\alpha^* = \beta^* = \gamma^* = 90°$

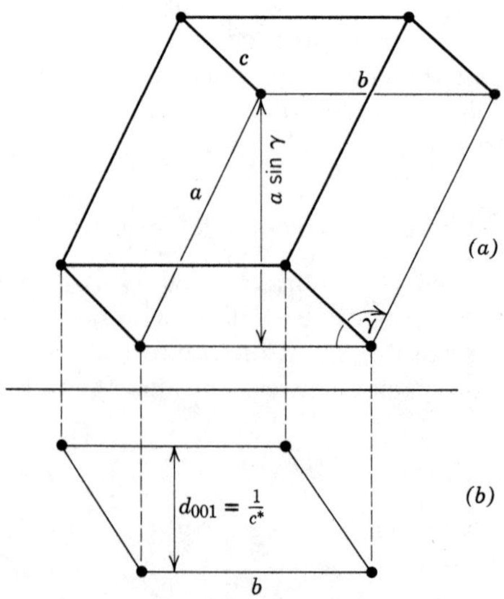

Fig. A6-1

These equations contain one unit reciprocal translation each and can, therefore, be used to calculate these constants when the direct elements are known

$$a^* = (bc \sin \alpha)/V \tag{4}$$

$$b^* = (ac \sin \beta)/V \tag{5}$$

$$c^* = (ab \sin \gamma)/V \tag{6}$$

The axial ratios of the reciprocal cell can now be expressed in terms of direct-cell elements

$$a^*:b^*:c^* = \frac{\sin \alpha}{a}:\frac{\sin \beta}{b}:\frac{\sin \gamma}{c} \tag{7}$$

The volume of the reciprocal cell leads to similar equations

$$V^* = \frac{b^*c^* \sin \alpha^*}{a} = \frac{a^*c^* \sin \beta^*}{b} = \frac{a^*b^* \sin \gamma^*}{c} \tag{8, 9, 10}$$

from which expressions may be obtained for the direct-unit translations

$$a = (b^*c^* \sin \alpha^*)/V^* \tag{11}$$

$$b = (a^*c^* \sin \beta^*)/V^* \tag{12}$$

$$c = (a^*b^* \sin \gamma^*)/V^* \tag{13}$$

and the axial ratios of the direct cell

$$a:b:c = \frac{\sin \alpha^*}{a^*}:\frac{\sin \beta^*}{b^*}:\frac{\sin \gamma^*}{c^*} \tag{14}$$

The Volume of a Triclinic Cell in Terms of Its Own Elements

The expressions for the volume of a triclinic cell developed above (Equations 1, 2, and 3), rely on the value of one reciprocal cell edge. It is often necessary to have an expression for the volume entirely in terms of its own elements.

In Fig. A6-2, OZ is the perpendicular distance (the interplanar spacing) between successive ab planes of the cell, and thus OX, OY, and OZ are mutually perpendicular; ε and δ are the angles, respectively, between the c edge and OX and OZ. The angle between the a edge and OX is evidently $(90 - \gamma)$.

From Fig. A6-2

$$V = abc \sin\gamma \cos\delta$$

Only δ needs to be expressed in more appropriate terms in this relation.

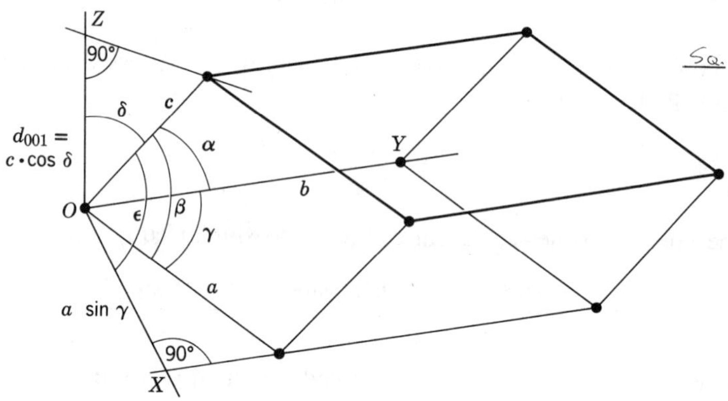

Fig. A6-2

The direction of the c edge, with respect to OX, OY, and OZ, is given by the cosine relation

$$\cos^2\varepsilon + \cos^2\alpha + \cos^2\delta = 1$$

from which

$$\cos\delta = (1 - \cos^2\alpha - \cos^2\varepsilon)^{1/2}$$

The angle between the a and c edges (namely β) can be expressed in terms of the angles that the edges make with OX, OY and OZ

$$\begin{aligned}\cos\beta &= \cos(90 - \gamma)\cos\varepsilon + \cos\gamma\cos\alpha + \cos 90 \cos\delta \\ &= \sin\gamma\cos\varepsilon + \cos\gamma\cos\alpha\end{aligned}$$

and then

$$\cos\varepsilon = \frac{1}{\sin\gamma}(\cos\beta - \cos\gamma\cos\alpha)$$

Substituting this value of $\cos \varepsilon$ in the equation for $\cos \delta$ derived above

$$\cos \delta = \left[1 - \cos^2 \alpha - \left(\frac{1}{\sin \gamma}[\cos \beta - \cos \gamma \cos \alpha]\right)^2\right]^{1/2}$$

and thus the volume can be expressed in terms of the cell elements as

$$\begin{aligned} V &= abc \sin \gamma \left[1 - \cos^2 \alpha - \left(\frac{1}{\sin \gamma}[\cos \beta - \cos \gamma \cos \alpha]\right)^2\right]^{1/2} \\ &= abc[\sin^2 \gamma - \sin^2 \gamma \cos^2 \alpha - \cos^2 \beta + 2 \cos \alpha \cos \beta \cos \gamma + \\ &\qquad + \cos^2 \gamma \cos^2 \alpha]^{1/2} \\ &= abc[\sin^2 \gamma - \cos^2 \beta - \cos^2 \alpha (\sin^2 \gamma + \cos^2 \gamma) + \\ &\qquad + 2 \cos \alpha \cos \beta \cos \gamma]^{1/2} \\ &= abc[1 - \cos^2 \gamma - \cos^2 \beta - \cos^2 \alpha + 2 \cos \alpha \cos \beta \cos \gamma]^{1/2} \qquad (15) \end{aligned}$$

because $(\sin^2 \gamma + \cos^2 \gamma) = 1$. Similarly, for the reciprocal cell

$$V^* = a^*b^*c^*[1 - \cos^2 \gamma - \cos^2 \beta - \cos^2 \alpha + 2 \cos \alpha \cos \beta \cos \gamma]^{1/2} \qquad (16)$$

A Unit Translation in Terms of Its Corresponding Reciprocal Constant

In addition to the expressions derived above for the unit translations (Equations 4 to 6 and 11 to 13), it is useful to have a relation between a unit translation and its corresponding reciprocal constant. Figures A6-3*a* and *b* are constructed to derive an expression involving a^* and a. Figure A6-3*a* shows a direct cell (solid circles) oriented with the c axis normal to the page. As a consequence, a^*, b^*, and the interaxial angle γ^* are in the plane of the page. Figure A6-3*b* is a section in the ac plane. The (100)* reciprocal planes, which are normal to the a axis, therefore appear as traces in this orientation. It follows that $OY = d^*_{100} = 1/a$. From Fig. A6-3*a*

$$OX = a^* \sin \gamma^*$$

and from Fig. A6-3*b*

$$OX = \frac{OY}{\sin(180 - \beta)} = \frac{OY}{\sin \beta} = \frac{1}{a \sin \beta}$$

Therefore,

$$a^* \sin \gamma^* = \frac{1}{a \sin \beta}$$

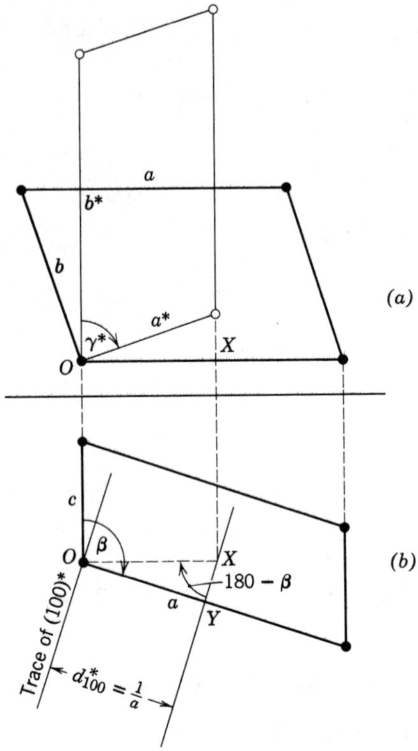

Fig. A6-3

and

$$a = \frac{1}{a^* \sin \beta \sin \gamma^*} \tag{17}$$

A similar relation involving $\sin \beta^*$ and $\sin \gamma$ can be derived between a and a^*. The direct cell of Fig. A6-4*a* (solid circles) is oriented with the a and b axes, and consequently the angle γ, in the plane of the page. Hence the reciprocal axis, c^*, is the normal to the page. Figure A6-4*b* is a section in the a^*c^* plane. The (100) planes, which are normal to the a^* axis, therefore appear as traces in this figure, and $OY = d_{100} = 1/a^*$. From Fig. A6-4*a*

$$OX = a \sin(180 - \gamma) = a \sin \gamma$$

and from Fig. A6-4*b*

$$OX = \frac{OY}{\sin \beta^*} = \frac{1}{a^* \sin \beta^*}$$

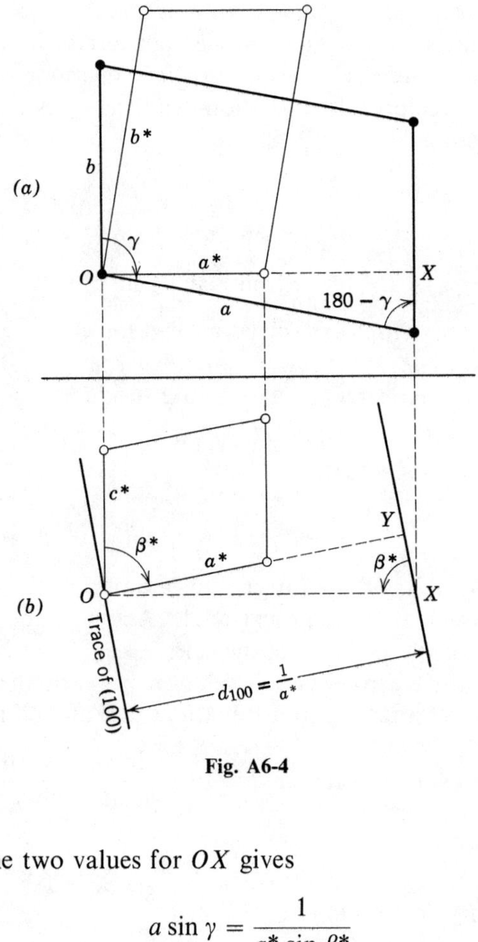

Fig. A6-4

Combining the two values for OX gives

$$a \sin \gamma = \frac{1}{a^* \sin \beta^*}$$

and

$$a = \frac{1}{a^* \sin \beta^* \sin \gamma} \tag{18}$$

The related expressions for b and c are

$$b = \frac{1}{b^* \sin \alpha^* \sin \gamma} = \frac{1}{b^* \sin \alpha \sin \gamma^*} \tag{19, 20}$$

$$c = \frac{1}{c^* \sin \alpha^* \sin \beta} = \frac{1}{c^* \sin \alpha \sin \beta^*} \tag{21, 22}$$

New expressions for the reciprocal volume can now be formulated by inserting these values for a, b, and c in the appropriate volume equations (8 to 10) derived above. This results in three relations each of which involves the three reciprocal translations and the sines of three angles, one being a direct angle

$$V^* = a^*b^*c^* \sin \alpha^* \sin \beta^* \sin \gamma \tag{23}$$

$$= a^*b^*c^* \sin \alpha^* \sin \beta \sin \gamma^* \tag{24}$$

$$= a^*b^*c^* \sin \alpha \sin \beta^* \sin \gamma^* \tag{25}$$

Similarly, if the expressions are solved instead for a^*, b^*, and c^* respectively and appropriate substitutions are made in the direct volume equations (1 to 3), three similar relations are produced

$$V = abc \sin \alpha \sin \beta \sin \gamma^* \tag{26}$$

$$= abc \sin \alpha \sin \beta^* \sin \gamma \tag{27}$$

$$= abc \sin \alpha^* \sin \beta \sin \gamma \tag{28}$$

Each relation contains three direct translations and the sines of three angles, one of which is a reciprocal angle. Actually, the relations are identical, except for the position of the star.

Finally, it is possible to derive a relation between the sines of the reciprocal and direct angles. From Equations (21) and (22)

$$\sin \alpha^* \sin \beta = \sin \alpha \sin \beta^* \quad \text{or} \quad \frac{\sin \alpha}{\sin \alpha^*} = \frac{\sin \beta}{\sin \beta^*}$$

and from (17) and (18)

$$\sin \beta^* \sin \gamma = \sin \beta \sin \gamma^* \quad \text{or} \quad \frac{\sin \beta}{\sin \beta^*} = \frac{\sin \gamma}{\sin \gamma^*}$$

and consequently

$$\frac{\sin \alpha}{\sin \alpha^*} = \frac{\sin \beta}{\sin \beta^*} = \frac{\sin \gamma}{\sin \gamma^*} \tag{29}$$

An Angle Entirely in Terms of the Angles of Its Reciprocal Cell

It is advantageous to have an expression for an interaxial angle entirely in terms of the interaxial angles of its reciprocal cell. The angles α, β, and γ between the axes of a triclinic cell and the dihedral angles A, B, C between the planes contained by these axes can be shown in spherical projection (Fig. A6-5). The relations between these six elements

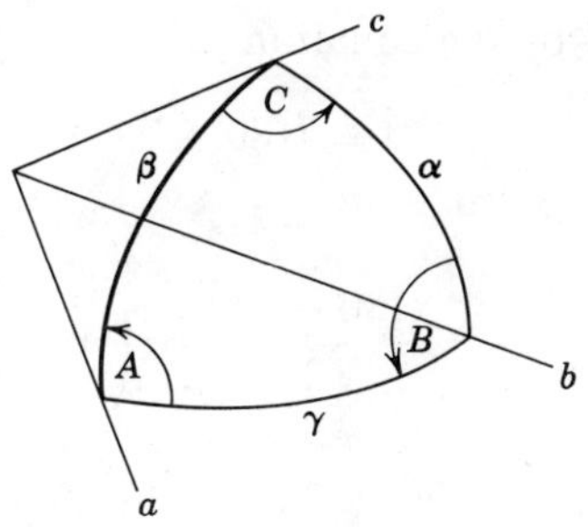

Fig. A6-5

of a spherical triangle are well known. For example

$$\tan\frac{A}{2} = \left[\frac{\sin(\sigma - \beta)\sin(\alpha - \gamma)}{\sin\sigma\sin(\sigma - \alpha)}\right]^{1/2} \tag{30}$$

where $\sigma = (\alpha + \beta + \gamma)/2$.

Figure A6-6 shows a direct triclinic cell with *a* normal to the page. Consequently, the (010) and (001) planes appear as traces in the diagram. The angle between these traces is the dihedral angle *A*. The reciprocal axes b^* and c^* are normal, respectively, to the *ac* and *ab* planes, hence they lie in the plane of the page. The angle α^* between these axes is

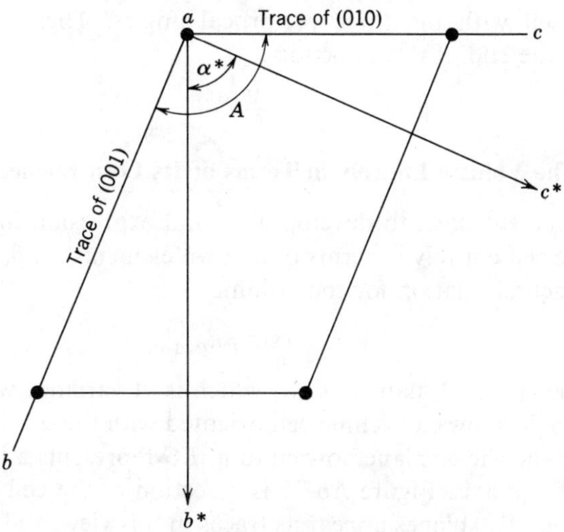

Fig. A6-6

evidently related to A by the equation

$$A = (180 - \alpha^*)$$

Therefore

$$\frac{A}{2} = \frac{(180 - \alpha^*)}{2} = \left(90 - \frac{\alpha^*}{2}\right)$$

and

$$\tan\frac{A}{2} = \left(90 - \frac{\alpha^*}{2}\right) = \cot\frac{\alpha^*}{2}$$

Thus from (30)

$$\cot\frac{\alpha^*}{2} = \left[\frac{\sin(\sigma - \beta)\sin(\alpha - \gamma)}{\sin\sigma\sin(\sigma - \alpha)}\right]^{1/2} \tag{31}$$

The related equations for the remaining two reciprocal angles are evidently

$$\cot\frac{\beta^*}{2} = \frac{\sin(\sigma - \alpha)\sin(\sigma - \gamma)}{\sin\sigma\sin(\sigma - \beta)} \tag{32}$$

$$\cot\frac{\gamma^*}{2} = \frac{\sin(\sigma - \alpha)\sin(\sigma - \beta)}{\sin\sigma\sin(\sigma - \gamma)} \tag{33}$$

Obviously, similar relations can be derived connecting a half-angle in the direct cell with the three reciprocal angles. These equations are included at the end of this appendix.

The Volume Entirely in Terms of Its Own Elements

It is convenient now, to develop a second expression for the volume of a triclinic cell entirely in terms of its own elements: α, β, γ, a, b, and c. The fundamental relation for the volume

$$V = ab\sin\gamma\, d_{001}$$

contains one constant, namely d_{001}, which is at variance with this.

Figure A6-7a shows a triclinic cell oriented with the a axis in the plane of the page, and the ac plane normal to it. EG represents a line in a plane normal to the a axis. Figure A6-7b is a section of the cell in this plane. The (001) and (010) planes appear as traces in this view, and consequently EG, which is constructed normal to the (001) traces, equals the spacing

d_{001}. From the triangle EFG

$$d_{001} = EF \sin A$$

Figure A6-7c is a section of the cell which contains the a and c axes and EF. From the triangle EHF

$$EF = c \sin \beta$$

The volume of the cell can now be expressed as

$$V = abc \sin \beta \sin \gamma \sin A$$

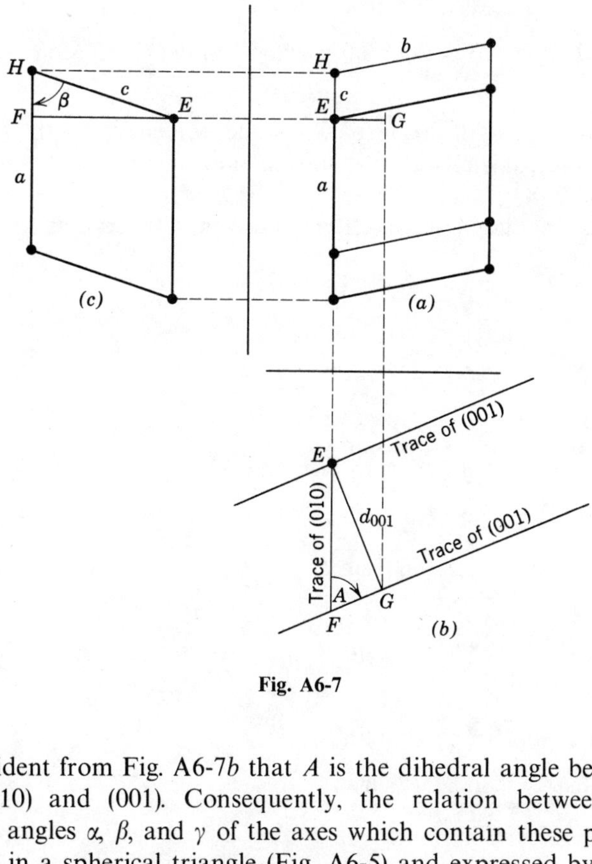

Fig. A6-7

It is evident from Fig. A6-7b that A is the dihedral angle between the planes (010) and (001). Consequently, the relation between A and interaxial angles α, β, and γ of the axes which contain these planes can be shown in a spherical triangle (Fig. A6-5) and expressed by the well-known formulas governing the half angles

$$\cos \frac{A}{2} = \left[\frac{\sin(\sigma - \beta) \sin(\sigma - \gamma)}{\sin \beta \sin \gamma} \right]^{1/2}$$

$$\sin\frac{A}{2} = \left[\frac{\sin\sigma\sin(\sigma-\alpha)}{\sin\beta\sin\gamma}\right]^{1/2}$$

where $\sigma = (\alpha + \beta + \gamma)/2$.

A fundamental formula in plane trigonometry states that

$$\sin A = 2\sin\frac{A}{2}\cos\frac{A}{2}$$

Consequently

$$\sin A = \frac{2[\sin\sigma\sin(\sigma-\alpha)\sin(\sigma-\beta)\sin(\sigma-\gamma)]^{1/2}}{\sin\beta\sin\gamma}$$

and

$$V = 2abc[\sin\sigma\sin(\sigma-\alpha)\sin(\sigma-\beta)\sin(\sigma-\gamma)]^{1/2} \tag{34}$$

The expression for the reciprocal volume is identical with this, except that all of the constants are the starred equivalents.

Equations Relating Triclinic-, Reciprocal-, and Direct-Lattice Constants

$$a = \frac{b^*c^*\sin\alpha^*}{V} \qquad a^* = \frac{bc\sin\alpha}{V}$$

$$b = \frac{a^*c^*\sin\beta^*}{V} \qquad b^* = \frac{ac\sin\beta}{V}$$

$$c = \frac{a^*b^*\sin\gamma^*}{V} \qquad c^* = \frac{ab\sin\gamma}{V}$$

$$a = \frac{1}{a^*\sin\beta^*\sin\gamma} = \frac{1}{a^*\sin\beta\sin\gamma^*}$$

$$b = \frac{1}{b^*\sin\alpha^*\sin\gamma} = \frac{1}{b^*\sin\alpha\sin\gamma^*}$$

$$c = \frac{1}{c^*\sin\alpha^*\sin\beta} = \frac{1}{c^*\sin\alpha\sin\beta^*}$$

$$a:b:c = \frac{\sin\alpha^*}{a^*} = \frac{\sin\beta^*}{b^*} = \frac{\sin\gamma^*}{c^*}$$

$$a^*:b^*:c^* = \frac{\sin\alpha}{a} = \frac{\sin\beta}{b} = \frac{\sin\gamma}{c}$$

$$\cot\frac{\alpha}{2} = \left[\frac{\sin(\sigma^*-\beta^*)\sin(\sigma^*-\gamma^*)}{\sin\sigma^*\sin(\sigma^*-\alpha^*)}\right]^{1/2}$$

$$\cot\frac{\beta}{2} = \left[\frac{\sin(\sigma^* - \alpha^*)\sin(\sigma^* - \gamma^*)}{\sin\sigma^*\sin(\sigma^* - \beta^*)}\right]^{1/2}$$

$$\cot\frac{\gamma}{2} = \left[\frac{\sin(\sigma^* - \alpha^*)\sin(\sigma^* - \beta^*)}{\sin\sigma^*\sin(\sigma^* - \gamma^*)}\right]^{1/2}$$

$$\cot\frac{\alpha^*}{2} = \left[\frac{\sin(\sigma - \beta)\sin(\sigma - \gamma)}{\sin\sigma\sin(\sigma - \alpha)}\right]^{1/2}$$

$$\cot\frac{\beta^*}{2} = \left[\frac{\sin(\sigma - \alpha)\sin(\sigma - \gamma)}{\sin\sigma\sin(\sigma - \beta)}\right]^{1/2}$$

$$\cot\frac{\gamma^*}{2} = \left[\frac{\sin(\sigma - \alpha)\sin(\sigma - \beta)}{\sin\sigma\sin(\sigma - \gamma)}\right]^{1/2}$$

$$\frac{\sin\alpha}{\sin\alpha^*} = \frac{\sin\beta}{\sin\beta^*} = \frac{\sin\gamma}{\sin\gamma^*}$$

$$V = \frac{bc\sin\alpha}{a^*} = \frac{ac\sin\beta}{b^*} = \frac{ab\sin\gamma}{c^*}$$

$$\begin{aligned} V &= abc\sin\alpha\sin\beta\sin\gamma^* \\ &= abc\sin\alpha\sin\beta^*\sin\gamma \\ &= abc\sin\alpha^*\sin\beta\sin\gamma \end{aligned}$$

$$V = abc(1 - \cos^2\alpha - \cos^2\beta - \cos^2\gamma + 2\cos\alpha\cos\beta\cos\gamma)^{1/2}$$

$$V = 2abc[\sin\sigma\sin(\sigma - \alpha)\sin(\sigma - \beta)\sin(\sigma - \gamma)]^{1/2}$$

$$V^* = \frac{b^*c^*\sin\alpha^*}{a}\;\frac{a^*c^*\sin\beta^*}{b}\;\frac{a^*b^*\sin\gamma^*}{c}$$

$$\begin{aligned} V^* &= a^*b^*c^*\sin\alpha^*\sin\beta^*\sin\gamma \\ &= a^*b^*c^*\sin\alpha^*\sin\beta\sin\gamma^* \\ &= a^*b^*c^*\sin\alpha\sin\beta^*\sin\gamma^* \end{aligned}$$

$$V^* = a^*b^*c^*(1 - \cos^2\alpha^* - \cos^2\beta^* - \cos^2\gamma^* + 2\cos\alpha^*\cos\beta^*\cos\gamma^*)^{1/2}$$

$$V^* = 2a^*b^*c^*[\sin\sigma^*\sin(\sigma^* - \alpha^*)\sin(\sigma^* - \beta^*)\sin(\sigma^* - \gamma^*)]^{1/2}$$

where $\sigma = \dfrac{\alpha + \beta + \gamma}{2}$ and $\sigma^* = \dfrac{\alpha^* + \beta^* + \gamma^*}{2}$

Index

Absorption coefficient, 40, 137, 138
Absorption edge, 35–40
Absorption factor, 71
Absorption of x-rays, 40
  in Debye-Scherrer method, 155–161
Amorphons, 8
Angstrom units, 5–7
ASTM index to powder data file, 133–136
ASTM x-ray powder data file, 130–137
  Fink index, 136
  index, 133–136
  Matthews index, 136, 137
Atomic scattering factor (f), 64–66
Avogadro's number, 1, 5

Back-reflection Laue method, 226, 232–235
Beam, divergence of, 155
  line focus, 42–45, 174
  spot focus, 42–45
Bernal chart for rotation photographs, 292
Bradley-Jay film mount, 106, 150, 151
Bragg angle (θ), 4, 58
Bragg equation for reflection, 4, 58
Buerger diffractometer, 343–345
Buerger precession method, *see* Precession method

Center of symmetry, 86, 101
Central rows, 304, 305
  diffraction in terms of, 305–324
Characteristic spectra, 29, 31–35

Collimator, parallel plate, 175
Continuous spectra, 29–31
Counters (x-ray), 177–196
  comparison of three, 185–196
  energy resolution, 186–188
  figure of merit, 193–196
  gas-filled detectors, 177–184
  Geiger, 177, 179, 180–183
  photoelectric detectors, 184, 185
  proportional, 183, 184
  pulse height discrimination, 188–193
  quantum counting efficiency, 186
  scintillation, 184, 185
Cryptocrystalline, 8
Crystal, 7, 8
Crystal classes, 26
Crystal holder, 209–212
Crystals, classification of, 8
  definition, 7, 8
  projection of morphological, 212–221
Crystallinity, 8
Crystalline, 8
Crystallite size, determination of, 147–149
Crystallography, 7
Crystal structure, definition, 8
  determination, 79–84
Cubic crystal analyzer, 120
Cylindrical coordinates of a reciprocal point, 289, 315, 316

Debye-Scherrer method, *see* Powder methods

De Jong-Bouman method, 250–252
Detectors, *see* Counters
Diffraction symmetry, 84–86
Diffraction tubes, commercial, 42–45
  gas, 40, 41
  hot cathode, 40–42
Diffraction of x-rays, 46–86
  by a plane, 50–52
  by a row, 48–50
  by a three-dimensional array, 52–55
  geometry of, 46–62
  intensity of, 62–79
  in terms of reciprocal lattice, 244–249
Diffractometer (powder), 172
  comparison with Debye-Scherrer method, 206, 207
  counting circuits, 196–201
  counting rate, measurement of, 199–202
  curved-sample device, 176–177
  data, factors affecting, 201–204
  goniometer, 172–177
  optical system of goniometer, 172–174
  pulse height analyzer, 188–196
  sample, 173–177, 204–206
  scaler, 196, 198, 199
  ratemeter, 196, 198
  x-ray counters, 177–196
Diffractometer (single crystal), 342–350

Einstein equation, 29
Equi-inclination diffractometer, 343–345
Equi-inclination Weissenberg method, 312, 313
Equivalent positions, 24–26
Eulerian cradle, 345–347
Excitation potential, 34, 35
Extinction, primary and secondary, 75–79
Extrapolation functions for lattice constants, 157–161, 163, 164, 334, 335

Film coordinates of a reflection, 307, 315–317
  on a rotation photograph, 290, 291
  on a Weissenberg photograph, 302, 303, 306
Film shrinkage, 150–155
Filters (x-ray), 38, 40
Fink index to ASTM x-ray powder file, 136
Flat-cone Weissenberg method, 310–312
Flat-film rotation method for orienting randomly set crystals, 351–362
Fluorescence, 35–38
Focal spot on diffraction tubes, 42–45
Fourier summation, 80–82
Friedel's law, 86, 101

Gas amplification, 178
Geiger counter, 177, 179, 180–183, 184, 186, 192, 196
Generators, x-ray, 45
Glancing (Bragg) angle (θ), 4, 58
Glide planes, 22, 23
  effect on reflections, 93–96
Gnomonic projection, 213, 221
  relation to reciprocal lattice, 243, 244
Goniometer, optical, 208–212
  x-ray, 172–177
Goniometer head, 209–212
Gratings, ruled, 6
Greninger chart, 230–232
Guinier-de Wolff focusing camera, 168, 169

Hermann-Manguin symbols, 24
High-temperature powder cameras, 171

Imperfect crystals, 75–78
Integrated intensity, measurement with diffractometer, 201
Integrated reflection, 73–75
  from real crystals, 75–79
  of Debye-Scherrer reflections, 109, 110
Intensity of diffraction, 62–73
  of Debye-Scherrer reflections, 107–111

kx Units, 6, 7
K series, 33, 34

Lattice dimensions with Debye-Scherrer method, source and correction of errors, 149–166
absorption, 155–161
beam divergence, 155
comparison method of minimizing errors, 165, 166
film shrinkage, 150–155
measurement errors, 162–165
refraction of x-rays, 160, 162
sample eccentricity, 155
Lattice line template for Weissenberg equi-inclination photographs, 327
Lattices, definition, 9
effect of centering on reflections, 87–93
interplanar spacing, 13, 14
rotation of points, rows, planes, 9, 11–13
three-dimensional (Bravais, space), 9–11
two-dimensional, 9
Laue equations for diffraction, 53
by a row, 50
in terms of reciprocal lattice, 249
Laue experiments, 2, 3
Laue groups, 18, 20, 21
Laue method, 222–237
back-reflection photographs, 226, 232–235
orienting crystals with, 236–237
transmission photographs, 226–230
Laue symmetry, 84–86, 100
Linear diffractometer, 347–350
Line breadth, 147–149
Line spectra, 29, 31–35
Lorentz factor, 72, 73, 107

Matthews coordinate index to ASTM x-ray powder file, 136, 137
Miller indices, 12, 13
Mixtures, identification of components with Debye-Scherrer method, 134–136
quantitative analysis of, 137–147
Monochromatic radiation, 34
Monochromator, 168, 169
Mosaic structure, 75, 78
Multiplicity factor, 108, 109

Nelson and Riley, correction of absorption errors, 157–161
Normal-beam Weissenberg method, 309, 310
Noncentral rows, 304, 305
diffraction in terms of, 324–330

Orienting a crystal, by optical methods, 208–212
with flat-film rotation photographs, 351–362
with Laue photographs, 236, 237
with oscillation photographs, 296–300
with precession photographs, 266–272
Orientation of randomly set crystal, determination by flat-film rotation method, 351–362
Laue method, 232–235
Orientation of triclinic cells, 26–28
Orientation-precession photographs, 266–272
Oscillation method, 282, 294–300
indexing oscillation films, 294, 295
orienting crystals with, 296–300

Parafocusing arrangement in powder diffractometer, 172–177
Parafocusing cameras, 166–169
Patterson summation, 82–84
Perfect crystals, 75–78
Phase angle, 66–69, 80–82, 84
Pinhole powder techniques, 169–171
transmission method, 169, 170
back-reflection method, 170, 171
Plateaus, of gas-filled detectors, 178–180
of scintillation counter, 185
Point groups, 14–18, 20, 21
Polarization factor, 64
Polycrystal methods, *see* Powder methods
Powder counter methods, *see* Diffractometer
Powder film methods, Debye-Scherrer technique, 105–166
advantages, 105
camera sizes, 106, 107
crystalline size determination, 147–149

Powder film methods, film arrangement, 105, 106, 150–155
film measurement, 116–118, 162–165
identification, 130–137
intensities of reflections, 107–111
intensity measurements, 110–111
indexing, 118–125
lattice dimensions, 124, 149–166
lattice type, 128, 129
Lorentz factor, 107
multiplicity factor, 108, 109
principle of, 103, 104
quantitative analysis of mixtures, 137–147
radiation, 114–116
relative integrated reflection, 109, 110
sample, 111, 112, 114
space group, 128, 129
unit cell contents, 125–128
Powder film methods, special techniques and apparatus, 166–171
Guinier-de Wolff focusing camera, 168, 169
high-temperature cameras, 171
parafocusing cameras, 166–169
pinhole techniques, 169–171
Precession method, 250–281
choice of precession axis, 274, 275
choice of wavelength, 264–266
determination of d^*, 261–264
determination of lattice constants, 278–281
motion of crystal, 253, 254
nomogram for setting instrument, 262
orientation error chart, 273
orientation photographs, 266–272
Primary extinction, 75–78
Proportional counter, 177–180, 182–196
Pulse height discrimination, 188–196

Ratemeter, 196, 198
Reciprocal lattice, 238–249
centered lattices, 242, 243
reflection of x-rays in terms of, 244–249
relation to direct lattice, 239, 240, 241–243, 391–403
Reciprocal point, cylindrical coordinates, 289
Recovery time, of Geiger counter, 182
of proportional counter, 183
of scintillation counter, 185
Reduced cell, 26
Reflection coordinates, 305–308, 315
Reflection of x-rays, 55–62
by a typical crystal structure, 60–62
conditions for reflection (Bragg law), 57, 58
integrated reflection, 73–75
in terms of reciprocal lattice, 244–249
notation, 59
Reflection sphere, 244–249
Refraction of x-rays, 160, 162
Rotation method, Bernal chart, 292
indexing rotation films, 288–293
lattice dimensions, 285–288
measurement of rotation films, 288

Sample eccentricity, effect in Debye-Scherrer method, 155, 156
Scale factor, 196, 198
Scaler, 196, 198, 199
Scaling circuit, 196
Scattering of x-rays, 46, 62–69
Schoenflies symbols, 24
Scintillation counter, 177, 182, 184–196
Screw axes, 18, 19, 22
effect on reflections, 96–98
Secondary extinction, 75–79
Slit, divergence and receiving, 174, 175
equatorial layer line, 302
Space groups, 18–24, 365–369
determination of, 98–102
determination tables, 372–384
Hermann-Mauguin symbols, 24
orthorhombic permutations, 25, 370, 371
Schoenflies symbols, 24
Space group symmetry elements, 18–19, 22–23
systematically absent reflections due to, 93–98
Spectra, characteristic, 29, 31–35
continuous, 29
electromagnetic, 29, 30

Sphere of reflection, 244–249
Stereographic projection, 214–220
  of back-reflection Laue photographs, 230, 231, 232–235
  of transmission Laue photographs, 226–230
Stereographic scales, 227, 231
  data for constructing, 388–390
Straumanis film mount, 150, 152–154
Structure amplitude, 67–69
Structure factor, 67–69
Symmetry, elements (operations), 9, 15
  of crystals, 20, 21
  recognition on randomly oriented crystals, 362–364
  axes, 14, 15, 18, 19, 22
  planes, 15, 18, 22, 23
Systematically absent reflections, 87–102
  due to lattice centering, 87–93
  due to space-group symmetry, 93–98

Takeoff angle, 43, 175
Temperature factor, 69, 70
Threshold voltage, 179, 180
Time constant of ratemeter, 198, 199
Transmission Laue method, 223–226
Triclinic cells, 26–28

Unit cell, 9
  calculation of cell contents, 125–128

Van Arkel film mount, 150–152

Wavelength, angstrom units, 6
  first absolute wavelength, 4, 5
  K series values, 34
  kx Units, 6
Wavelength angstrom units, Siegbahn, kx values, 6, 7
  units, 5, 6
  x units, 6
Weissenberg method, scales for measurement of films, 317–319
  chart for deriving instrument settings, 314
  diffraction symmetry, 330–332
  equi-inclination, 312, 313
  flat-cone, 310–312
  lattice constants, 332–341
  lattice line template, 327
  measurement of films, 319–322, 324
  normal beam, 309, 310
  reconstruction of reciprocal lattice, 321, 322, 328–330
White radiation, 29–31
Wilson film mount, 150, 155
Wulff net, 217–220

x Units, 6
X-ray, detectors, *see* Counters
  diffraction tubes, 40–45
  generators, 45
  spectrometer, 4
  wavelengths, 34
X-rays, characteristic (line), 31–35
  continuous (white), 30, 31
  fluorescent, 35–39
  generation of, 29–35
  reflection of, 55–62
  scattering by an atom, 64–67
  scattering by an electron, 62–64
  scattering by a unit cell, 67–69

Zone, axis, 208